# Finite Simple Groups

## An Introduction to Their Classification

**THE UNIVERSITY SERIES IN MATHEMATICS**

Series Editor: Joseph J. Kohn
*Princeton University*

**INTRODUCTION TO PSEUDODIFFERENTIAL AND FOURIER INTEGRAL OPERATORS**
François Treves
VOLUME 1: PSEUDODIFFERENTIAL OPERATORS
VOLUME 2: FOURIER INTEGRAL OPERATORS

**A SCRAPBOOK OF COMPLEX CURVE THEORY**
C. Herbert Clemens

**FINITE SIMPLE GROUPS: An Introduction to Their Classification**
Daniel Gorenstein

# Finite Simple Groups

## An Introduction to Their Classification

**Daniel Gorenstein**

*Rutgers, The State University of New Jersey*
*New Brunswick, New Jersey*

PLENUM PRESS • NEW YORK AND LONDON

Library of Congress Cataloging in Publication Data

Gorenstein, Daniel.
Finite simple groups.

(The University series in mathematics)
Bibliography: p.
Includes index.
1. Finite simple groups. I. Title. II. Series: University series in mathematics (Plenum Press)

QA171.G6417 512'.2 81-23414
ISBN 0-306-40779-5 AACR2




A Division of Plenum Publishing Corporation
233 Spring Street, New York, N.Y. 10013



Printed in the United States of America

**In memory of**
**Richard and Ilse Brauer**

**He, for his pioneering studies**
**of finite simple groups**

**She, for her steadfast support**
**on the long journey**

# Acknowledgments

I have earlier acknowledged the considerable help of J. L. Alperin, Michael Aschbacher, Robert Gilman, George Glauberman, Robert Griess, David Hunt, Richard Lyons, Michael O'Nan, and Charles Sims for the American Mathematical Society *Bulletin* article on simple groups, of which this book is an expanded version. I should like to add here my special thanks to Richard Lyons, who, from his deep knowledge of simple group theory, clarified for me many of the points covered in the text, and also to Enrico Bombieri, Paul Fong, George Glauberman, Morton Harris, and Stephen Smith, who read the manuscript and made many valuable comments on it.

Daniel Gorenstein

# Contents

# Introduction

In February 1981, the classification of the finite simple groups (D1)* was completed,[†, ‡] representing one of the most remarkable achievements in the history or mathematics. Involving the combined efforts of several hundred mathematicians from around the world over a period of 30 years, the full proof covered something between 5,000 and 10,000 journal pages, spread over 300 to 500 individual papers.

The single result that, more than any other, opened up the field and foreshadowed the vastness of the full classification proof was the celebrated theorem of Walter Feit and John Thompson in 1962, which stated that every finite group of odd order (D2) is solvable (D3)—a statement expressible in a single line, yet its proof required a full 255-page issue of the *Pacific Journal of Mathematics* [93].

Soon thereafter, in 1965, came the first new *sporadic* simple group in over 100 years, the Zvonimir Janko group $J_1$, to further stimulate the

*To make the book as self-contained as possible, we are including definitions of various terms as they occur in the text. However, in order not to disrupt the continuity of the discussion, we have placed them at the end of the Introduction. We denote these definitions by (D1), (D2), (D3), etc.

† The classification theorem asserts that an arbitrary finite simple group is necessarily isomorphic to one of the groups on a specified list of simple groups. (See Section 2.11 for the detailed list.)

‡ The final mathematical step of the classification was carried out by Simon Norton of the University of Cambridge, in England, establishing the "uniqueness" of the Fischer–Griess sporadic simple group $F_1$. Griess had earlier constructed $F_1$ in terms of complex matrices of degree 196,883. Moreover, Thompson had shown that there existed at most one simple group of "type $F_1$" which could be so represented by complex matrices of this degree. What Norton did was to prove that any group of type $F_1$ could, in fact, be represented by such matrices. (All these results are described more fully in Section 2.10.)

We also note that as of February 1981 several manuscripts (including Norton's) concerning the classification were still in preparation.

interest of the mathematical community in finite simple groups [187]. The sporadic groups acquired their name because they are not members of any infinite family of finite simple groups. Emile Mathieu, in 1861, had discovered five such groups [210–212], yet $J_1$ remained undetected for a full century, despite the fact that it has only 175,560 elements (a very small number by the standards of simple group theory). Then in rapid succession over the next ten years, 20 more sporadic groups were discovered, the largest of these the group $F_1$ of Bernd Fischer and Robert Griess (recently constructed by Griess [152]) of order 808, 017, 424, 794, 512, 875, 886, 459, 904, 961, 710, 757, 005, 754, 268, 000, 000, 000 (approximately $10^{54}$), and because of its size, originally dubbed the "monster." An additional intriguing aspect of these new sporadic groups is the fact that several have depended upon computer calculations for their construction.

The pioneer in the field was Richard Brauer, who began to study simple groups in the late 1940s. He was the first to see the intimate and fundamental relationship between the structure of a group and the *centralizers* (D4) of its *involutions* (elements of order 2; D5), obtaining both quantitative and qualitative connections. As an example of the first, he showed that there are a finite number of simple groups with a specified centralizer of an involution [46]. As an example of the second, he proved that if the centralizer of an involution in a simple group $G$ is isomorphic to the general linear group $GL(2, q)$ (D6) over the finite field with $q$ elements, $q$ odd, then either $G$ is isomorphic to the three-dimensional projective special linear group $L_3(q)$ (D7), or else $q = 3$ and $G$ is isomorphic to the smallest Mathieu group $M_{11}$ of order $8 \cdot 9 \cdot 10 \cdot 11$ [40, 42]. This last result, which Brauer announced in his address at the International Congress of Mathematicians in Amsterdam in 1954, represented the starting point for the classification of simple groups in terms of the structure of the centralizers of involutions. Moreover, it foreshadowed the fascinating fact that conclusions of general classification theorems would necessarily include sporadic simple groups as exceptional cases.

In the early years, Brauer had been essentially a lone figure working on simple groups, although Claude Chevalley's seminal paper of 1955 on the finite groups of Lie type [66] had considerable impact on the field. By the late 1950s two disciples of Brauer, Michio Suzuki and Feit, had joined the battle. However, it was the Feit–Thompson theorem that provided the primary impetus for the great expansion of the study of simple groups. The field literally exploded in the 1960s, with a large number of talented young mathematicians attracted to the subject, primarily in the United States, England, Germany, and Japan. For the next fifteen years, the papers came pouring out—long, long papers: Thompson's classification of *minimal* simple groups (i.e., simple groups in which all proper subgroups are solvable), 410 pages in six parts, the first appearing in 1968 and the last in 1974 [289]; John Walter's classification of simple groups with abelian Sylow 2-subgroups, 109 pages of the *Annals of Mathematics* in 1969 [314]; the Alperin–Brauer–Gorenstein classification of simple groups with quasi-dihedral or wreathed Sylow 2-subgroups (see D7), 261 pages of the *Transactions of the American Mathematical Society* in 1970 [3]; the Gorenstein–Harada classification of simple groups whose 2-subgroups are generated by at most 4 elements, a 461-page *Memoir of the American Mathematical Society* in 1971 [136], to name but a few. Even near the end, we find Michael Aschbacher's fundamental "classical involution" theorem, 115 pages of the *Annals of Mathematics* in 1977 [13].

Furthermore, the search for new simple groups was keeping pace with this effort, with a discovery rate of roughly one per year. The phenomenon can be compared with elementary particle theory, in which one must scan a large horizon with the aid of one's intuition and theoretical knowledge in the hope of distinguishing a new particle. If Janko's group $J_1$ has been constructed from the centralizer of one of its involutions [isomorphic to $Z_2 \times L_2(4)$], then examine other likely candidates as potential centralizers of involutions in a new simple group. If Janko's second group $J_2$ turns out to be a *transitive rank* 3 permutation group (D8), conduct a more general investigation of such permutation groups. If the automorphism group of the remarkable 24-dimensional Euclidean lattice of John Leech yields John Conway's three simple groups .1, .2, .3, then look for other integral Euclidean lattices that may have a "large" automorphism group. Any plausible direction is worth considering; just keep in mind that the probability of success is very low. In the end, several sporadic groups were discovered by both the centralizer-of-involution and rank 3 permutation group approach, but unfortunately the study of integral lattices yielded no further new groups.

Another aspect of sporadic group theory makes the analogy with elementary particle theory even more apt. In a number of cases (primarily but not exclusively those in which computer calculations were ultimately required), "discovery" did not include the actual construction of a group—all that was established was strong evidence for the existence of a simple group $G$ satisfying some specified set of conditions $X$. The operative metamathematical principle is this: if the investigation of an arbitrary group $G$ having property $X$ does not lead to a contradiction but rather to a "compatible" internal subgroup structure, then there exists an actual group with property $X$. In all cases, the principle has been vindicated; however, the interval between discovery and construction has varied from a few months to several years. Although the major credit has usually gone to the initial discoverer, existence and uniqueness were often established by others, or at least with someone else's assistance.

The excitement generated by the discovery (and construction) of these new simple groups was intense. Moreover, there was a long period in which it was felt that there might well exist infinitely many sporadic groups (from the point of view of classification at least, this was a disturbing thought, since this possibility would very likely have precluded the achievement of a complete classification of simple groups. A certain haphazard, almost random quality accompanied the search; some of the groups seemed literally plucked from thin air. I have always felt great admiration for the remarkable intuition of these indefatigable explorers.

It is essential to distinguish between the notion of *discovery* (including construction) and *classification*. One can search for a new simple group in any direction, and discovery is its own reward, requiring no further theoretical justification. However, in contrast, the solution of a general classification problem must be systematic and all-inclusive—*every* simple group with the specified property must be determined. In particular, the analysis must uncover every sporadic group satisfying the given conditions, previously discovered or not. For example, Fischer's first three sporadic groups, $M(22)$, $M(23)$, and $M(24)$, were discovered and constructed in the process of proving just such a classification theorem [97, 98]. Likewise, some years earlier, Suzuki's exceptional family of groups of Lie type of characteristic 2 was discovered in the process of classifying groups in which the centralizer of every involution has order a power of 2 [276, 278].

Whatever one's attitude toward the possible number of sporadic groups, it was certainly true that a complete classification of the finite simple groups was regarded at that time as very remote, for the steady stream of developments was producing as much turmoil as light. The chaotic state of affairs was well expressed in the verses of a song entitled "A Simple Song" (to be sung to the tune of "Sweet Betsy from Pike")), published in the *American Mathematical Monthly* (1973, p. 1028), and summed up in its final stanza:

> No doubt you noted the last lines don't rhyme.
> Well, that is, quite simply, a sign of the time.
> There's chaos, not order, among simple groups;
> And maybe we'd better go back to the loops.

I believe I have the distinction of being the original optimist regarding a possible classification of the simple groups. Even as early as 1968, in the final section of my book *Finite Groups* [130], I had placed great emphasis on

Thompson's classification of minimal simple groups, a magnificent result in which Thompson showed the fundamental significance of the "local" methods of the odd-order paper for the study of simple groups. In my comments, I suggested that his techniques might well be applicable to much broader classification problems. Over the next few years my thoughts in this direction continued to evolve, and gradually I developed a global picture of how it might be possible to carry through a complete classification. At J. L. Alperin's urging, I presented these ideas at a group theory conference at the University of Chicago in 1972, and in four lectures I laid out a *sixteen*-step program for classifying the finite simple groups [132; Appendix]. The program was met with considerable skepticism. I doubt that I made any converts at that time—the pessimists were still strongly in the ascendancy.

However, in the next few years substantial progress was made on some of the individual portions of the program: the complete classification of "nonconnected" simple groups, the first inroads into the "$B$-conjecture," and a deeper understanding of the structure of centralizers of involutions in groups of "component" type. Aschbacher, who had entered the field during this period, came on now like a whirlwind, moving directly to a leadership position and sweeping aside all obstacles, as he proved one astonishing result after another. Within five years, the program, which at its formulation had been a far-off dream, began to take on a sense of immediacy, with genuine prospects for fulfillment.

Hardly surprisingly, individual steps of the program had to be modified along the way—in 1972, the key notions of "tightly embedded subgroup," "Aschbacher block," and George Glauberman's entire theory of "pushing up" had not even existed. In addition, the program overemphasized the role of the prime 3 in the analysis of groups of "characteristic 2 type." But more significantly, in 1972 I had not appreciated the far-reaching impact that Fischer's "internal geometric" approach would have for the study of simple groups. Despite these shortcomings, the overall program remained largely intact, so that at all times we had a way of measuring the extent of our achievements and could describe quite accurately the steps remaining to complete the classification.

The turning point undoubtedly occurred at the 1976 summer conference in Duluth, Minnesota. The theorems presented there were so strong that the audience was unable to avoid the conclusion that the full classification could not be far off. From that point on, the practicing finite group theorists became increasingly convinced that the "end was near"—at first within five years, then within two years, and finally momentarily. Residual skepticism was confined largely to the general mathematical community, which quite reasonably would not accept at face value the assertion that the classification theorem was "almost proved."

But the classification of the finite simple groups is not an ordinary theorem: it makes more sense to think of it as an entire field of mathematics, in which by some accident the central questions become incorporated into the statement of a single theorem. Indeed, as in any other broad field of mathematics, later results depend upon earlier theorems. Thus, all the major results after 1963 depend crucially upon the solvability of groups of odd order, and none provides an alternate proof. As one can well imagine, the logical interconnections between the several hundred papers making up the classification proof are often very subtle, and it is no easy task to present a completely detailed flow diagram.

Because of the excessive lengths of the papers and the specialized techniques developed for the study of simple groups, the classification proof has remained quite inaccessible to non–finite group theorists. Not because of lack of interest: indeed, many mathematicians have followed the developments rather closely, especially those related to the sporadic groups and the groups of Lie type. But very few have managed to penetrate beneath these "boundary" aspects of the subject to the core of the classification proof. Even within the field, many finite group theorists, working in a specialized area of the subject, have had similar difficulties developing a global picture of the full classification proof.

I hope that this detailed outline of the classification theorem will represent a first step toward correcting this situation by illuminating the broad features of the subject of simple groups—the known simple groups themselves, the techniques underlying the classification, and the major components of the classification proof.

The concept of a group is so central to mathematics, it is difficult to imagine that ideas that have been so fruitful for the study of simple groups will find no further mathematical applicability. The "signalizer functor" theorem, the "classical involution" theorem, the "$B$-property" of finite groups, the "root involution" theorem, the "$C(G; T)$" theorem—to name but a few of the basic results—have such conceptually natural statements, one would certainly hope that analogues of at least some of these theorems exist for some families of rings or algebras. Thus, I have had in mind the subsidiary objective of enabling (primarily) algebraists, number theorists, and geometers to consider the central ideas of finite simple group theory in relation to their own fields.

Although this book is clearly aimed at a mathematical audience, portions of it should be of interest to physicists, crystallographers, and perhaps other theoretical scientists as well—especially the descriptions of the known finite simple groups: the construction of the groups of Lie type from their associated Lie algebras, the origins and definitions of each of the sporadic groups, the list of orders of the known simple groups, many of their basic properties, etc.

For expository purposes it has seemed best to split the total endeavor into two parts. In this book we present a global picture of the classification proof (Chapter 1), a detailed description of the known simple groups, including the sporadic groups (Chapters 2, 3), and a long discussion of the major techniques underlying the proof of the classification theorem (Chapter 4). Its contents represent an updated and considerably expanded version of an article written for the Brauer memorial issue of the *Bulletin of the American Mathematical Society*, published in January 1979 [132].

We hope that the material covered here, which can be regarded as preparatory to the classification proof itself, will stimulate the reader to pursue the more detailed outline, which will form the contents of two further books. However, by itself, this book should provide the reader with considerable insight into simple group theory: in particular, an overall picture of the fundamental four-part subdivision of the classification proof, the group-theoretic origins and definitions of each of the known simple groups, and a good feeling for the methods that have been developed for the study of simple groups.

I have attempted to make the book as self-contained as reasonably possible, so that it will be accessible to anyone with a sound mathematical background and a modest knowledge of abstract algebra. In particular, I have included the definition of essentially every term used in the text (even such basic notions as simple group and Sylow's theorem). Furthermore, even though the organizational structure follows a precise logical pattern, I have tried to make the individual sections more or less independent, so that a selective reading of the text is possible. Moreover, I have included only a few proofs (and outlines of proofs); these have been selected either because of the intrinsic importance of the result or as illustrations of a group-theoretic technique in action. However, I do not mean to imply that the book will necessarily make easy reading, for finite simple group theory involves many deep and difficult concepts whose applicability is often very elaborate.

There is one aspect of the classification "proof" that we must mention before concluding this introduction. Indeed, many of the papers on simple groups are known to contain a considerable number of "local" errors. The fact that it seems beyond human capacity to present a closely reasoned argument of several hundred pages with absolute accuracy may provide the explanation, but this explanation does not eliminate doubt concerning the validity of the proof. Most of these errors, when uncovered, can be fixed up "on the spot." But many of the arguments are ad hoc, so how can one be certain that the "sieve" has not let slip a configuration leading to yet another simple group? The prevailing opinion among finite group theorists is that the overall proof is accurate and that with so many individuals working on simple groups during the past fifteen years, often from such differing perspectives, every significant configuration has loomed into view sufficiently often and could not have been overlooked.

However, clearly the first task of the post-classification era must involve a reexamination of the proof to eliminate these local errors. That reexamination will have two other objectives as well. First, because the full proof evolved over a thirty-year period, some of the early papers were written without the benefit of subsequent developments. Second, because of the lengths of most of the major papers, prior results were usually quoted wherever possible, even when a slight additional argument might have avoided a particular reference.

Thus the local errors will most likely be corrected as part of the broader task of "revising" the existing classification proof in an attempt to discover its "essential" core. Helmut Bender actually began this effort ten years ago, making significant simplifications in the local group-theoretic portions of the odd-order paper [28]. The "Bender method," as his approach came to be called, subsequently developed into a standard technique, finding application in several classification problems (see Sections 4.3, 4.8). But only very recently, with the full classification proof nearly in view, have finite group theorists begun to systematically consider a global reexamination.

The outline I shall present is meant to be a historical summary of the original classification proof. Thus, apart from Bender's work, which has already become an integral part of the proof, I shall avoid discussion in the body of the text of any recently achieved revisions. In the final chapter of the sequel, I shall briefly describe some of the revisionist game plans for improving the classification proof.

Finally, except for a few passing remarks, I shall not discuss the remarkable, recently discovered connections between the Fischer–Griess group $F_1$ and classical elliptic function theory. These connections have their origin in the serendipitous observation of John McKay that the coefficient of $q$ in the expansion at infinity of the elliptic modular function $J(q)$ is 196,884, while the minimal degree of a faithful irreducible complex representation of $F_1$ is 196,883. Although considerable "numerological" interconnections have since been uncovered [70, 195, 298], the deeper explanation of this relationship remains a mystery. However, because approximately 20 of the 26 sporadic groups are embedded one way or the other in $F_1$, there is a distinct possibility that there may ultimately exist a unified, coherent description of most of the sporadic groups. Although these developments are not needed for the classification theorem per se, they do indicate that an interest in the finite simple groups will persist long after the classification.

## Definitions for the Introduction

**D1.** A group $G$ is *simple* if its only normal subgroups are the entire group $G$ and the trivial subgroup consisting of the identity element 1 of $G$. In general, a subgroup $X$ of a group $G$ is *normal* if $g^{-1}xg \in X$ for every $x \in X$ and $g \in G$. When this is the case, the set of (right) cosets of $X$ in $G$ form a group, called the *factor* or *quotient* group of $G$ by $X$ and denoted by $G/X$. Multiplication in $G/X$ is given by the rule $(Xg)\cdot(Xg') = X(gg')$ for $g, g' \in G$. Here, for a given $g \in G$, the *coset* $Xg$ denotes the set of elements $xg$ with $x$ ranging over $X$. The mapping $\phi$: $g \mapsto Xg$ is a homomorphism of $G$ onto $G/X$. Every homomorphic image of $G$ is easily shown to arise in this fashion. Thus a group $G$ is simple if and only if its only homomorphic images are itself and the trivial group.

**D2.** The *order* of a group is the number of its elements.

**D3.** A group is *solvable* if it possesses a normal series with abelian (i.e., commutative) factors. A *normal series* of a group $G$ is a chain of subgroups $G = G_1, G_2, \dots, G_n = 1$ with each $G_i$ normal in $G_{i-1}$, $2 \leq i \leq n$. The quotient groups $G_{i-1}/G_i$ are called the *factors* of the series.

**D4.** The *centralizer* of a subset $X$ of a group is the set of elements of $G$ that commute elementwise with $X$, i.e., the set of $g \in G$ such that $g^{-1}xg = x$ for all $x \in X$. The *center* of $G$ is the set of elements of $G$ whose centralizer is $G$ itself—i.e., the set of $x \in G$ that commute with every element of $G$.

**D5.** The *order* of an element $x$ of a group $G$ is the order of the *cyclic* subgroup that it generates, i.e., the subgroup $\{x^i | i \in \mathbb{Z}\}$.

**D6.** $GL(n, q)$ is the group of all nonsingular $n \times n$ matrices with entries in the finite field $GF(q)$ with $q$ elements. It has a normal subgroup $SL(n, q)$, the *special linear* group, consisting of those matrices of determinant 1. The factor group of $SL(n, q)$ by its subgroup of scalar matrices (of determinant 1) is the *projective special linear* group and is denoted by both $PSL(n, q)$ and $L_n(q)$. It is known to be simple when $n \geq 3$ or $n = 2$ and $q \geq 4$.

**D7.** A 2-group $S$ is *quasi-dihedral* if $S$ is generated by elements $x, y$ subject to the relations $x^2 = y^{2^n} = 1$, $x^{-1}yx = y^{-1+2^{n-1}}$, $n \geq 3$; and $S$ is *wreathed* if $S$ is generated by elements $x, y, z$, subject to the relations $x^{2^n} = y^{2^n} = z^2 = 1$, $xy = yx$, $z^{-1}xz = y$, $z^{-1}yz = x$, $n \geq 2$.

**D8.** The group of all permutations of a (finite) set $\Omega$ (i.e., one-to-one transformations of $\Omega$ onto itself) under the natural operation of composition is called the *symmetric* group on $\Omega$. Any subgroup $X$ of the symmetric group is called a *permutation* group (on $\Omega$). The cardinality of $\Omega$ is called the *degree* of $X$. $X$ is *k-fold transitive* on $\Omega$ if any two ordered $k$-tuples of distinct elements of $\Omega$ can be transformed into each other by an element of $X$. One writes *transitive*, *doubly transitive*, *triply transitive*, etc., for 1-fold, 2-fold, 3-fold transitive, etc. $X$ has (permutation) *rank* $r$ if $X$ is transitive on $\Omega$ and the subgroup of $X$ fixing a point of $\Omega$ has exactly $r$ orbits on $\Omega$. Thus double transitivity is equivalent to permutation rank 2.

# 1

# Local Analysis and the Four Phases of the Classification

## 1.1. From Character Theory to Local Analysis

It seems best to begin with an explanation of the historical origins and general meaning of the primary underlying method of the classification proof: *local group-theoretic analysis*. This was not always the principal approach to the study of simple groups, for Brauer's methods were almost entirely *representation-* and *character-theoretic* (D1).* In the middle 1930s he had introduced and developed the concept of *modular* characters (see D1) of a finite group. He soon realized the power of these ideas, which played an instrumental role in his proof of a conjecture of Artin on $L$-series in algebraic number fields, and he saw how they could be applied to obtain deep results concerning the structure of simple groups.† From the middle 1940s until his death, Brauer systematically developed the general theory of modular characters and *blocks* of irreducible characters (D2), with increasingly significant applications to simple group theory.

These methods were especially suited for investigating "small" simple groups: complex linear groups of transformations of low dimension, alternating groups of low degree (D3), groups with very restricted Sylow 2-subgroups [e.g., quaternion, dihedral, quasi-dihedral, wreathed, abelian, etc. (D4)]. This was certainly fortunate, since at the outset of the study of simple groups, it was obviously most natural to focus on the smallest ones.

*Again we include some definitions, denoted (D1), (D2), etc., which we have placed at the end of the section (see the footnote on page 1). We follow the same procedure in Sections 1.2 and 1.3.

†Throughout the text the term *simple* group will always refer to a *nonabelian* simple group.

Indeed, the methods were so effective that in the early years there was a strong conviction that character theory would remain a principal tool—perhaps even *the* most essential—for investigating simple groups (even though it was also recognized that larger-rank situations would involve considerable computational difficulties).

However, even in treating small groups, the method had drawbacks, for the way it worked was this: if one had rather precise information about the structure of some subgroup $H$ of $G$ (such as the centralizer of an involution), one could relate the characters of $H$ to those of $G$ and use this connection to obtain conclusions about the structure of $G$. This was the thrust of the Brauer method.

The difficulties arise if one asks a broad enough question, for then one cannot assert *a priori* that any critical subgroup $H$ of $G$ has a restricted shape. I should like to illustrate this point by considering a specific classification problem, namely, the determination of all simple groups $G$ of order $p^a q^b r^c$, $p, q, r$ primes with $p<q<r$. In view of the classical Burnside theorem that all groups of order $p^a q^b$ are solvable [see Theorem 4.130], this is a problem of natural interest. Among the known simple groups, there are exactly *eight* whose orders have this form (see D5):

$$A_5,\quad A_6,\quad L_2(7),\quad L_2(8),\quad L_2(17),\quad L_3(3),\quad U_3(3),\quad \text{and}\quad U_4(2).$$

Each of these can certainly be considered to be a "small" group (the largest order is, in fact, 25,920). (Note that in these groups $p=2$, $q=3$, and $r=5$, 7, 13, or 17.)

The obvious conjecture is that an arbitrary simple group of order $p^a q^b r^c$ is necessarily isomorphic to one of these eight groups.

For brevity, let us call these eight groups *$K_3$-groups* and an arbitrary group of order $p^a q^b r^c$ a $(p, q, r)$-group, so that our conjecture assumes the form:

*A simple $(p, q, r)$-group is necessarily a $K_3$-group.*

Of course, until shown otherwise, our conjecture may be false. As in every general classification problem, it is therefore natural to focus attention on a *minimal* counterexample $G$ to the conjecture, i.e., a simple $(p, q, r)$-group $G$ of *least* order which is not a $K_3$-group. Clearly, to establish the conjecture, we must show that no such group $G$ exists—equivalently, that $G$ must, in fact, be a $K_3$-group. The advantage of considering a minimal counterexample is that if $H$ is any proper subgroup of $G$, then the (nonsolvable) composition factors of $H$ (D6) are simple $(p, q, r)$-groups of *lower* order than $G$ and so are necessarily $K_3$-groups. In particular, this suggests extending the notion of a $K_3$-group to include any finite group whose nonsolvable composition factors are $K_3$-groups. Using this terminology, our conjecture thus reduces to verification of the following assertion:

*If $G$ is a simple $(p, q, r)$-group, each of whose proper subgroups is a $K_3$-group, then $G$ is a $K_3$-group.*

This is the typical situation one encounters at the outset of a general classification problem. We are interested in the circumstances under which character theory can be used to determine the structure of $G$. It turns out that there are two specific sets of conditions in which this is the case.

(a) The centralizer of some involution of $G$ is isomorphic to (or at least "closely approximates") the centralizer of an involution in one of the eight simple $K_3$-groups.
(b) The largest prime $r$ occurs only to the first power in the order of $G$ [or, more generally, $G$ has cyclic Sylow $r$-subgroups (D9)].

If (a) holds, one takes as critical subgroup $H$ the centralizer of the given involution of $G$ and investigates the characters of $H$ and $G$, using block theory for the prime 2. If (b) holds, one takes $H$ to be the normalizer (D8) in $G$ of a Sylow $r$-subgroup and investigates the characters of $H$ and $G$, using block theory for the prime $r$.

We can view conditions (a) and (b) as saying that $G$ "internally resembles" one of the simple $K_3$-groups. Thus character theory enables one to prove the following assertion:

*If a simple $(p, q, r)$-group internally resembles a simple $K_3$-group, then (up to isomorphism) $G$ is a $K_3$-group.*

This indicates the limitation (as well as the power) of character theory, for why should our minimal counterexample $G$ necessarily internally resemble a $K_3$-group? *A priori*, why cannot its involutions have centralizers of arbitrarily high order or its Sylow $r$-subgroup an arbitrarily complex structure? Yet unless one can establish such an internal resemblance of $G$ to one of our $K_3$-groups, there will clearly be no hope of ever proving that $G$ is, in fact, isomorphic to a $K_3$-group. Unfortunately, it seems to be beyond the scope of character theory to deal effectively with these aspects of the classification.

Our discussion suggests that a successful attack on a general classification problem requires techniques capable of forcing the internal structure of the simple group under investigation to resemble that of a *known* simple group. Only after such internal resemblance has been established—in the passage from "resemblance" to "isomorphism"—is character theory potentially a useful tool. However, as the classification of simple groups progressed, it has, in fact, turned out that this passage from resemblance to isomorphism has in most situations been achieved without recourse to character theory. Indeed, in the end, the applicability of character theory has been confined precisely to the "small" simple groups.

Our aim is surely not to minimize the importance of character theory, for its role in the solution of the small classification problems and in the analysis of certain sporadic groups has been crucial. Rather, our purpose is to explain the need for other techniques for analyzing the *internal subgroup structure* of the general finite simple group. Such techniques were indeed developed over the last twenty years, forming the basis for what has come to be called *local group-theoretic analysis* (for brevity, *local analysis*). It is this, the major portion of simple group theory, that has remained a mysterious subject to the outside mathematical world and which will occupy most of our attention.

Likewise, the example of simple $(p, q, r)$-groups was designed to show that despite the smallness of the simple $K_3$-groups, their full classification involves the same broad considerations that must be faced in classifying all finite simple groups. To reinforce this point, we note that as an independent effort, simple $(p, q, r)$-groups have never been fully classified; that classification is obtained only as a corollary of the complete classification of all finite simple groups! (Thompson's classification of the minimal simple groups [284] shows that only the four triples $(p, q, r)$ associated with the simple $K_3$-groups need to be considered, namely, $p=2$, $q=3$, and $r=5$, 7, 13, or 17. Simple $(2,3,13)$-groups were determined by Kenneth Klinger [197] and simple $(2,3,17)$-groups by Geoffrey Mason [209], the answer being, respectively, $L_3(3)$ and $L_2(17)$. However, simple $(2,3,5)$- and $(2,3,7)$-groups were never fully treated as independent problems.)

The origins of local analysis are contained in the proof of the solvability of groups of odd order, a further indication of the quintessential significance of the Feit–Thompson theorem. [Intimations of the method appear in Thompson's doctoral thesis, in which he proved the celebrated conjecture of Georg Frobenius that a finite group which admits an automorphism of prime period leaving only the identity element fixed is necessarily nilpotent (D9)].

When Feit and Thompson began their attack on the general odd-order problem, they had a conceptual framework for a proof, based on a special case previously successfully treated by Suzuki, who had studied simple groups of odd order in which the centralizer of every nonidentity element is assumed to be abelian. In this situation, it was very easy to determine the structure of *every* maximal subgroup $M$ of $G$ (D10).

1. $M=AK$, where $K$ is an abelian normal subgroup of $M$, $A$ is cyclic, $A\neq 1$, and every nontrivial element of $A$ induces by conjugation an automorphism of $K$ leaving only the identity fixed ($M$ is an example of a *Frobenius* group with *kernel* $K$ and *complement* $A$).
2. $K$ has order relatively prime to its index in $G$. One says that $K$ is a *Hall* $\pi$-subgroup of $G$ (after Philip Hall) for the set of primes $\pi$ dividing the order of $K$.
3. If $g\in G$, then $g^{-1}Kg\cap K=1$ or $K$ [$K$ is said to be a *trivial intersection* set (T.I. set) in $G$].

These are precisely the sort of tight conditions amenable to character-theoretic analysis. Indeed, at this point Suzuki applied the theory of *exceptional* characters (D11), developed by Brauer and himself, to relate certain characters of $M$ to those of $G$. This connection applies to every maximal subgroup of $G$. Choosing representatives $M_1, M_2,\ldots,M_n$ for the distinct conjugacy classes of maximal subgroups and applying the procedure for each $i$, $1\leqslant i\leqslant n$, Suzuki was able to obtain information about *all* the characters of $G$. At this point he was able to derive a contradiction by a delicate arithmetical analysis of the character values of the elements of $G$.

Suzuki's proof was highly original; in particular, it was the first example of a classification theorem that required an analysis of *every* proper subgroup of $G$. In other words, the first step consisted in obtaining a complete description of the full lattice of proper subgroups of $G$. This is the meaning of "internal resemblance" for this problem. The fact that the passage from "internal resemblance" to "isomorphism" led to a contradiction rather than to an actual simple group should not cloud the basic two-part division of Suzuki's proof.

Feit and Thompson, together with Marshall Hall, Jr., first extended the Suzuki argument to the case in which the centralizer of every nonidentity element is assumed to be nilpotent (abelian being a very special case of nilpotent). Again they were able to determine the lattice of proper subgroups of $G$ (but with a more complicated argument than in the Suzuki case), again proving that every maximal subgroup of $G$ was a Frobenius group with kernel a (nilpotent) Hall $\pi$-subgroup and a T.I. set in $G$. Again this put them in a position to apply exceptional character theory to reach an arithmetic contradiction as before (although somewhat more difficult calculations were involved).

Thus it was evident to Feit and Thompson, when they approached the general odd-order problem, that their first task was to pin down as precisely as possible the structure and embedding of the maximal subgroups of a minimal counterexample $G$. But now there were no special assumptions on centralizers of elements: indeed, these centralizers could now be arbitrary *solvable* groups (of odd order). This aspect of the problem turned out to be extremely formidable; in the process of carrying through the analysis, Thompson was forced to develop totally new techniques, which have come to play a central role in the study of simple groups. In particular, he established the first so-called "transitivity" theorems and "uniqueness" theorem from which the entire basic "signalizer functor method" later evolved. Thompson's approach looked very elementary: just take normalizers (and centralizers) of various nontrivial subgroups of prime power order and analyze their relationships. These normalizers, being proper subgroups of $G$, were therefore solvable and so their general structure was given by P. Hall's classical theory of solvable groups, which deals primarily with the so-called *extended* Sylow theorems [156–159] (also cf. [130, Chapter 6)]. Despite its apparent simplicity, Thompson's analysis required the most elaborate and ingenious study of the full subgroup structure of $G$ to achieve its objective.

However, he was now no longer able to show that every maximal subgroup of $G$ was a Frobenius group—this time several "approximate" Frobenius structures were also possible. Furthermore, only a weakened form of the T.I. property could now be established. The resulting structure—looser than in the prior special cases—created considerable obstacles for the applicability of exceptional character theory. To overcome them, Feit was forced to develop an elaborate extension of the entire Brauer–Suzuki theory of exceptional characters. Unfortunately, an even greater obstacle awaited Feit and Thompson, for one of the final configurations of maximal subgroups completely eluded the hoped-for arithmetic contradiction. It was a full year before Thompson found a way of eliminating this last configuration—by a remarkable analysis of generators and relations arising from the critical subgroups of the configuration. Ultimately the problem reduced to a question concerning the number of solutions of a certain equation with coefficients in a finite field; and its resolution led to a final contradiction. (A more detailed outline of the proof will be given in the sequel.)

Some time after the Feit–Thompson theorem, Alperin coined the term *local* subgroup for the normalizer in $G$ of a nonidentity subgroup of prime power order $p^n$ ( *p-local* if one wishes to distinguish the prime $p$). Gradually the term *local group-theoretic analysis* came to refer to the portion of the study of simple groups that deals with the determination of the local subgroup structure of $G$, primarily to obtain information about the structure and embedding in $G$ of (a) maximal subgroups, (b) centralizers of involutions, and (c) centralizers of elements of odd prime order.

To say that a simple group $G$ "internally resembles" a known simple group $G^*$, we mean that at least certain subgroups of $G$ of type (a), (b), or (c) have a similar shape to the corresponding subgroups of $G^*$. In the odd-order problem the critical set of subgroups was the complete set of maximal subgroups. However, as we shall see later, the structure and embedding of an appropriate *single* centralizer of an involution or element of odd prime order will often suffice for the purposes of internal resemblance.

The *techniques* of local analysis are the methods used to attain these objectives. Even though the odd-order theorem, by showing that a simple group must have even order, strikingly reinforced Brauer's contention that the structure of a simple group is intimately connected with its involutions, its proof could not directly furnish techniques for studying centralizers of involutions, since a group of odd order has no involutions.

Such new insights were provided by Thompson's classification of minimal simple groups—a natural next problem for him to consider, since the odd-order theorem can be properly viewed as the classification of minimal simple groups of *odd* order. Since by definition every proper subgroup of a minimal simple group is solvable, so obviously is each of its local subgroups. But, by its very nature, local analysis involves only interconnections between the local subgroups of $G$; so with a little extra effort, Thompson was able to extend his arguments to the broader class of simple groups, each of whose local subgroups is solvable (for brevity, such a group is called an *N-group*). [For example, $L_2(25)$ is an $N$-group but is not minimal simple since it contains the simple group $L_2(5)$ of order 60 as a subgroup.]

The final result, as in the case of Brauer's theorem mentioned in the Introduction, is typical of classification theorems: there is a "generic" solution–in this case, certain families of groups of Lie type of Lie rank 1 (these being the "minimal" such groups); and, apart from this, a few individual additional groups that satisfy the given hypothesis. The precise statement of Thompson's theorem is as follows:

THEOREM. *If G is a simple N-group, then G is isomorphic to one of the following groups*:

(1) $L_2(q)$, $q>3$.
(2) $Sz(q)$, $q=2^{2n+1}$, $n\geqslant 1$.
(3) $L_3(3)$, $U_3(3)$, ${}^2F_4(2)'$, $A_7$, *or* $M_{11}$.

Here, $Sz(q)$ denotes the family of simple groups discovered by Suzuki and related to the 4-dimensional symplectic groups over $GF(q)$; and ${}^2F_4(2)'$ denotes the derived group of ${}^2F_4(2)$, the smallest member of the family of simple groups discovered by Rimhak Ree, related to the exceptional groups $F_4(q)$, $q=2^{2n+1}$, $n\geqslant 0$, of Lie type. [The group ${}^2F_4(2)$ itself is not simple, but its derived group, of index 2, is simple. See Chapter 2 for detailed definitions.]

Again, Thompson's strategy was to show first that an arbitrary simple $N$-group $G$ internally resembles one of the listed groups and then to prove in each case that this resemblance leads to actual isomorphism. Here one could see for the first time how the local analytic ideas of the odd-order paper could be used and further developed to yield information about the structure of the centralizers of involutions and maximal 2-local subgroups of $G$. In particular, the fundamental concept of a "strongly embedded" subgroup, one of the most crucial tools of simple group theory, has its origins in the $N$-group analysis (see Section 4.2). (Strong intimations of the idea appear in earlier works of Feit and Suzuki.) A fuller discussion of the $N$-group paper and its significance for the study of simple groups will be given in the sequel.

In the analysis of $N$-groups (as in groups of odd order), every critical subgroup is solvable. However, in more general classification problems, local subgroups may well be nonsolvable. The first theorem involving nonsolvable local analysis was the determination of all simple groups with dihedral Sylow 2-subgroups, which Walter and I obtained in the early 1960s [140, 141]. (Thompson uses the dihedral result in his $N$-group analysis.)

THEOREM. *If G is a simple group with dihedral Sylow* 2-*subgroups, then* $G\cong L_2(q)$, *q odd*, $q>3$, *or* $A_7$.

To carry through the analysis, Walter and I were forced to use many specific properties of the groups $L_2(q)$, $q$ odd, and $A_7$, inasmuch as these groups could occur as composition factors of local subgroups in a minimal counterexample. Thus our proof strongly indicated that any broad extension of Thompson's local analytic methods would require the prior verification of many properties of the known simple groups. And so it has turned out, for underlying the complete classification of simple groups is an elaborately worked-out theory of *K-groups*—finite groups whose composition factors are among the known simple groups.† (See Sections 4.14, 4.15).

Let us finally sum up the broad features of the classification of simple groups. Just as in the case of simple $(p, q, r)$-groups, considered above, we can formulate the general problem as follows:

*If G is a finite simple group, each of whose proper subgroups is a K-group, prove that G is a K-group.*

Again, at the outset, the local subgroups of $G$ may have arbitrarily complex structure. As we have repeatedly emphasized, the goal of the first part of the local analysis is to prove, to the contrary, that some portion of the local structure of $G$ is very restricted and, in fact, resembles that of the corresponding portion of some known simple group $G^*$. Once internal resemblance is established, there follows the task of turning this resemblance into an isomorphism of $G$ with $G^*$.

To fully understand the nature of the classification proof, it is best to divide the passage from resemblance to isomorphism itself into two distinct parts. The first part of that analysis involves a refinement of the initial internal resemblance to an isomorphism (in some cases, only to an "approximate isomorphism") of some portion of the subgroup structure of $G$ with that of the corresponding portion of that of some known simple group $G^*$. (It is in this phase of the overall proof that character theory may be needed.) There thus remains the problem of turning this "internal isomorphism" into an actual isomorphism of $G$ with $G^*$.

Clearly to make this final deduction, $G \cong G^*$, we must not only know that the group $G^*$ exists, but must also be able to "identify" $G^*$ among all finite groups from the given set of *internal* conditions specified by the previous phase of the proof. Thus ultimately local analysis (or local analysis plus character theory) reduces the classification problem to questions concerning the *existence* and *uniqueness* of the known simple groups. Whatever the stipulated form of these internal conditions, they can always be equivalently expressed in terms of a set of defining relations among a set of

†Of course, now that the classification has been completed, the distinction between "finite group" and "$K$-group" disappears. However, to discuss the classification proof itself, it is clearly preferable to separate these two notions.

generators of the group $G^*$.

This division should also help to explain why the discovery of many of the sporadic groups has occurred in two stages. The "evidence" stage of the analysis starts from an internal resemblance and turns it into an internal isomorphism of the given group $G$ with an as yet unknown simple group $G^*$. The ensuing "construction" stage involves a proof of the existence (and usually also uniqueness) of a group $G^*$ satisfying the given set of internal conditions.

Summarizing, we see that the proof of a general classification theorem occurs in three stages:

A. Internal resemblance of $G$ with some "known" simple group $G^*$.
B. Internal isomorphism of $G$ with $G^*$.
C. Actual isomorphism of $G$ with $G^*$.

We emphasize that this last chapter of simple group theory, involving the existence and uniqueness of the known simple groups, is not directly part of local group-theoretic analysis, but is rather to be regarded as the external framework that delineates its boundaries.

A final comment about the odd-order theorem. Despite the fact that there are no (nonabelian) simple groups of odd order, the Feit–Thompson proof of this assertion has the same three-part division as any general classification theorem concerning simple groups (but with the final generator–relation stage leading to a contradiction rather than to an isomorphism). This is yet further evidence of its fundamental significance for the classification of simple groups.

This discussion should help to place our overall organizational structure in perspective. For before one can take up the classification itself, one must have a precise description of the known simple groups, characterizations of them by internal conditions, and an understanding of the major methods and results of local analysis needed for the proof.

## Definitions for Section 1.1

**D1.** A *representation* of a group $G$ *over* a field $F$ is a homomorphism $\phi$ of $G$ into the group of nonsingular linear transformations of some (finite-dimensional) vector space $V$ over $F$. The *degree* $n$ of $\phi$, denoted by $\deg(\phi)$ is the dimension of $V$ over $F$. $\phi$ is *ordinary* if $F$ is the field $\mathbb{C}$ of complex numbers (more generally, if $F$ is any algebraically closed field of characteristic 0); and $\phi$ is *modular* if $F$ has prime characteristic $p>0$. With respect to a basis $(v)=v_1, v_2, \ldots, v_n$ of $V$, the linear transformations $\phi(g)$ for $g\in G$ can be expressed in terms of $n\times n$ matrices $\phi(g)_{(v)}$ with entries in $F$. In the complex case, the function $\chi$, which assigns to each $g\in G$ the *trace* of the matrix $\phi(g)_{(v)}$, is called an (ordinary) *character* of $G$ (and also of $\phi$). Since similar matrices have the same traces, $\chi(g)$ is completely determined from $\phi(g)$, independent of the particular basis $(v)$. Deg$(\phi)$ is also called the *degree* of $\chi$ and denoted by deg$(\chi)$. *Modular* characters are more difficult to describe, but they are certain analogously defined *complex-valued* functions on $G$ associated with its modular representations (cf. [84, 88, 126, 182]).

**D2.** A representation $\phi$ of $G$ on a vector space $V$ over a field $F$ is *reducible* if $\phi(G)$ leaves invariant a subspace $W$ of $V$ with $0<W<V$; otherwise $\phi$ is *irreducible*. If $F=\mathbb{C}$, the character $\chi$ of $\phi$ is correspondingly said to be *reducible* or *irreducible*. A group $G$ has only a finite number of distinct irreducible characters (equal to the number of distinct conjugacy classes of elements of $G$; cf. [130, Theorem 3.6.14]). For a fixed prime $p$ these irreducible characters are partitioned into subsets, called *blocks*; two characters $\chi_1$ and $\chi_2$ lie in the same block if and only if for all $g\in G$ the character values $\chi_1(g)$, $\chi_2(g)$ satisfy a suitable congruence modulo the prime $p$ (cf. [84, 88, 126, 182]).

**D3.** If $\Sigma_n$ is the symmetric group on the set $\Omega=\{1,2,\dots,n\}$, every element $x$ of $\Sigma_n$ can be written as a product of *transpositions* [i.e., permutations $(ij)$ interchanging a pair of letters $i$, $j$ and fixing the remaining $n-2$ letters]. Moreover, the number of transpositions in any two such representations of $x$ has the same parity. According as this common parity is 0 or 1, $x$ is called *even* or *odd*. The set of even permutations forms a subgroup of index 2 in $\Sigma$, called the *alternating* group and denoted by $A_n$. As is well known, $A_n$ is simple for $n\geqslant 5$.

**D4.** A 2-group $S$ is (generalized) *quaternion* or *dihedral*, respectively if $S$ is generated by elements $x, y$ subject to the relations $x^{-1}yx=y^{-1}$ and correspondingly $x^4=y^{2^n}=1$, $x^2=y^{2^{n-1}}$, $n\geqslant 2$, or $x^2=y^{2^n}=1$, $n\geqslant 1$.

**D5.** Here, for $n\geqslant 3$, $U_n(q)=PSU(n,q)=SU(n,q)/(\text{mod scalars})$ is the *projective special unitary* group over $GF(q)$ [see Section 2.1 for the definition of $SU(n,q)$]. $U_n(q)$ is simple for $q\geqslant 3$ or $n\geqslant 4$.

**D6.** A normal series $G=G_1>G_2>\cdots>G_n=1$ of the group $G$ is called a *composition* series if each $G_i$ is maximal subject to being normal in $G_{i-1}$. In this case the *composition* factors $G_{i-1}/G_i$, $1\leqslant i\leqslant n-1$, are necessarily simple. A theorem of Schreier asserts that every normal series (with distinct members) can be "refined" to a composition series [130, Theorem 1.2.7]. Moreover, by the well-known Jordan–Hölder theorem any two composition series of $G$ have the same "length" and the corresponding composition factors are (up to permutation) isomorphic in pairs [130, Theorem 1.2.8]. As a result, one speaks of *the* composition factors of $G$. (Thus $G$ is solvable if and only if its composition factors are all of prime order.)

**D7.** For any prime $p$, if $p^a$ is the exact power of $p$ dividing the order of a group $G$, Sylow's theorem asserts that $G$ contains a subgroup of order $p^a$ (called a *Sylow* $p$-subgroup of $G$), any two Sylow $p$-subgroups of $G$ are conjugate, and the number of Sylow $p$-subgroups of $G$ is congruent to 1 modulo $p$ [130, Theorem 1.2.9]. (Two subgroups or subsets $X$, $Y$ of $G$ are *conjugate* if $g^{-1}Xg=Y$ for some $g\in G$.)

**D8.** If $X$ is a subgroup of a group $G$, the *normalizer* of $X$ in $G$ is the set of $g\in G$ such that $g^{-1}Xg=X$, i.e., $g^{-1}xg\in X$ for all $x\in X$. (Thus $X$ is normal in $G$ if and only if its normalizer is $G$ itself.)

**D9.** A group $G$ is *nilpotent* if $G$ has a normal series $G=G_1\geqslant G_2\geqslant\cdots\geqslant G_n=1$ such that each $G_i$ is normal in $G$ and $G_{i-1}/G_i$ is contained in the center of the group $G/G_i$, $2\leqslant i\leqslant n$. A nilpotent group is necessarily solvable, but not conversely. Indeed, it is easily shown that $G$ is nilpotent if and only if $G$ is the direct product of its Sylow subgroups [130, Theorem 2.3.5]. In particular, $G$ is nilpotent if it has prime power order.

**D10.** A subgroup $M$ of a group $G$ is *maximal* if $M < G$ and whenever $M \leqslant X \leqslant G$ for some subgroup $X$ of $G$, then necessarily $X = M$ or $X = G$.

**D11.** If $X$ is a subgroup of a group $G$, there is a natural way of "inducing" a character $\chi$ of $X$ to obtain a character $\chi^*$ of $G$, which is obtained by inducing the representation $\phi$ of $X$ corresponding to $\chi$ to a representation $\phi^*$ of $G$ and defining $\chi^*$ to be the character of $\phi^*$. If $g_1, g_2, \ldots, g_n$ are a set of coset representatives of $X$ in $G$, one defines $\phi^*(g)$ for $g \in G$ to be the $n \times n$ block matrix whose $ij$th-block is the matrix $\phi(g_i g g_j^{-1})$, where by convention $\phi(g')$ is the 0 matrix for all elements of $G$ not in $X$.

When $X$ is a T.I. set in $G$, there is a close relationship between the values of $\chi$ and $\chi^*$ on the elements of $X$; in this case the characters of $G$ which arise from this process are called *exceptional*. Brauer and Suzuki developed the general theory and showed how to apply it to particular group-theoretic situations (cf. [277]).

## 1.2. Internal Geometric Analysis

To the nonexpert, the name Bernd Fischer is known solely for its connection with a number of sporadic simple groups, but to the practitioner, he is recognized as the founder of *internal geometric analysis*. His work has given rise to five new simple groups; supplemented by the work of his disciple Franz Timmesfeld, it has also provided a fundamental general technique for studying simple groups, which can reasonably be regarded as second in importance for the classification only to local group-theoretic analysis.

In contrast to local analysis, Fischer's work appears as a personal creation, and aspects of its subsequent development have an almost magical quality. The odd-order and $N$-group theorems give one the feeling that Thompson's "moves" were literally forced upon him—that his brilliance lay in the very inevitability of the techniques he developed to study the internal subgroup structure of simple groups. Indeed, the details of the 255-page odd-order theorem have been thoroughly analyzed and dissected in the almost twenty years since it was first proved and a great many improvements made in various portions of the argument. Yet the updated proof which Glauberman and David Sibley have been writing during the past few years and which includes all of Bender's simplifications maintains the identical conceptual framework as the original Feit–Thompson argument (and still requires approximately 150 pages). Likewise, in studying the early version of the $N$-group paper, the inexorable quality of the proof had a profound impact on my own understanding of simple groups. In particular, it convinced me that the ideas contained in it, if fully exploited, could provide a possible basis for a complete classification of simple groups.

On the other hand, in Fischer's case, originality begins with the very question he raised [96, 97, 99, 100]:

*Which finite groups can be generated by a conjugacy class* (D12) *of involutions, the product of any two of which has order* 1, 2, *or* 3?

The question is natural enough since every symmetric group has this property: indeed, each is generated by its transpositions $(ij)$, and the product $(ij)(mn)$ of the two transpositions $(ij),(mn)$ has order 1, 2, or 3 according as $\{i, j\} \cap \{m, n\}$ consists of 2, 0, or 1 element, respectively [in the last case, if, say, $j = m$, then $(ij)(mn)$ is the 3-cycle $(ijn)$].

However, there have been many equally "natural" group-theoretic questions which have turned out to be either totally intractable, manageable only for small groups, or of little or no significance for the classification of simple groups. Respective examples are the following: (a) Can all simple groups be generated by an element of order 2 and an element of odd prime order? (b) Which simple groups have an irreducible complex representation of degree $n$? (c) Can the known simple groups be characterized in terms of their character tables (D13)?

What distinguished Fischer's question from the preceding was the fact that it was amenable to a complete solution, involving powerful techniques capable of deep generalization. A more detailed discussion of Fischer's contributions will be given in Chapter 2, but I should like to describe the general features of his theory—in particular, to justify the term "geometric" in its title. Let then $G$ be an arbitrary group generated by a conjugacy class $D$ of involutions, the product of any two of which has order 1, 2, or 3. Fischer calls $D$ a class of 3-*transpositions*. There is a natural way to associate a "geometry" with $G$ on which $G$ "acts as a group of automorphisms." Indeed, consider the *graph* $\Gamma$ whose vertices are the elements of the class $D$ with two elements $x, y \in D$ connected by an edge if and only if they commute. Since $D$ is a conjugacy class in $G$, $g^{-1}xg \in D$ for any $x \in D$ and $g \in G$. Moreover, if $x, y \in D$ commute, then so do the elements $g^{-1}xg$ and $g^{-1}yg$. Thus, under conjugation the elements of $G$ permute the vertices of $\Gamma$ and preserve the given incidence relation on $\Gamma$, so by definition $G$ acts as a group of automorphisms of $\Gamma$.

Clearly then the structure of $G$ is intimately related to the geometry of $\Gamma$. Of course, this same construction can be carried out for any group $G$ and any conjugacy class or union of conjugacy classes. However, what gives the geometry here its rich shape is Fischer's basic hypothesis on the order of the products of pairs of elements of $D$. As a first step in his analysis, Fischer investigates the "root" subsets of $D$—maximal subsets $E$ of *mutually* commuting elements of $D$, i.e., subgraphs $\Gamma_E$ of maximal size with the property that each pair of vertices is connected by an edge—and studies the *stabilizer* in $G$ of $\Gamma_E$, equivalently, the normalizer $N$ of $E$ in $G$. Eventually, he pins down all allowable possibilities for $N$; in particular, he shows that $N$ acts doubly transitively on the set $E$. But Fischer's deepest hold on the geometry arises from the remarkable fact that his hypothesis carries over to certain subgroups of $G$ related to $D$, thus permitting him to use induction in the analysis: namely, for $x \in D$, let $D_x$ be the subset of elements of $D$ commuting with $x$, but distinct from $x$, i.e., the set of vertices of $\Gamma$ *connected* to $x$. Fischer proves that the subgroup $D_x^*$ of $G$ generated by the elements of $D_x$ is itself a 3-transposition group with respect to the set $D_x$, i.e., $D_x$ is a conjugacy class of involutions of $D_x^*$ (obviously the condition on products carries over by restriction to $D_x$). We hope this brief discussion gives some flavor of Fischer's internal geometric analysis.

Fischer's influence on Timmesfeld was direct, but his work also had a strong, though more indirect, impact on Aschbacher. Indeed, one of Aschbacher's first major group-theoretic results was an extension of Fischer's classification of 3-transposition groups to *odd* transposition groups, i.e., groups generated by a conjugacy class of involutions, the product of any two of which has order 1, 2, or some odd number. But beyond this, his fundamental "classical involution" theorem [13], for which he won the Cole prize in algebra in 1980, can be viewed as a study of the geometry associated with a conjugacy class of classical involutions (D14).

In addition, Aschbacher realized that the key results on small groups—the classification of groups with a "proper 2-generated core" and of groups with a "nonconnected Sylow 2-subgroup"—could be rephrased in terms of the connectivity of a suitable graph $\Gamma$, whose vertices are the *Klein four subgroups* of the given group $G$ (i.e., subgroups isomorphic to $Z_2 \times Z_2$), again with two vertices connected by an edge if and only if they commute (elementwise), thus conceptualizing this basic area of simple group theory. We say that $G$ is *connected* or *nonconnected* according as the corresponding graph is connected or nonconnected. Essentially, the meaning of the term "small" group, which we have introduced here solely for its suggestive qualities, is one whose associated graph is disconnected. This will be described in more detail in Section 1.5.

## Definitions for Section 1.2

**D12.** In any group $G$, the relation of conjugacy partitions the elements into disjoint classes, called *conjugacy classes*. (Thus $x, y \in G$ lie in the same conjugacy class if and only if $g^{-1}xg = y$ for some $g \in G$.)

**D13.** For any group $G$, the number $r$ of irreducible complex characters of $G$ is equal to the number of conjugacy classes of $G$ [130, Theorem 3.6.14]. Since conjugate matrices have the same traces, the characters of $G$ are constant on conjugacy classes. Thus if $\chi_i$, $1 \leqslant i \leqslant r$, denote the distinct irreducible complex characters of $G$ and if $g_j$, $1 \leqslant j \leqslant r$, are representatives of the distinct conjugacy classes of $G$, then the $r \times r$ matrix with entry $\chi_i(g_j)$ in the $ij$th position provides a complete description of the characters of $G$. It is called the *character table* of $G$.

**D14.** Let $t$ be an involution of a group $G$, let $C$ be its centralizer in $G$ and set $\bar{C} = C/O(C)$, where $O$ is the unique largest normal subgroup of $C$ of odd order. $t$ is called *classical* provided $\bar{C}$ contains a subgroup $\bar{K} \cong SL(2, q)$ for some odd prime power $q$ such that

(a) $\bar{t} \in \bar{K}$ [$SL(2, q)$, $q$ odd, has a unique involution: namely, $\begin{pmatrix} -1 & 0 \\ 0 & -1 \end{pmatrix}$]; and

(b) $\bar{K}$ is a "component" of $\bar{C}$, i.e., the smallest normal subgroup of $\bar{C}$ containing $\bar{K}$ is the product of pairwise centralizing subgroups, each isomorphic to $\bar{K}$.

Nearly all groups of Lie type over fields of odd characteristic contain a classical involution. Aschbacher's marvelous theorem is essentially a converse, giving a complete determination of all simple groups possessing such an involution: apart from a few exceptional cases, they are indeed just the groups of Lie type of odd characteristic. Thus his theorem "characterizes" the groups of Lie type of odd characteristic from the structure of the centralizer of a *single* involution. Aschbacher's proof will be outlined in the sequel.

## 1.3. Why the Extreme Length?

An often expressed feeling in the general mathematical community is that the present approach to the classification of simple groups must be wrong—no single theorem can possibly require a 10,000-page proof! Since most of the known simple groups are finite analogues of Lie groups, one should be able to build a geometry from suitable internal properties of a simple group $G$ and then determine $G$ from its action on this geometry. This is the most frequently suggested alternate approach. If successful, it would reduce the classification to the presumably more tractable problem of classifying the corresponding geometries. After all, this is exactly how the classification of complex simple Lie algebras proceeds (and with it the classification of simple Lie groups): the entire problem is reduced to the analysis of connected "Dynkin diagrams" (see Section 2.1)—equivalently to the solution of a certain question about sets of vectors in Euclidean $n$-space.

Just as there are five exceptional complex simple Lie algebras and Lie groups—$G_2$, $F_4$, $E_6$, $E_7$, $E_8$—so one could visualize certain exceptional geometries arising from the finite simple groups, each of which would lead to one of the sporadic groups.

The suggestion is so plausible, why has it not been possible to proceed along such lines to study simple groups? Let me attempt an answer. Consider a minimal counterexample $G$ to the proposed classification theorem, so that every proper subgroup of $G$ is a $K$-group. On the other hand, let $G^*$ be a completely arbitrary $K$-group, as far from being simple as you can conceive. Is there any *a priori* reason why our minimal counterexample $G$ should not have the identical lattice of *proper* subgroups as $G^*$? If this is so, then the *internal* structure of $G$ bears no resemblance to that of any simple group but rather is equivalent to that of an *arbitrary* $K$-group. But then if we attempted to construct a geometry directly from $G$, the resulting geometry would be identical to one constructed from $G^*$. Thus, if the suggested approach were to be followed at the very outset, it would necessarily generate as many distinct geometries as there are finite $K$-groups. The classification of such geometries would appear to be a hopeless task.

The implication of this discussion is the following: one cannot expect a geometric approach to be effective until *after* one uses the *simplicity* of $G$ to show that its internal structure is, in fact, much more restricted than that of the general $K$-group $G^*$—indeed, until one shows that its internal structure "resembles" that of a *simple* $K$-group. It is precisely this resemblance that local group-theoretic analysis has as its primary objective. Once $G$ has been shown to resemble internally a simple $K$-group $G^*$, the proof that $G$ is actually isomorphic to $G^*$ can be regarded as involving the construction of a "geometry."

By far the largest portion of the classification of simple groups is taken up with this resemblance phase of the proof. The elaborateness of the task has no analogue in the classification of complex simple Lie algebras, since there the nondegeneracy of the Killing form is such a powerful criterion for semisimplicity that it quickly reduces the analysis to the geometric problem just described. I believe it is the very brevity of this reduction that makes one so totally unprepared for the degree of complexity of the corresponding problem for simple groups. I might add that the situation in the as yet uncompleted classification of finite-dimensional simple Lie algebras over algebraically closed fields of characteristic $p > 0$, in terms of the difficulty of carrying out the reduction, seems closer to that of simple groups than to that of complex simple Lie algebras.

If, as we are suggesting, the local analytic approach to the determination of the simple groups is the only available one, then the question of the length of the classification proof reduces to the amount of work involved in settling the following two problems:

A. Proving that $G$ resembles internally a simple $K$-group $G^*$.
B. Proving that whenever $G$ resembles such a group $G^*$, then $G$ must be isomorphic to $G^*$.

Clearly, answers to A and B in turn involve the following two subsidiary questions:

I. How uniform are the internal structures of simple $K$-groups? In other words, how many distinct "types" of simple $K$-groups must one consider?
II. For those groups $G$ that will end up isomorphic to a simple $K$-group $G^*$ of a given type, can the analysis under A and B, respectively, be carried out uniformly for each of the given types?

At the minimum it seems necessary to consider the following families of $K$-groups to be of distinct types (the known simple groups will be described in detail in Chapter 2).

1. The groups of prime order.
2. The alternating groups.
3. The groups of Lie type over $GF(q)$, $q$ odd.
4. The groups of Lie type over $GF(q)$, $q$ even.
5. The 26 sporadic groups.

The distinction between 3 and 4 is an essential one, for in a group $G^*$ of Lie type defined over $GF(q)$, an involution of $G^*$ corresponds to a *semisimple* element or a *unipotent* element according as $q$ is odd or even. Correspondingly the centralizers of these involutions have totally distinct structures (see Section 1.5).

Moreover, in treating many situations, one must further subdivide the groups of Lie types as follows:

a. The classical groups: linear, symplectic, orthogonal.
b. The exceptional groups: $G_2$, $F_4$, $E_6$, $E_7$, $E_8$.
c. The "algebraic twisted" groups: unitary groups, "triality" $D_4$, twisted $E_6$ (also finite analogues of real orthogonal groups).
d. The "exceptional" twisted groups: Suzuki groups, Ree groups of characteristic 3, Ree groups of characteristic 2.

Some collections of sporadic groups arise from a single context (the first two Mathieu groups $M_{11}$, $M_{12}$; the remaining three Mathieu groups

$M_{22}, M_{23}, M_{24}$; Janko's second and third groups $J_2, J_3$; the three Conway groups .1,.2,.3; the three Fischer groups $M(22)$, $M(23)$, $M(24)'$; the group $F_1$ and its three subgroups, the Fischer group $F_2$, Thompson's group $F_3$, and Koichiro Harada's group $F_5$), but from the point of view of their classification by internal properties, the 26 sporadic groups are best considered to be of distinct types.

The above subdivision gives rise to roughly 40 distinct types of simple $K$-groups (over half representing individual groups). This means that at various stages in carrying through the analysis of A and B above, up to 40 distinct cases may have to be treated.

In addition, many other cases unrelated to any known simple group must often be considered. For example, the final configuration of the odd-order paper is as "tight" as any in which $G$ has an internal structure resembling a simple $K$-group $G^*$. Thus it is not so farfetched to say in this case that the internal structure of $G$ resembles that of a *nonexistent* simple group. The point is that the analysis one is often forced to take to eliminate a particular configuration is in spirit not unlike that which one carries out in "real" cases, in showing either that $G$ internally resembles some simple $K$-group $G^*$ or, at a later stage, that $G$ is isomorphic to $G^*$.

As already indicated, special methods are required for investigating small simple groups. But beyond that, the analysis in these cases is extremely complicated and each specific problem involves its own special configurations and its own ad hoc arguments. For example, the determination of the simple groups with quasi-dihedral or wreathed Sylow 2-subgroups takes over 300 pages. Yet apart from the actual result itself and a few general ideas of the proof, no specific proposition or lemma has ever been utilized in any other classification problem. The complete analysis of small groups may run to almost 3,000 journal pages, and, in addition, involves considerable computer time for the construction and proof of uniqueness of certain sporadic groups.

Here then is an "insider's" explanation of the enormous length of the classification proof. The presently contemplated revisionist approach to the classification of simple groups, if successful, may well succeed in condensing the proof by a factor of, say, *three*; however, a tenfold reduction would certainly require some totally new ideas.

## 1.4. Some Standard Terminology and Results

By now the reader undoubtedly realizes that even the most superficial discussion of simple groups requires a great many basic group-theoretic concepts. I want to conclude this introduction by making more precise the major subdivisions of the classification proof; but to do so, I need many additional standard definitions and terminology. Rather than introduce these in a further sequence of footnotes, it seems preferable to review them systematically at this time along with some equally standard basic results that we shall need throughout the book. I apologize for the length of the list, but it is impossible to discuss the division meaningfully without a reasonable vocabulary. The reader unfamiliar with these terms will probably do best to assimilate this material as it comes up in the text.

We let $X$ be an arbitrary finite group.

$X^{\#}$ denotes the set of nonidentity elements of $X$.

$Y \leqslant X$ means $Y$ is a subgroup of $X$.

$Y < X$ means $Y$ is a proper subgroup of $X$, i.e., $Y \leqslant X$ and $Y \neq X$.

$Y \lhd X$ means $Y$ is a normal subgroup of $X$.

$Y \lhd \lhd X$ means $Y$ is a *subnormal* subgroup of $X$, i.e., $Y = Y_1 \lhd Y_2 \lhd Y_3 \lhd \cdots \lhd Y_n = X$ for suitable subgroups $Y_i$ of $X$, $1 \leqslant i \leqslant n$.

$Y$ char $X$ means $Y$ is a *characteristic* subgroup of $X$, i.e., $Y$ is invariant under all automorphisms of $X$.

$Y$ is a *section* of $X$ means $Y = A/B$, where $A, B \leqslant X$, $B \lhd A$.

$Y$ is *involved* in $X$ means $Y$ is isomorphic to a section of $X$.

$Y$ *covers* a section $A/B$ of $X$ means $Y \leqslant X$ and $A = B(Y \cap A)$.

$|S|$ denotes the cardinality of the set $S$. In particular, $|Y|$ is the order of any subset $Y$ of $X$.

$|X:Y|$ denotes the *index* of the subgroup $Y$ of $X$, i.e., $|X|/|Y|$. (By Lagrange's theorem, $|Y|$ divides $|X|$).

$\langle Y \rangle$ denotes the unique smallest subgroup of $X$ containing the subset $Y$ of $X$. $\langle Y \rangle$ is the intersection of all subgroups of $X$ containing $Y$. If $\langle Y \rangle = X$, $Y$ is said to *generate* $X$ or to be a *set of generators* of $X$.

$C_X(Y)$ and $N_X(Y)$ denote, respectively, the centralizer and normalizer of the subset $Y$ of $X$.

$\langle Y^X \rangle$ denotes the *normal closure* in $X$ of the subset $Y$ of $X$, i.e., the subgroup generated by all conjugates of $Y$ in $X$.

$Z(X)$ denotes the center of $X$.

$Syl_p(X)$ denotes the set of Sylow $p$-subgroups of $X$ for the prime $p$.

$Sol(X)$ denotes the unique largest normal solvable subgroup of $X$.

$F(X)$ denotes the *Fitting* subgroup of $X$, the unique largest normal nilpotent subgroup of $X$. $F(X)$ is the join of all normal nilpotent subgroups of $X$.

$\phi(X)$ denotes the *Frattini* subgroup of $X$, which is the intersection of the maximal subgroups of $X$.

$\mathrm{Aut}(X)$ denotes the automorphism group of $X$.

1 denotes the identity subgroup and element of $X$ as well as the trivial group with one element.

$X$ *acts* on the group $Y$ means that there is a homomorphism of $X$ into $\mathrm{Aut}(Y)$. In particular, if $Y \lhd X$, then $X$ acts on $Y$ by conjugation.

$AB = \{ab \mid a \in A, b \in B\}$ for any subsets $A$, $B$ of $X$. The notation extends in a natural way to any finite number of subsets of $X$.

$A^B = \{b^{-1}ab \mid a \in A, b \in B\}$ for any subsets $A$, $B$ of $X$.

$[a, b] = a^{-1}b^{-1}ab = a^{-1}a^b$ for any elements $a, b \in X$.

$[A, B] = \langle [a, b] \| a \in A, b \in B \rangle$ for any subsets $A$, $B$ of $X$.

$[a, b, c] = [[a, b], c]$ for any elements $a, b, c \in X$.

Inductively, one defines commutators of arbitrary length, likewise for sequences of subgroups.

$X' = [X, X]$ denotes the commutator subgroup of $X$. If $X = X'$, $x$ is *perfect*.

The chain of subgroups $X, X', (X')', \ldots$ is called the *derived series* of $X$. It is easily shown that $X$ is solvable if and only if its derived series terminates in the identity.

The chain of subgroups $X, X', [X', X], [[X', X], X], \ldots$ is called the *lower central series* of $X$. Likewise, it is easily shown that $X$ is nilpotent if and only if its lower central series terminates in the identity.

If $X$ is nilpotent, the *class* of $X$ is one less than the number of terms in its lower central series. Thus an abelian group is nilpotent of class 1.

$Z_n$ denotes the cyclic group of order $n$.
$Q_{2^n}$ denotes a (generalized) quaternion group of order $2^n$, $n \geqslant 3$.
$D_{2n}$ denotes a dihedral group of order $2n$, $n \geqslant 2$.

A group of order 2 power of a prime $p$ is called a *p-group*. An abelian $p$-group $X$ is *homocyclic* if it is the direct product of cyclic subgroups of the same order $p^m$. If $m = 1$, $X$ is called *elementary*. We write $E_{p^n}$ for an elementary abelian $p$-group of rank $n$. Note that $E_{p^n}$ can be identified with an $n$-dimensional vector space over the prime field $GF(p)$. Note also that, for example, $E_8$ and $Z_2 \times Z_2 \times Z_2$ are isomorphic groups.

A group isomorphic to $Z_2 \times Z_2$ is called a *four group*.

If $X$ is a $p$-group, the *rank* of $X$, denoted by $m(X)$ (not to be confused with the (permutation) rank of a permutation group), is the maximum rank of an elementary abelian subgroup of $X$.

$X$ is a *central* product of subgroups $A$, $B$ if $X=AB$ and $[A, B]=1$. The product is *direct* if, in addition, $A\cap B=1$. In the latter case, we write $X=A\times B$. On the other hand, if $A\cap B\neq 1$ [note that $A\cap B\leq Z(A)\cap Z(B)$], we write $X=A*B$. These concepts and notation extend in a natural way to finite products of subgroups of $X$.

If $A$ is a group and $B$ a subgroup of Aut($A$), the *semidirect product* or *split extension* of $B$ by $A$ is a group of the form $A^*B^*$ with $A^*\cong A$, $B^*\cong B$, $A^*\cap B^*=1$, $A^*\lhd A^*B^*$, And the action of $B^*$ on $A^*$ by conjugation is determined by the action of $B$ as a group of automorphisms of $A$. One identifies $A^*$ with $A$, $B^*$ with $B$ and writes $A\cdot B$ (or simply $AB$) for the semidirect product.

Thus, if $X=AB$ with $A\lhd X$, and $A\cap B=1$, then $X$ is a *split* extension of $A$ by $B$. Also, $B$ is called a *complement* of $A$ and vice versa. On the other hand, if $A\cap B\neq 1$, we say that the extension is *nonsplit*.

To *adjoin* an automorphism $\beta$ to a group $A$ or to *extend* $A$ by $\beta$ means simply to form the group $A\cdot\langle\beta\rangle$.

There are a number of other conventions for designating group extensions. For example, one often writes $A/B$ for a group $X$ having a normal subgroup isomorphic to $B$ with corresponding factor group isomorphic to $A$. This notation is intended to be solely schematic and is usually restricted to the case in which $A$ and $B$ are known groups, themselves having a somewhat involved expression. For example, one might write $A_5/(Q_8*D_8)$. Of course, the symbol $A/B$ customarily denotes the quotient of a group $A$ by a normal subgroup $B$. It will be clear from the context which meaning of the slash is intended.

Perfect central extensions of a known group $X$ by an element of prime order are usually designated by $\hat{X}$. Thus $\hat{A}_5$ and $A_5/Z_2$ (nonsplit) represent the same groups.

Sometimes, for brevity, one may write $(X)\cdot 2$ for an extension of the group $X$ by $Z_2$ (which may or may not split over $X$). When one uses this notation, it is assumed that some specified action of the element of order 2 on $X$ is given.

The *wreath product* of a group $A$ by a group $B$, denoted by $A\int B$, although conceptually clear, is cumbersome to express. One forms the direct product $A^*$ of $n=|B|$ copies of $A$, indexed from 1 to $n$, and identifies $B$ in a natural way with a subgroup of $\Sigma_n$.[†] The permutation action of $B$ on the set

[†]If $X=\{x_1, x_2,\ldots,x_n\}$ is a group, then for $x\in X$, the map $\phi_x$: $x_i\mapsto x_ix$, $1\leq i\leq n$, is a permutation of the set $\{x_1, x_2,\ldots,x_n\}$. If we identify this set with $\{1,2,\ldots,n\}$, then the map $\theta$: $x\mapsto\phi_x$ for all $x\in X$, is an isomorphism of $X$ with a subgroup of $\Sigma_n$. (This is Cayley's theorem.) Using $\theta$, we can identify $X$ with a subgroup of $\Sigma_n$.

$\{1,2,\ldots,n\}$ is then used to induce an automorphism action of $B$ on $A^*$, namely, for $(a_1, a_2,\ldots,a_n)\in A^*$, $a_i\in A$, $1\leqslant i\leqslant n$, and $b\in B$, one sets

$$(a_1, a_2,\ldots, a_n)^b=(a_1', a_2',\ldots, a_n'),$$

$$\text{where } a_i'=a_j \text{ if } i=j^b,\ 1\leqslant i,j\leqslant n. \tag{1.1}$$

Finally, using this action, one defines the wreath product $A\int B$ to be the semidirect product $A^*\cdot B$.

For example, if $|B|=2$, then $A\int B$ consists of the direct product of two copies of $A$, interchanged under conjugation by the involution of $B$. In this terminology, our earlier definition of a wreathed 2-group is just $Z_{2^m}\int Z_2$ for some integer $m\geqslant 2$.

If $\pi$ is a set of primes, $\pi'$ denotes the complementary set of primes.

$O_\pi(X)$ denotes the unique largest normal subgroup of $X$ whose order is divisible only by primes in the set $\pi$. It is the join of all normal subgroups of $X$ with this property.

$O^\pi(X)$ denotes the unique smallest normal subgroup $Y$ of $X$ such that $X/Y$ has order divisible only by primes in $\pi$. It is the intersection of all normal subgroups of $X$ such that the corresponding factor group has this property.

$O_{p'}(X)$ is the *$p'$-core* of $X$, the unique largest normal subgroup of $X$ of order relatively prime to $p$, $p$ a prime.

$O(X)=O_{2'}(X)$ is the *core* of $X$, the unique largest normal subgroup of $X$ of odd order.

$X$ has a *normal $p$-complement* provided $X/O_{p'}(X)$ is a $p$-group [whence $X=O_{p'}(X)P$ for any $P\in Syl_p(X)$].

If $O_{p'}(X)=1$, $X$ is *$p$-constrained,* $p$ any prime, provided $C_X(O_p(X))\leqslant O_p(X)$. More generally, $X$ is $p$-constrained if $X/O_{p'}(X)$ is. [Note that $O_{p'}(X/O_{p'}(X))=1$.]

The following additional terminology, although not so standard, will be very useful.

$\mathcal{I}_p(X)$ denotes the set of elements of $X$ of order $p$, $p$ a prime.

$\mathcal{I}(X)=\mathcal{I}_2(X)$.

$y\in\mathcal{I}_p(X)$ is *$p$-central* if $y$ lies in the center of some Sylow $p$-subgroup of $X$.

Further notation will be introduced as we go along.

We conclude this section with some further terminology and basic facts about representations of groups, permutation groups, generators and relations, $p$-groups, and the transfer homomorphism.

D1 and D2 at the end of Section 1.1 have already given some definitions concerning representations. To add to this list, let $\phi$ be a representation of $X$ on a vector space $V$ over a field $F$, so that $\phi(X) \leqslant GL(V; F)$, the group of all nonsingular linear transformations of $V$ over $F$.

$\ker(\phi)$, the *kernel* of $\phi$, is the set of $x \in X$ such that $\phi(x)$ is the identity linear transformation on $V$. Clearly, $\ker(\phi) \lhd X$.

$\phi$ is *faithful* if $\ker(\phi) = 1$, equivalently, if $\phi(X) \cong X$.

$\phi$ induces a (*factor group*) representation $\phi^*$ of $X/\ker(\phi)$ on $V$ by the rule $\phi^*(\ker(\phi)x) = \phi(x)$ for all $x \in X$. The representation $\phi^*$ is faithful.

If $W$ is a subspace of $V$ invariant under $\phi(X)$, then $\phi$ induces (by *restriction*) a representation $\phi|_W$ of $X$ on $W$. Likewise $\phi$ induces a (*quotient*) representation $\phi_{V/W}$ of $X$ on $V/W$ by composing $\phi$ with the natural homomorphism of $V$ onto $V/W$.

We often say that $\phi$ is the "sum" of $\phi|_W$ and $\phi_{V/W}$ and that $\phi|_W$ and $\phi_{V/W}$ are *constituents* of $\phi$. Repeating this decomposition process, we can always represent $\phi$ as a sum if *irreducible constituents*. [This corresponds to a chain of $\phi(X)$-invariant subspaces $0 = W_0 < W_1 < W_2 < \cdots < W_n = V$ such that $\phi(X)$ is irreducible on $W_i/W_{i-1}$, $1 \leqslant i \leqslant n$.]

$\phi$ is *completely reducible* if $V$ is the direct sum of $\phi(X)$-invariant subspaces $V_1, V_2, \ldots, V_m$ such that $\phi|_{V_i}$ is irreducible, for all $i$, $1 \leqslant i \leqslant m$.

Not every representation of $X$ over a field $F$ is necessarily completely reducible. Maschke's well-known theorem [130, Theorem 3.3.1] gives an important sufficient condition on $F$ for this to be true.

THEOREM 1.1. *If $F$ is a field of characteristic 0 or of characteristic relatively prime to $|X|$, then every representation of $X$ on a vector space $V$ over $F$ is completely reducible.*

$\phi$ is *linear* if $V$ has dimension 1 over $F$. In this case $\phi$ is simply a homomorphism of $X$ into $F$. Clearly linear representations are always irreducible.

It is well known that irreducible representations of abelian groups over any algebraically closed field are necessarily linear. (This is a consequence of *Schur's lemma*, which asserts that if a group $X$ possesses a faithful irreducible representation, then the center $Z(X)$ of $X$ must be cyclic [130, Theorem 3.5.3]). As a corollary one has

THEOREM 1.2. *If $\phi$ is an irreducible representation of an abelian group $X$ over a field $F$, then $X/\ker(\phi)$ is cyclic.*

Distinct representations of $X$ may be "essentially" the same, in the sense that one can be obtained from the other by a suitable change of notation. Formally, we say that two representations $\phi_1$ and $\phi_2$ of $X$ on vector spaces $V_1$ and $V_2$ over a field $F$ are *equivalent* provided there is a vector space isomorphism $\psi$ of $V_1$ onto $V_2$ such that for each $g \in G$ and $v \in V_1$,

$$\phi_2(g)(\psi(v)) = \psi(\phi_1(g)(v)).$$

Clearly $\phi_1(g)$ and $\phi_2(g)$ have the same matrices with respect to a basis of $V_1$ and its image under $\psi$, respectively. In particular, it follows in the case $F = \mathbb{C}$ that equivalent complex representations of $X$ have the same characters.

The theory of complex representations and characters of finite groups was fully developed at the turn of the century by Frobenius and William Burnside. We state here only a few of the many known properties (see [130, Chapter 4]).

THEOREM 1.3. (i) *$X$ has only a finite number of inequivalent irreducible complex representations and hence only a finite number $n$ of distinct irreducible characters.*

(ii) *The integer $n$ is equal to the number of distinct conjugacy classes of $X$.*

(iii) *If $d_1, d_2, \ldots, d_n$ denote the degrees of the distinct irreducible characters of $X$, then*

$$|X| = d_1^2 + d_2^2 + \cdots + d_n^2.$$

PROPOSITION 1.4. (i) *If $\chi$ is a character of $X$, then*

$$\ker(\chi) = \{x \in X \mid \chi(x) = \chi(1) = \deg(\chi)\}.$$

(ii) *Two irreducible complex representations of $X$ are equivalent if and only if their characters are identical (i.e., have the same values for each $x \in X$).*

If $\phi$ is a representation of $X$ on a vector space $V$ over a field $F$, then the map

$$\phi^*: x \mapsto \phi(x^{-1})^t \qquad \text{for } x \in X$$

(where $t$ denotes transpose) is also a representation of $X$ on $V$, called the *dual* of $\phi$.

If $F=\mathbb{C}$, the character $\chi^*$ of $\phi^*$ is likewise called the *dual* of the character $\chi$ of $\phi$. It is immediate that

$$\chi^*(x)=\overline{\chi(x)},$$

where the bar denotes complex conjugation. In particular, $\chi$ is *self-dual* (i.e., $\chi^*=\chi$) if and only if $\chi$ is real-valued.

Clifford's well-known theorem [130, Theorem 3.4.1] gives a precise description of the decomposition of an irreducible representation with respect to a normal subgroup of $X$.

THEOREM 1.5. *Let $\phi$ be an irreducible representation of $X$ on a vector space $V$ over a field $F$ and let $Y$ be a normal subgroup of $X$. Then the following conditions hold*:

(i) *$V$ is the direct sum of $\phi(Y)$-invariant subspaces $V_i$ of $V$, $i\leqslant i\leqslant n$.*

(ii) *For each $i$, $V_i$ is the direct sum of irreducible $\phi(Y)$-invariant subspaces $V_{ij}$, $1\leqslant j\leqslant t$, where $t$ is independent of $i$ and the representations of $Y$ on $V_{ij}$ and $V_{i'j'}$ are equivalent if and only if $i=i'$ (in particular, $V_{i1},V_{i2},\ldots,V_{it}$ determine equivalent irreducible representations of $Y$.*

(iii) *As a group of linear transformations of $V$, $\phi(X)$ permutes the subspaces $V_1,V_2,\ldots,V_n$ among themselves and this action is transitive [i.e., for any pair $i,i'$, there is some $x\in X$ (which depends on $i$ and $i'$) such that $\phi(x)(V_i)=V_{i'}$].*

We assume the reader is familiar with the notion of the *tensor product* of two vector spaces $V$, $W$ over a field $F$, customarily denoted by $V\otimes_F W$. The vector space $V\otimes_F W$ is generated by elements of the form $v\otimes w$, $v\in V$, $w\in W$, which satisfy suitable bilinearity relations. (In Sections 2.1 and 4.14, we shall also need the notion of the tensor product of two modules over the integers $\mathbb{Z}$.)

If $\phi$ and $\psi$ denote representations of $X$ on $V$ and $W$, respectively, we obtain a representation $\phi\otimes\psi$ of $X$ in the obvious way, namely, for $x\in X$, $v\in V$, and $w\in W$, by defining

$$\phi\otimes\psi(x)(v\otimes w)=\phi(x)(v)\otimes\psi(x)(w).$$

$\phi\otimes\psi$ is called the *tensor product* of the representations $\phi$ and $\psi$. In particular, we can define the tensor *square* $\phi\otimes\phi$ of a representation $\phi$ of $X$ on a vector $V$ over $F$; likewise, we define the tensor *cube* as $(\phi\otimes\phi)\otimes\phi$; etc.

[It is easily shown that $(\phi\otimes\phi)\otimes\phi$ and $\phi\otimes(\phi\otimes\phi)$ determine equivalent representations of $X$, so that the notion of tensor power is well defined.]

There exist two other products of representations closely related to tensor products. Consider the tensor product $\phi\otimes\phi$ of $X$ on $V\otimes V$, where $\phi$ is a representation of $X$ on the vector space $V$ over $F$. If $W$ denotes the subspace of $V\otimes V$ generated by the vectors $v\otimes v$, $v\in V$, it is immediate that $\phi\otimes\phi(X)$ leaves $W$ invariant. Hence there is an induced factor representation of $X$ on the vector space $V\otimes V/W$. It is called the *exterior* or *wedge* product of $\phi$ and is denoted by $\phi\wedge\phi$ with $V\otimes V/W$ likewise denoted by $V\wedge V$.

Observe that $\phi\otimes\phi(X)$ also leaves invariant the subspace $U$ of $V$ generated by the vectors $v\otimes w-w\otimes v$ for $v$, $w\in V$. Moreover, the kernel of the homomorphism $\psi$ of $V\otimes V$ onto $U$ given by $\psi$: $v\otimes w\mapsto v\otimes w-w\otimes v$ is precisely the subspace $W$. Hence the wedge product $\phi\wedge\phi$ is equivalent to the restriction of the representation $\phi\otimes\phi$ to the subspace $U$.

Similarly there is an induced representation of $X$ on $V\otimes V/U$, called the *symmetric square* of $\phi$ and denoted by $\phi^2$. If $F$ is not of characteristic 2 (in particular, if $F=\mathbb{C}$), one checks that $\phi^2$ is equivalent to the restriction of $\phi\otimes\phi$ to the subspace $W$. Moreover, in this case, one has

$$V\otimes V=W\oplus U,$$

so that $\phi\otimes\phi$ decomposes as the sum of $\phi^2$ and $\phi\wedge\phi$.

The higher exterior and symmetric powers of $\phi$ are similarly defined.

Next we consider permutation groups.

$\Sigma(\Omega)$ denotes the symmetric group of all permutations of the (finite) set $\Omega$. Thus $\Sigma_n=\Sigma(\Omega)$ for $\Omega=\{1,2,\ldots,n\}$.

$A(\Omega)$ (resp. $A_n$) denotes the alternating subgroup of even permutations of $\Sigma(\Omega)$ (resp. $\Sigma_n$).

$X$ is a *permutation* group on $\Omega$ if $X\leqslant\Sigma(\Omega)$. $|\Omega|$ is the *degree* of $X$. [Cf. D7 of the Introduction, where we have also defined transitivity and $k$-fold transitivity as well as the (permutation) rank of $X$.]

If $X\leqslant\Sigma(\Omega)$, then for any $\Lambda\subset\Omega$ the $\Lambda$-*point stabilizer* $X_\Lambda$ of $X$ is the subset of permutations of $X$ fixing each point of $\Lambda$. It is clearly a subgroup of $X$. If $X$ is $k$-fold transitive, then for any subsets $\Lambda_1$, $\Lambda_2$ of $\Omega$ of cardinality $k$, $X_{\Lambda_1}$ and $X_{\Lambda_2}$ are conjugate subgroups of $X$. Because of this, we speak simply of the $k$-*point stabilizer* of $X$ (when $X$ is $k$-fold transitive). It is determined up to conjugacy.

If $X\leqslant\Sigma(\Omega)$ is transitive on $\Omega$, it is immediate that the one-point stabilizer has index $|\Omega|$ in $X$.

$X \leqslant \Sigma(\Omega)$ is *primitive* if $X$ is transitive on $\Omega$ and its stabilizer is a maximal subgroup of $X$.

A homomorphism $\phi$ of $X$ into $\Sigma(\Omega)$ is called a *permutation representation* of $X$ on $\Omega$. $|\Omega|$ is the *degree* of $\phi$. We say that $\phi$ is *k-fold transitive* if $\phi(X)$ is $k$-fold transitive on $\Omega$. Also clearly $X/\ker(\phi) \leqslant \Sigma(\Omega)$.

Two permutation representations $\phi$ and $\phi'$ of $X$ on sets $\Omega$, $\Omega'$ are *equivalent* if there exists a one-to-one mapping $\theta$ of $\Omega$ on $\Omega'$ such that $\phi' = \theta^{-1}\phi\theta$.

For $Y \leqslant X$, let $\Omega_Y = \{Y_1, Y_2, \ldots, Y_n\}$ be the set of right cosets of $Y$ in $X$. For $x \in X$, the mapping $\phi_x$: $Y_i \mapsto Y_i x$, $1 \leqslant i \leqslant n$, is a permutation of $\Omega_Y$ and the mapping $\phi_Y$: $x \mapsto \phi_x$ for $x \in X$ defines a transitive permutation representation of $X$ on $\Omega_Y$ having $Y$ as a one-point stabilizer. Moreover, $\ker(\phi_Y) = \bigcap_{x \in X} Y^x$.

Similarly, let $\Omega'_Y = \{Y'_1, Y'_2, \ldots, Y'_n\}$ be the set of distinct conjugates of $Y$ in $X$. Defining $\phi'_x$: $Y_i \mapsto Y_i^x$, $1 \leqslant i \leqslant n$, for $x \in X$ and $\phi'_Y$: $x \mapsto \phi'_x$ for $x \in X$, we see that $\phi'_Y$ is a transitive permutation representation of $X$ on $\Omega'_Y$. Moreover, $\phi'_Y$ is equivalent to $\phi_{N_X(Y)}$. In particular, $\phi'_Y$ is equivalent to $\phi_Y$ if $Y = N_X(Y)$.

The significance of these particular permutation representations can be seen from the following elementary result [130, Theorem 2.7.1].

Proposition 1.6. *Every transitive permutation representation of a group $X$ is equivalent to the transitive permutation representation on the right cosets of some subgroup of $X$.*

Finally one has the following equally basic fact [130, Theorem 2.7.3].

Proposition 1.7. *The permutation representation of a group $X$ on the right cosets of $Y \leqslant X$ is doubly transitive if and only if $X = Y \cup YxY$ for any $x \in X - Y$.*

Now for generators and relations. If $Y$ is a set of generators of $X$, a *word* in $Y$ is a finite formal product $abc\cdots$ of elements $a, b, c, \ldots$ of $Y$ or their inverses. A word $W$ is called a *relator* if it represents the identity element of $X$. The statement $W=1$ is called a *relation*. If $P_i$, $1 \leqslant i \leqslant n$, are relators (in $Y$), a word $W$ in $Y$ is *derivable* from the $P_i$, if the following operations, applied a finite number of times, transform $W$ into the empty word:

a. Insertion of some $P_i$ or $P_i^{-1}$ between any two symbols of $W$ or before $W$ or after $W$.
b. Deletion of some $P_i$ or $P_i^{-1}$ if it forms a block of consecutive symbols of $W$.

Clearly in such a case $W$ itself is also a relator.

If every relator (in $Y$) is derivable from the relators $P_i$, $1 \leqslant i \leqslant n$, together with the trivial relators $\{aa^{-1} | a \in Y\}$, then the $P_i$ are called a *complete set of defining relators* and the equalities $P_i = 1$ a *complete set of defining relations* for $X$. The set of generators $Y$ together with a complete set of defining relators or relations is called a *presentation* of $X$.

The importance of presentations is that they "characterize" the group $X$. Indeed, if $X$, $X^*$ are groups with presentations $(Y, \{P_i, 1 \leqslant i \leqslant n\})$ and $(Y^*, \{P_i^*, 1 \leqslant i \leqslant n^*\})$ and if there exists a one-to-one mapping $\theta$ from $Y$ to $Y^*$ (whence $|Y| = |Y^*|$) which induces in the obvious way a one-to-one map from the set $\{P_i, 1 \leqslant i \leqslant n\}$ onto the set $\{P_i^*, 1 \leqslant i \leqslant n^*\}$ (whence $n = n^*$), then the natural extension of $\theta$ to the set of all words in $Y$ onto all words in $Y^*$ induces an isomorphism of $X$ on $X^*$.

Next, some elementary facts about $p$-groups.

PROPOSITION 1.8. *An abelian p-group X is the direct product of cyclic groups and the number of nontrivial factors n is independent of the decomposition.*

The integer $n$ is called the *rank* of $X$.

PROPOSITION 1.9. *If $X$ is a p-group and $Y$ is a nontrivial normal subgroup of $X$, then $Y \cap Z(X) \neq 1$. In particular, $Z(X) \neq 1$.*

This is often used in the following way: if the $p$-group $A$ acts on the nontrivial $p$-group $X$, then $C_X(A) \neq 1$. Indeed, the semidirect product of $X$ by $A$ is a $p$-group with normal subgroup $X$, so the result follows from the proposition.

Note that it also implies that a group of order $p^2$ is necessarily abelian.

The following property provides a basic characterization of nilpotent groups [130, Theorem 2.3.4].

THEOREM 1.10. *$X$ is nilpotent if and only if $Y < N_X(Y)$ for every proper subgroup $Y$ of $X$.*

The theorem has a simple corollary which is very important in local analysis.

COROLLARY 1.11. *Let $P$ be a p-subgroup of $X$, $p$ a prime. If $P \in Syl_p(N_X(P))$, then $P \in Syl_p(X)$.*

Indeed,* let $R \in Syl_p(X)$ with $P \leq R$. If $P < R$, then as $R$ is nilpotent, $N_R(P) > P$ by the theorem, in which case $P \notin Syl_p(N_X(P))$. Hence, if $P \in Syl_p(N_X(P))$, we must have $P = R$.

If $X$ is a $p$-group, $\Omega_i(X)$ denotes the subgroup of $X$ generated by all elements of $X$ of order at most $p^i$. Clearly $\Omega_i(X)$ is characteristic in $X$.

If $p^n$ is the maximum order of an element of $X$, then $X$ has *exponent* $p^n$. Thus if $X$ is abelian of exponent $p$, then $X$ is elementary.

The following properties of $p$-groups are very old (see [130, Chapter 5]).

PROPOSITION 1.12. *If $X$ is a $p$-group, then*

(i) *$X/\phi(X)$ is elementary abelian. (In particular, $X' \leq \phi(X)$.)*
(ii) *If $Y \leq X$ and $Y\phi(X) = X$, then $Y = X$.*
(iii) *If $\alpha$ is an automorphism of $X$ of order prime to $p$ and $\alpha$ acts trivially on $X/\phi(X)$, then $\alpha$ is the identity automorphism.*

A $p$-group $X$ is *special* if either $X$ is elementary abelian or $X' = \phi(X) = Z(X)$ is elementary. In particular, $X$ has class 1 or 2. Moreover, $X$ is called *extra-special* if $X$ is nonabelian special and $|X'| = p$.

Extra-special $p$-groups are important in simple group theory, in large measure because of the following result of P. Hall [130, Theorem 5.4.9]; but they also arise as minimal cases in representation theory (see Section 4.1).

THEOREM 1.13. *If a $p$-group $X$ has no noncyclic characteristic abelian subgroups, then*

$$X = A * B,$$

*where $A$ is extra-special (or $A = 1$) and either $B$ is cyclic or $p = 2$ and $B$ is dihedral, quasi-dihedral, or quaternion.*

The structure of extra-special $p$-groups can be precisely determined (cf. [130, Section 5.5]).

PROPOSITION 1.14. *If $X$ is an extra-special $p$-group, then*

$$X = A_1 * A_2 * \cdots * A_n,$$

*where each $A_i$ is extra-special of order $p^3$.*

*In keeping with the informality of the exposition, we shall normally use expressions such as "Indeed" or "We outline the proof" as indications of an argument to follow, rather than the customary "*Proof*," deviating from this practice only in the few situations in which a lengthy proof is being presented in its entirety.

The integer $n$ is called the *width* of $X$. For a given $p$, there are two nonisomorphic extra-special $p$-groups of order $p^3$. If $p$ is odd, one of these is of exponent $p$ and the other has a maximal cyclic subgroup of order $p^2$; if $p=2$, one is quaternion of order 8 and the other dihedral of order 8. Moreover, it is easily checked that $Q_8 * Q_8 \cong D_8 * D_8$.

We introduce the symbol $(Q_8)^k$ [resp.$(D_8)^k$] to denote the extra-special 2-group of width $k$, each of whose factors is $Q_8$ (resp. $D_8$). In view of this discussion, one can easily prove

PROPOSITION 1.15. *If $X$ is an extra-special 2-group of width $n$, then $X \cong (Q_8)^n$ or $(Q_8)^{n-1} * D_8$. Moreover, the latter two groups are not isomorphic.*

Finally, a $p$-group $X$ with no noncyclic characteristic abelian subgroups is said to be of *symplectic* type.

The following considerations will explain this choice of terminology. Suppose $X$ is extra-special and set $Z(X)=\langle z\rangle$, $\bar{X}=X/Z(X)$. Thus $\bar{X}$ is a vector space over $GF(p)$ and $|z|=p$. We define a map

$$\rho\colon \bar{X}\times\bar{X}\mapsto GF(p).$$

For $\bar{x}, \bar{y}\in\bar{X}$, let $x$, $y$ be representatives of $\bar{x}$, $\bar{y}$ in $X$. Then

$$[x, y]=z^i$$

for some $i$, depending on $\bar{x}$, $\bar{y}$ but not on the choice of the representatives $x$, $y$. Since $|z|=p$, we can view $i$ as an element of $GF(p)$. We set $\rho(\bar{x}, \bar{y})=i$. It is now easily verified that $\rho$ is a nondegenerate *alternating bilinear* form on $\bar{X}$. Since the symplectic group on a vector space is defined to be the group of linear transformations preserving a nondegenerate alternating bilinear form, this suggests the term "symplectic" for $p$-groups of this general form.

P. Hall's theorem (Theorem 1.13) constitutes one of the few cases in which it has been possible to completely determine the structure of all $p$-groups satisfying some specified condition. A second example, important for the classification, is that of 2-groups of "maximal class."

It is clear from the definition of class, together with the fact that a group of order $p^2$ is necessarily abelian, that the class of a $p$-group $P$ of order $p^n$ is at most $n-1$. If equality holds, we say that $P$ is of *maximal class*.

The following criterion for a $p$-group to have this property is easily proved by induction on $|X|$.

PROPOSITION 1.16. *If a p-group X contains a subgroup A of order* $p^2$ *such that* $A = C_X(A)$, *then X is of maximal class.*

For 2-groups, we have the following complete classification [130, Theorem 5.4.5].

THEOREM 1.17. *If X is a (nonabelian) 2-group of maximal class, then X is either quaternion, dihedral, or quasi-dihedral.*

Some further properties of low-rank $p$-groups are given in the next section.

Although transfer can be defined more generally, we limit ourselves to the case of a Sylow $p$-subgroup $P$ of the group $X$. Let $x_1, x_2, \ldots, x_n$ be a set of representatives of the distinct right cosets of $P$ in $X$. For each $x \in X$, we can write

$$x_i x = p_i(x) x_{\pi(i)},$$

for suitable $p_i(x) \in P$ and $\pi(i) = j$, where $1 \leqslant j \leqslant n$ and $j$ is determined from the condition that $x_i x$ is an element of the coset $Px_j$.

With this terminology, let $A$ be a subgroup of $P$ containing $P'$ and set

$$\tau_A(x) = (p_1(x) p_2(x) \cdots p_n(x)) A,$$

so that $\tau_A(x)$ is an element of the abelian factor group $P/A$.

PROPOSITION 1.18. (i) *For* $x \in X$, $\tau_A(x)$ *is determined independently of the set of coset representatives of P in X.*

(ii) $\tau_A$ *is a homomorphism of X into the abelian group* $P/A$.

$\tau_A$ is called the *transfer* (or *transfer homomorphism*) of $X$ into $P/A$ (cf. [130, Section 7.3]).

If $Y$ is the kernel of $\tau_A$, then clearly $X/Y$ is isomorphic to the image of $\tau_A(X)$ of $X$ in $P/A$ and so $X/Y$ is an abelian $p$-group. In particular, if $\tau_A(X) \neq 1$, it follows that $X$ has a normal subgroup of index $p$ and so is not simple. Thus we can view the transfer as a tool for showing that a given group $X$ is not simple.

It is not difficult to prove that in a group $X$ there is a *uniquely* determined normal subgroup $Y$ of $X$ which is *minimal* subject to the condition that $X/Y$ is an abelian $p$-group and, further, that $X/Y \cong P/P \cap X'$. (Since $P' \leqslant X'$, the group $P/P \cap X'$ is certainly an abelian $p$-group.) The subgroup $P \cap X'$ is called the *focal* subgroup of $X$ (for the prime $p$). Clearly then $X$ possesses a normal subgroup of index $p$ if and only if the focal subgroup is proper in $P$.

Thus we have two criteria for $X$ to have a normal subgroup of index $p$. That these are directly related is shown by the following result, which also gives an effective internal description of the focal subgroup [130, Theorem 7.3.4].

THEOREM 1.19. *Let $X$ be a group, $p$ a prime, and $P$ a Sylow $p$-subgroup of $X$. Then we have*

(i) *The transfer $\tau_{P \cap X'}$ maps $X$ onto $P/P \cap X'$.*
(ii) *$P \cap X'$ is generated by the set of elements $y^{-1}y^x$, where $y$ and $y^x \in P$ and $x \in X$.*

A number of classical results have been proved by means of the transfer map. We mention two, due, respectively, to Frobenius and Burnside, which give criteria for a group to possess a normal $p$-complement [130, Theorems 7.4.2, 7.4.5].

THEOREM 1.20. *Let $X$ be a group and $P$ a Sylow $p$-subgroup of $X$. Then $X$ has a normal $p$-complement under either of the following conditions*:

(i) *For every $Q \leqslant P$, $N_X(Q)/C_X(Q)$ is a $p$-group.*
(ii) *$P$ is abelian and $P \leqslant Z(N_X(P))$.*

Using transfer, Thompson has established a very useful criterion for a group to have a normal subgroup of index 2, known as the "Thompson transfer lemma" [287, Lemma 5.38].

PROPOSITION 1.21. *Let $X$ be a group, $P$ a Sylow 2-subgroup of $X$, and $Q$ a maximal subgroup of $P$ such that $P - Q$ contains an involution. If $u$ is an involution of $P - Q$, then one of the following holds*:

(i) *$X$ has a normal subgroup of index 2 not containing $u$.*
(ii) *$u$ is conjugate in $X$ to an element of $Q$.*

One considers the transfer homomorphism $\tau_Q$ of $X$ into $P/Q$. (As $P/Q$ is abelian, $Q \geqslant P'$, so $\tau_Q$ is well defined.) If $u$ is not $X$-conjugate to an element of $Q$, a direct calculation shows that $\tau_Q(u) \neq 1$. Set $Y = \ker(\tau_Q)$. Since $Q \leqslant Y$ and $u \notin Y$, it follows that $Q \in Syl_2(Y)$. But $\ker(\tau_Q)$ also clearly contains all elements of $X$ of odd order, so $Y$ is, in fact, a normal subgroup of $X$ of index 2 (with $u \notin Y$).

Finally some notational conventions. Throughout the paper the letter $G$ will be reserved for arbitrary simple groups and more generally, for any group to be investigated by local analytic methods. For $x \in \mathcal{I}_p(G)$, $p$ any prime, we shall for simplicity write $C_x$ for $C_G(x)$. However, we shall not use this contraction in any other connection.

We shall use the letter $G^*$ for $K$-groups, and usually the letter $X$ will denote an arbitrary finite group. We also adopt the bar convention for homomorphic images $\bar{X}$ of $X$—i.e., $\bar{Y}$ will denote the image in $\bar{X}$ of any subgroup or subset $Y$ of $X$.

## 1.5. The Shape of the Proof

With the aid of certain of the terms and ideas discussed in the preceding section, we can now describe the breakup of the analysis of simple groups. We first define some key general concepts that were first introduced by Walter and me in [143] and [144]. An improved approach to some of these ideas was given by Bender in [27]. Again let $X$ be an arbitrary group.

DEFINITION 1.22. $X$ is *quasisimple* if $X$ is perfect and $X/Z(X)$ is simple.

For example, if $X = SL(2, q)$, $q$ odd, $q > 3$, then $X$ is quasisimple, but not simple, $Z(X)$ being of order 2 and generated by $\left(\begin{smallmatrix} -1 & 0 \\ 0 & -1 \end{smallmatrix}\right)$. [$SL(2,3)$ is solvable and hence is not perfect.]

If $X$ is quasisimple, one speaks of $X$ as being a *covering* group of $X/Z(X)$. The possible covering groups of every simple $K$-group have been computed. This will be discussed in more detail in Section 4.15. Thus all quasisimple $K$-groups have been determined. Of course, with the classification theorem, we, in fact, now know all quasisimple groups. Issai Schur [245] showed that a simple (and, more generally, perfect) group $X$ possesses a "universal" covering group $\hat{X}$ with the property that every covering group of $X$ is a homomorphic image of $\hat{X}$. $Z(\hat{X})$ is called the *Schur multiplier* of $X$.

Thus the problem of determining the possible covering groups of $X$ reduces to computing the center of $\hat{X}$ (cf. Section 4.15).

For example, $A_n$ has $Z_2$ as its Schur multiplier except for $n=6$ or 7, in which cases the Schur multiplier is cyclic of order 6. Thus $A_6$ and $A_7$ have three nontrivial covering groups, whose centers have orders 2, 3, and 6, respectively.

DEFINITION 1.23. $X$ is *semisimple* if either $X$ is a central product of quasisimple groups or $X=1$.

If $X\neq 1$ is semisimple, it is easily shown that the quasisimple factors of $X$ are precisely the set of normal subgroups of $X$ which are minimal subject to being nonsolvable and, in particular, are uniquely determined by $X$. They are called the *components* of $X$. (One includes the trivial group in the definition solely for convenience.)

The following is easily proved.

PROPOSITION 1.24. *A finite group $X$ possesses a unique maximal normal semisimple subgroup. It is called the layer of $X$ and is denoted by $L(X)$.*

It is immediate that $L(X)$ centralizes $Sol(X)$ and hence $F(X)$.

The next concept, due to Bender, is fundamental and has its origin in a basic property of solvable groups, known as *Fitting's* theorem [130, Theorem 6.1.3].

PROPOSITION 1.25. *If $X$ is a solvable group, then*

$$C_X(F(X))\leqslant F(X).$$

*In particular, $X$ is $p$-constrained for every prime $p$.*

That the second assertion follows from the first is immediate from the definition of $p$-constraint and the fact that $F(X)=O_p(X)$ if $O_{p'}(X)=1$.

It is this property of solvable groups which Bender generalized.

DEFINITION 1.26. $F^*(X)=L(X)F(X)$ is called the *generalized* Fitting subgroup of $X$.

Justification for this terminology is given by the following general result of Bender [27]. (The special case in which $O_{p'}(X)=1$ is proved in [143] and [144].)

PROPOSITION 1.27. *For any group $X$, $C_X(F^*(X)) \leqslant F^*(X)$. Moreover, if $F(X) \leqslant Y \lhd X$ and $C_X(Y) \leqslant Y$, then $F^*(X) \leqslant Y$.*

Thus $F^*(X)$ is the unique normal subgroup of $X$ which is minimal subject to containing $F(X)$ and to containing its own centralizer in $X$. For example, if $F^* = F^*(X)$ is simple, the proposition shows that $X \leqslant \mathrm{Aut}(F^*)$. More generally if $F^*$ is quasisimple, it yields that $X/Z(F^*) \leqslant \mathrm{Aut}(F^*/Z(F^*))$. Likewise, as an immediate corollary of the proposition, we have

COROLLARY 1.28. *For any group $X$, $C_X(F(X)) \leqslant F(X)$ if and only if $L(X)=1$. In particular, $X$ is $p$-constrained for any prime $p$ if and only if $L(X/O_{p'}(X))=1$.*

These concepts enable us to describe in group-theoretic terms the general structure of the centralizers of involution in simple $K$-groups. Ree [240], Nagayoshi Iwahori [184], Nicholas Burgoyne and C. Williamson [54, 55], and Roger Carter [60, 61] have described these centralizers in the groups of Lie type of odd characteristic (the third pair of authors covered more generally the centralizers of "semisimple" elements in groups of Lie type of arbitrary characteristic). A detailed statement of their results will be given in Chapter 4. On the other hand, the general shape of such centralizers in groups of Lie type of characteristic 2 follows from a theorem of Armand Borel and Jacques Tits [37] which gives the basic general structure of $p$-local subgroups in any group of Lie type of characteristic $p$ (see Theorems 1.41 and 1.42 below).

We list here only the following general properties.

PROPOSITION 1.29. *Let $G^*$ be of Lie type of characteristic $p$, $t^* \in \mathscr{I}(G^*)$, and set $C^* = C_{G^*}(t^*)$. Then we have*

(i) *If $p$ is odd, then $C^*/L(C^*)$ is solvable. In particular, either $L(C^*) \neq 1$ or $C^*$ is solvable.*
(ii) *If $p=2$, then $F^*(C^*)=O_2(C^*)$ (equivalently, $C^*$ is 2-constrained with trivial core).*

It will be instructive for the reader to compute $C^*$ when $G^*=SL(m,q)$ in the following two cases:

$$q \text{ odd and } t^* = \begin{bmatrix} -1 & & & & & & & \\ & -1 & & & & 0 & & \\ & & \ddots & & & & & \\ & & & -1 & & & & \\ & & & & 1 & & & \\ & & 0 & & & \ddots & & \\ & & & & & & & 1 \end{bmatrix} \tag{1.1}$$

with $k$ entries $-1$, $k$ even.

$$q \text{ even and } t^* = \begin{bmatrix} 1 & 0 & \cdot & \cdot & \cdot & 0 \\ 0 & 1 & & & & \vdots \\ \vdots & & \ddots & & & \vdots \\ 0 & & & \ddots & & \\ & & & & \ddots & 0 \\ 1 & 0 & \cdot & \cdot & 0 & 1 \end{bmatrix}. \tag{1.2}$$

In the first case, except for special values of $q$ and $k$,

$$L(C^*)=SL(k,q)\times SL(m-k,q).$$

In the second case, $F^*(C^*)$ is a special 2-group of order $q^{2n+1}$ with (elementary) center of order $q$.

The proposition shows a fundamental dichotomy in the structure of the centralizers of involutions in groups of Lie type of characteristic $p$ according as $p$ is odd or even. For odd $p$, their layers are always nontrivial (except in some small degenerate cases), while for $p=2$, these layers are always trivial and, in fact, their generalized Fitting subgroups are always 2-groups.

If $G^*=A_n$ and $t^*\in\mathcal{I}(G^*)$ is a product of $k$ disjoint transpositions, one easily checks that $L(C^*)$ is isomorphic to the alternating group on the remaining $n-2k$ letters provided $n-2k\geqslant 5$; in the contrary case, $L(C^*)=1$. In particular, if $n\geqslant 9$ and $t^*$ is a "short" involution (i.e., $k=2$), then

$L(C^*) \neq 1$, while if $t^*$ is a "long" involution [i.e., $k=(n-\delta)/2$, where $n-\delta$ is even and $0 \leq \delta \leq 3$], then $L(C^*)=1$.

Likewise in the sporadic groups, the centralizer of involutions can have trivial or nontrivial layers.

This dichotomy in the layers of centralizers of involutions is basic for the study of simple groups $G$. However, it is not $L(C_G(t))$, $t \in \mathcal{I}(G)$, which is initially important, but rather $L(C_G(t)/O(C_G(t)))$. In fact, as we shall see later, a central problem that must be resolved is the precise relationship between these two layers.

In view of the preceding discussion, it is therefore natural to divide all simple groups into two categories, as follows:

Definition 1.30. A group $X$ is said to be of *component* type if $L(C_X(t))/O(C_X(t)) \neq 1$ for some involution $t$ of $X$, and is said to be of *noncomponent* type if $L(C_X(t))/O(C_X(t))=1$ for every involution $t$ of $X$.

We turn now to the notion of connectivity which we mentioned at the end of Section 1.2. For later purposes, we define it for an arbitrary prime $p$.

Definition 1.31. A group $X$ is said to be *connected for the prime p* provided that for any two noncyclic elementary abelian $p$-subgroups $A$, $B$ of $X$, there exists a sequence $A=A_1, A_2, \ldots, A_n=B$ of noncyclic elementary abelian $p$-subgroups $A_i$ of $X$, $1 \leq i \leq n$, such that $A_i$ centralizes $A_{i+1}$, $1 \leq i \leq n-1$.

The connectivity of $X$ for the prime $p$ is equivalent to the connectivity (in the ordinary sense) of the associated graph $\Gamma(p)$ whose vertices are the subgroups of $X$ isomorphic to $Z_p \times Z_p$, with two vertices connected by an edge if and only if they commute. Thus the previously defined notion of the connectivity of $X$ is equivalent to the connectivity of $X$ for the prime 2.

It is the case $p=2$ in which we are interested here. The first major step in the classification proof is the determination of all nonconnected simple groups, thus reducing the classification problem to the connected case. The significance of connectedness is that, in its presence, certain general lines of argument are available, which have powerful consequences for the structure of the cores of centralizers of involutions in simple groups. We shall give an illustration shortly.

Unfortunately, nonconnectedness is a difficult problem to treat directly because the condition is not inductive to sections. As a consequence, the problem has been treated in two parts, the division based upon the following facts. A simple group $G$ may be connected, but a Sylow 2-subgroup of $G$ may be nonconnected. Alternatively, $G$ may be nonconnected but have a connected Sylow 2-subgroup. Janko's second and third groups $J_2$, $J_3$ are examples of the first kind (as shown by David Goldschmidt), while the groups $L_2(2^n)$ are examples of the second. This suggests the following subdivision:

A. Determine all nonconnected simple groups having a connected Sylow 2-subgroup.
B. Determine all simple groups having a nonconnected Sylow 2-subgroup.

We should next like to reformulate these problems in the terms in which they have actually been studied. Problem A is directly related to the notion of a *proper 2-generated core*, introduced in [142].

DEFINITION 1.32. Let $G$ be a group, $p$ a prime, $P$ a Sylow $p$-subgroup of $G$, and $k$ a positive integer with $k \leqslant m(P)$. Define

$$\Gamma_{P,k}(G) = \langle N_G(Q) | Q \leqslant P, m(Q) \geqslant k \rangle .$$

$\Gamma_{P,k}(G)$ is called the *k-generated p-core* of $G$ and is determined by $P$ up to conjugacy in $G$. If $p=2$, we speak of the *k-generated core*.

PROPOSITION 1.33. *Let $G$ be a group having a connected Sylow 2-subgroup. Then $G$ is disconnected if and only if the 2-generated core of $G$ is a proper subgroup.*

Indeed, let $\Gamma_0$ be a connected component of the associated graph $\Gamma$ of $G$; let $T$ be a vertex of $\Gamma_0$ and $S$ a Sylow 2-subgroup of $G$ containing the four group $T$. Every element $g$ of $G$ induces by conjugation a permutation of $\Gamma$, and the image of $\Gamma_0$ under $g$ is the connected component of $\Gamma$ containing $T^g$. We denote it by $\Gamma_0^g$. Clearly $G$ is connected if and only if $G$ leaves $\Gamma_0$ invariant. Likewise $\Gamma_0^g = \Gamma_0$ if and only if $\Gamma_0 \cap \Gamma_0^g \neq \varnothing$.

Set $H = \Gamma_{S,2}(G)$. We shall argue that $H$ acts on $\Gamma_0$ and that every vertex of $\Gamma_0$ is a four subgroup of $H$, which will immediately yield that $H < G$ if and only if $\Gamma_0 < \Gamma$ and hence if and only if $G$ is disconnected.

Note that as $S$ is assumed to be connected, its subgraph is connected, so the vertices of $\Gamma_0$ include every four subgroup of $S$. Now let $R \leqslant S$ with $m(R) \geqslant 2$ and let $U$ be any four subgroup of $R$. Then for $x \in N_G(R)$, $U$ and $U^x$ $(\leqslant R)$ are two four subgroups of $S$. Hence $U$ and $U^x \in \Gamma_0$. But as

$U \in \Gamma_0$, $U^x \in \Gamma_0^x$, so $U^x \in \Gamma_0 \cap \Gamma_0^x$, forcing $\Gamma_0 = \Gamma_0^x$. Thus $x$ leaves $\Gamma_0$ invariant, so $N_G(R)$ leaves $\Gamma_0$ invariant. We conclude that $H = \Gamma_{S,2}(G) = \langle N_G(R) | R \leqslant S, m(R) \geqslant 2 \rangle$ leaves $\Gamma_0$ invariant.

Now let $V$ be an arbitrary vertex of $\Gamma_0$. Let $T = T_1, T_2, \ldots, T_n = V$ be a chain of vertices of $\Gamma_0$ connecting $T$ to $V$. We have that $T = T_1 \leqslant H$. If $T_i \leqslant H$, then $N_G(T_i) \leqslant H$ by definition of $H$, so $T_{i+1} \leqslant H$ as $T_{i+1}$ centralizes $T_i$. We thus conclude inductively that $V \leqslant H$. This establishes the proposition.

Aschbacher has determined all finite groups having a proper 2-generated core [8], generalizing Bender's fundamental classification of groups possessing a "strongly embedded" subgroup (equivalently those having a proper 1-generated core) [26]. These results will be discussed in Section 4.2.

On the other hand, a solution of Problem B has been obtained as a corollary of a more general theorem classifying all groups of "sectional 2-rank at most 4." Because of this, groups of sectional 2-rank at most 4 are considered to be part of the category of "small" simple groups. To explain this concept, we introduce some terminology, involving various extensions of the notion of the rank of an abelian $p$-group.

DEFINITION 1.34. If $X$ is a $p$-group, $p$ a prime, the *normal* rank of $X$, denoted by $n(X)$, is the maximum rank of an abelian normal subgroup of $X$; and the *sectional rank* of $X$, denoted by $r(X)$ is the maximum rank of an abelian section of $X$.

Clearly $n(X) \leqslant m(X) \leqslant r(X)$. If $Y \leqslant X$, then $m(Y) \leqslant m(X)$ and $r(Y) \leqslant r(X)$, but the inequality need not hold for $n(Y)$. Moreover, if $S$ is a section of $X$, then $r(S) \leqslant r(X)$, but the inequality need not hold for either $m(X)$ or $n(X)$. Thus only the sectional rank is "inductive" to both subgroups *and* homomorphic images. It is for this reason that the concept of sectional rank is important in the study of simple groups.

There are two major results about $p$-groups of very low rank or normal rank (cf. [130, Chapter 5]).

PROPOSITION 1.35. *If $X$ is a $p$-group, we have*

(i) *If $m(X) = 1$, then either $X$ is cyclic or $p = 2$ and $X$ is quaternion.*
(ii) *If $n(X) = 1$ and $m(X) > 1$, then $p = 2$ and $X$ is either quaternion, dihedral of order at least* 16, *or quasi-dihedral.*

In particular, $r(X) \leqslant 2$ in either case.

Results of Norman Blackburn [34, 35] and Anne MacWilliams [207] give deeper properties of $p$-groups of normal rank 2.

THEOREM 1.36. *If $X$ is a $p$-group with $n(X)=2$, then we have*

(i) *If $p$ is odd, then $m(X)=2$.*
(ii) *If $p=2$, then $r(X)\leqslant 4$.*

Thus if $X$ is a 2-group with $n(X)\leqslant 2$, then $r(X)\leqslant 4$. It is this assertion which is critical for nonconnectivity, since one can easily establish the following result:

PROPOSITION 1.37. *If $X$ is a 2-group with $n(X)\geqslant 3$, then $X$ is connected.*

Indeed, let $E$ be an elementary normal subgroup of $X$ of rank at least 3. Since the notion of connectivity is clearly transitive, it suffices to show that any four subgroup $A$ of $X$ is connected to $E$. Since $E\lhd X$, $E\cap Z(X)\neq 1$ by Proposition 1.9, so $A$ centralizes a subgroup $E_0$ of $E$ of order 2. Let $E_0\leqslant A_0\leqslant\langle E_0, A\rangle$ with $A_0\cong Z_2\times Z_2$. Then $[A, A_0]=1$, so $A$ and $A_0$ are connected and hence we need only show that $A_0$ is connected to $E$. Hence we can assume without loss that $A=A_0$. Since $A_0\cap E\neq 1$, it follows that $A\cap E\neq 1$.

We can thus write $A=\langle e, a\rangle$, where $e\in E$. We need only show that $a$ centralizes a four subgroup $F$ of $E$, for then $F$ will centralize $A=\langle e, a\rangle$, whence $[A, F]=1$ and $[F, E]=1$, in which case $A$ will be connected to $E$ by definition of this term. But we can regard $E$ as a vector space over $Z_2$ and $a$ as a linear transformation of $E$ of period 1 or 2 (as $a^2=1$). Since $\dim(E)\geqslant 3$ (as $E$ has rank at least 3), we see that $E$ decomposes as $E_1\times E_2$ with $E_1$, $E_2$ nontrivial $a$-invariant subspaces of $E$. Then $E_i\lhd E_i\langle a\rangle$ and so $a$ centralizes some $e_i\in E_i^{\#}$, $i=1,2$ (again by Proposition 1.9). Thus $F=\langle e_1, e_2\rangle\cong Z_2\times Z_2$ centralizes $a$, as required.

We extend these notions of rank to arbitrary groups, as follows.

DEFINITION 1.38. For any group $X$, the *$p$-rank*, *normal $p$-rank*, and *sectional $p$-rank*, denoted by $m_p(X)$, $n_p(X)$, $r_p(X)$, respectively, are defined to be $m(P)$, $n(P)$, $r(P)$, respectively, for any $P\in Syl_p(X)$.

In this terminology, the preceding three results have the following corollary.

COROLLARY 1.39. *If $X$ is a group with $r_2(X)\geqslant 5$, then a Sylow 2-subgroup of $X$ is connected.*

Thus a simple group of sectional 2-rank at least 5 always has a connected Sylow 2-subgroup. The advantage of studying groups $G$ of sectional 2-rank of most 4 (instead of the more restricted problem of groups with a nonconnected Sylow 2-subgroup) lies in the fact that this condition is satisfied by *every* section of $G$ and so one can proceed inductively to try to classify all such groups. This has indeed been carried out by Harada and me [136]. Our proof will be discussed in the sequel, along with the prior classification theorems concerning groups of low 2-rank, which are needed for the argument.

We conclude this discussion with the statement of a particular case of a theorem of Walter's and mine [131, 142], which will give some indication of the force of connectedness. (Its proof utilizes several general local group-theoretic techniques, which will be discussed later.)

THEOREM 1.40. *Let $G$ be a group of noncomponent type of 2-rank at least 3 with $O(G)=1$. If $G$ is connected or if $G$ has a connected Sylow 2-subgroup, then $F^*(H)$ is a 2-group for any 2-local subgroup $H$ of $G$ [equivalently, $O(H)=1$ and $H$ is 2-constrained]. In particular, this holds for the centralizer of every involution of $G$.*

There are also notions of "smallness' for connected simple groups of noncomponent type. Such groups satisfy the conclusion of Theorem 1.40. To put this in its proper context, it will be helpful first to say a few words about the groups of Lie type.

Chevalley [66] gave the first systematic treatment of the finite analogues of complex Lie groups. Because of this, we denote by $Chev(p)$ the set of groups of Lie type defined over a finite field of characteristic $p$. Their structure is well understood in many aspects and will be described in more detail in Chapter 2.

Let $X \in Chev(p)$ for some prime $p$. If $P \in Syl_p(X)$, then $B=N_X(P)$ is called a *Borel* subgroup of $X$. By Sylow's theorem, all Borel subgroups of $X$ are conjugate. By the Schur–Zassenhaus theorem [130, Theorem 6.2.1],* $B$ splits over $P$, i.e., $B=CP$, where $C$ is a subgroup of $B$ such that $C \cap P=1$. $C$ is called a *Cartan* subgroup of $X$. All Cartan subgroups of $X$ are conjugate. It is known that $C$ is always abelian; and except possibly when $p=2$ and $X$ is defined over the prime field $GF(2)$ [also in the case $X=L_2(3)$], $C$ is

*Let $A$ be a normal Hall subgroup of the group $Y$, i.e., $A$ and $Y/A$ are of coprime orders. The Schur–Zassenhaus theorem asserts: (i) $A$ has a complement $B$ in $Y$ (thus $Y=AB$ with $A \cap B=1$); and (ii) if $A$ or $Y/A$ is solvable, then any two complements of $A$ are conjugate in $Y$.

nontrivial. Any proper subgroup of $X$ containing $B$ is called a *parabolic* subgroup of $X$. It is known that every parabolic subgroup $J$ of $X$ is $p$-constrained with trivial $p'$-core; that is, $F^*(J)$ is a $p$-group for every parabolic subgroup $J$ of $X$.

The Borel–Tits theorem, mentioned earlier, gives a purely group-theoretic characterization of the maximal parabolics.

THEOREM 1.41. *If $X \in Chev(p)$, then every maximal p-local subgroup of $X$ is a parabolic subgroup.*

Thus $F^*(H)$ is a $p$-group for every maximal $p$-local subgroup $H$ of $X$. Combining this with an easy group-theoretic argument, one can obtain the following extension of the theorem.

THEOREM 1.42. *If $X \in Chev(p)$, then $F^*(H)$ is a p-group for every p-local subgroup $H$ of $X$.*

In view of the theorem, it is natural to make the following definition.

DEFINITION 1.43. A group $X$ is said to be of *characteristic p type* if $F^*(H)$ is a $p$-group for every $p$-local subgroup $H$ of $X$.

Thus Theorem 1.40 asserts that a connected simple group of noncomponent type and 2-rank at least 3 must be of characteristic 2 type.

In the study of simple groups of component type, the role of centralizers of involutions is much more dominant than in groups of noncomponent type (and, in particular, groups of characteristic 2 type). This is because the "prototypes" of such groups are the groups $G^* \in Chev(p)$, $p$ odd, as previously noted, and in such a group $G^*$, involutions correspond to "semisimple" elements (in the Lie sense), in particular, those involutions lying in a Cartan subgroup of $G^*$.

It is also possible to determine $G$ from the centralizers of its involutions when $G$ internally resembles a group $G^* \in Chev(2)$. In particular, this has been done by Suzuki for most of the classical groups [282–284]. Thus, theoretically at least, if one could show that the centralizer of an involution in a simple group $G$ of characteristic 2 type resembled that in some simple $K$-group $G^*$, one should then be able to show that $G$ and $G^*$ are isomorphic. However, in most situations in such a group $G$, it can be very difficult to force $O_2(C_G(t))$, $t \in \mathcal{I}(G)$, to resemble $O_2(C_{G^*}(t^*))$ for some simple $K$-group $G^*$ and some $t^* \in \mathcal{I}(G^*)$.

Rather, the crucial objects of investigation in simple groups $G$ of characteristic 2 type are the *centralizers of elements $x$ of odd prime order $p$ which lie in a 2-local subgroup of $G$*. In trying to show that $G$ resembles internally a simple group $G^* \in Chev(2)$, the ultimate aim is to choose $p$ to be a prime that divides the order of the Cartan subgroup of the corresponding group $G^*$ and $x$ to correspond to an element of such a Cartan subgroup. Once this is achieved, $x$ will again correspond to a "semisimple" element of $G^*$ and the proof that $G$ and $G^*$ are isomorphic follows the same general pattern as the corresponding argument for centralizers of involutions in groups of component type. [In those cases in which $G^*$ has a trivial Cartan subgroup—i.e., when $G^*$ is a Chevalley group defined over $GF(2)$—one works with the prime $p=3$. This latter case is also closely connected with the "$F^*$ of symplectic type" problem, which will be described in a moment.]

Just as special methods are required for studying centralizers of involutions in groups of low 2-rank, the same is true when analyzing centralizers of elements of odd prime order in groups of characteristic 2 type which have "small" $p$-rank of all odd primes $p$. However, in this case the notion of smallness is a "relativized" one, which we proceed to define.

DEFINITION 1.44. Let $X$ be a group of characteristic 2 type. For any odd prime $p$, the *2-local $p$-rank* of $X$, denoted by $m_{2,p}(X)$, is the maximum of $m_p(H)$, taken over all 2-local subgroups $H$ of $X$. Furthermore, we set

$$e(X) = \max\{m_{2,p}(X) \mid p \text{ any odd prime}\}.$$

We call $e(X)$ the *odd-2-local rank* of $X$.

If $e(X)=0$, then every 2-local subgroup of $X$ is a 2-group, so by Frobenius's theorem [Theorem 1.20(i)] $X$ has normal 2-complement. Thus $e(G) \geq 1$ in any simple group $G$ of characteristic 2 type.

"Smallness" for odd primes in simple groups $G$ of characteristic 2 type means precisely that $e(G)=1$ or 2. Correspondingly we call $G$ a *thin* or *quasithin* group. Centralizers of elements of odd prime order dominate the analysis when $e(G) \geq 3$. In contrast, the analysis of thin and quasithin groups focuses directly on the structure of the maximal 2-local subgroups of $G$. [On the other hand, there are strong similarities in later stages of the arguments in both the $e(G) \geq 3$ and $e(G) \leq 2$ problems.]

There is yet a second notion of "smallness" connected with groups of characteristic 2 type which has no counterpart in the study of groups of component type. In Thompson's classification of $N$-groups, the proof divided into four major subcases:

I. $e(G) \geqslant 3$.
II. $e(G)=2$.
III. $e(G)=1$.
IV. $O_2(C_G(t))$ is of symplectic type for some $t \in \mathcal{I}(G)$.

This last possibility (IV) cut across all three of the preceding cases; in its presence, certain general lines of argument collapse. To handle it, Thompson was forced to produce a completely independent argument.

Since, as noted earlier, the analysis of arbitrary simple groups $G$ of characteristic 2 type can be viewed in many ways as a direct generalization of Thompson's $N$-group proof, one could expect the $O_2$ symplectic type case (i.e., case IV) to remain an important special problem. Moreover, it corresponds to a "real-life" situation in many of the groups $G^*$ of Lie type defined over the prime field $GF(2)$. For example, if $G^* = SL(m,2)$ and one takes $t^*$ as in (1.2), then

$$F^*(C^*) = O_2(C^*) \cong (D_8)^{m-2} \text{ is extra-special.}$$

This will explain the following terminology.

DEFINITION 1.45. A group $X$ is said to be of *GF(2)-type* if for some involution $t$ of $X$, $F^*(C_X(t))$ is a 2-group of symplectic type.

Although our primary interest is in the case in which $X$ is of characteristic 2 type, note that the definition requires 2-constraint only for $C_X(t)$ itself.

Remarkably, more than half of the sporadic groups are of $GF(2)$-type. It is for this reason and partly for other aspects of their internal structure that sporadic groups are often considered to be pathologies of groups of Lie type over $GF(2)$.

## 1.6. The Four Phases of the Classification

Summarizing the preceding discussion, we see that the classification of simple groups divides into the classification of simple groups in each of the following four categories:

A. Nonconnected simple groups.
B. Connected simple groups of component type.
C. Small simple groups of characteristic 2 type.
D. Simple groups of characteristic 2 type of large odd 2-local rank.

Specifically, category A refers to the determination of all simple groups having either a proper 2-generated core or sectional 2-rank at most 4; B to the determination of all connected simple groups of component type of 2-rank at least 3; C to the determination of all thin and quasithin simple groups of characteristic 2 type and of all simple groups of type $GF(2)$; and D to the determination of all simple groups of characteristic 2 type with $e(G) \geqslant 3$ and not of $GF(2)$-type.

In the sequel to the present volume, we shall give a detailed outline of the principal results within each of these four phases of the classification proof. However, at this point, the reader should at least have some idea of the global strategy behind the determination of the finite simple groups.

## 1.7. Consequences of the Classification

I think it is appropriate to conclude this introductory chapter with some comments concerning the implications of the classification theorem for group theory as well as related areas of mathematics and also to describe its likely impact on future group-theoretic research.

First, a number of long-standing questions are now reduced simply to checking an appropriate property of the known simple groups. The most celebrated of these is the Schreier conjecture.

THEOREM 1.46. *The outer automorphism group of every simple group is solvable.*

Although the following generational assertion has not yet been checked for every known group, one very likely has:

THEOREM 1.47. *Every simple group can be generated by two elements.*

The well-known fixed-point free automorphism problem is now settled.

THEOREM 1.48. *A finite group admitting an automorphism leaving only the identity element fixed is necessarily solvable.*

So likewise are a number of long-standing permutation-theoretic problems. We mention three.

THEOREM 1.49. *If $G$ is a nonsolvable doubly transitive permutation group of degree $n$ which contains an $n$-cycle, then one of the following holds:*

(i) $G \cong A_n$ *or* $\Sigma_n$.
(ii) $n=11$ *and* $G \cong L_2(11)$ *or* $M_{11}$.
(iii) $n=23$ *and* $G \cong M_{23}$.
(iv) $n=(q^k-1)/(q-1)$ *for some prime power $q$ and $G$ is isomorphic to a subgroup of* $\mathrm{Aut}(L_k(q))$ *containing* $L_k(q)$.

THEOREM 1.50. *The alternating groups and the Mathieu groups (excluding $M_{22}$) are the only simple quadruply transitive permutation groups.*

THEOREM 1.51 (WIELANDT'S PROBLEM). *If $G$ is a primitive, but not doubly transitive permutation group on $2p$ letters, $p$ a prime, then $p=5$ and $G \cong A_5$ or $\Sigma_5$.*

There are many character-theoretic consequences of the classification. We mention only the Alperin–McKay conjecture.

THEOREM 1.52. *For any group $G$ and any prime $p$, the number of irreducible characters of $G$ of degree coprime to $p$ equals the number of irreducible characters of $N_G(P)$ of degree coprime to $p$, where $P$ is a Sylow $p$-subgroup of $G$.*

Examining the known simple groups, one can obviously determine a variety of their properties, which now automatically become properties of all simple groups (the reader will find many such listed in Sections 4.14 and 4.15). For example, one has

THEOREM 1.53. *If $G$ is a simple group whose order is not divisible by $p^2$, $p$ a prime, then the order of* $\mathrm{Aut}(G)$ *is also not divisible by $p^2$.*

Feit lists a great many further consequences of the classification in his article [90]. We would like to mention two of these, related to polynomial equations and Galois groups.

A polynomial $f(x)$ in $K[x]$, $K$ a field, $x$ an indeterminate, is said to be *indecomposable* (over $K$) if whenever $f(x)=f_1(f_2(x))$ for suitable polynomials $f_1(x), f_2(x) \in K[x]$, then necessarily $f_1(x)$ or $f_2(x)$ has degree 1.

THEOREM 1.54. *Let $f(x)$ and $g(x)$ be two indecomposable polynomials in* $\mathbb{C}[x]$. *If $f(x)-g(y)$ factors in $\mathbb{C}[x, y]$, then one of the following holds*:

(i) $g(x)=f(ax+b)$ *for suitable* $a, b\in\mathbb{C}$.
(ii) $f(x)$ *and* $g(x)$ *have equal degree* 7, 11, 13, 15, 21, *or* 31.

*Moreover, there exist indecomposable polynomials $f(x)$, $g(x)$ of each of the degrees listed in* (ii) *such that $f(x)-g(y)$ factors.*

The following result about Galois groups over the rational field $\mathbb{Q}$ is a consequence of Theorem 1.49.

THEOREM 1.55. *Let $p$ be a prime and let $f(x)$ be an irreducible polynomial over $\mathbb{Q}$ of the form $f(x)=x^p+ax^k+b$, where $1\leqslant k\leqslant p-1$. If $G$ denotes the Galois group of $f(x)$ over $\mathbb{Q}$, then one of the following holds*:

(i) *$G$ is solvable.*
(ii) $G\cong A_p$ *or* $\Sigma_p$.
(iii) $p=7$ *and* $G\cong L_2(7)$.
(iv) $p=11$ *and* $G\cong L_2(11)$ *or* $M_{11}$.
(v) *$p=1+2^n$ is a Fermat prime exceeding 5 and $G$ is isomorphic to a subgroup of* $\mathrm{Aut}(L_2(2^n))$ *containing* $L_2(2^n)$.

We should like also to mention a recent result of Burton Fein, William Kantor, and Murray Schacher [85] concerning the nonexistence of "global fields" $L$ and $K$, with $L$ containing $K$ properly, whose "relative Brauer group" $B(L/K)$ is finite. The theorem is proved by a reduction to the following purely group-theoretic assertion, which in turn depends upon the full classification theorem.

THEOREM 1.56. *Let $G$ be a transitive permutation group on a set $\Omega$ with $|\Omega|>1$. Then there exists a prime $p$ and an element $x\in G$ of order a power of $p$ such that $x$ acts without fixed points on $\Omega$.*

Finally it is evident from equally recent work of Gary Seitz [248, 249] that one can utilize the classification theorem to establish further properties of the finite groups of Lie type, for example, to determine all subgroups containing a "maximal torus".

Thus the full import of the classification of the simple groups on finite group theory, number theory, finite geometry, and infinite groups remains to be explored. (For some interesting problem areas concerning infinite groups, see Gilbert Baumslag's article [24].)

## 1.8. The Future of Finite Group Theory

During the last few years of the classification proof, the idea spread that its completion would somehow coincide with the end of the subject of finite group theory itself. The prevalence of this view was undoubtedly fostered by the unusually wide (for mathematics) press coverage of simple groups and many of the comments (including certainly my own) by finite group theorists. Indeed, the headline of an article in the *New York Times* Week in Review of June 22, 1980, read "A School of Theorists Works Itself Out of A Job."

Mathematicians generally agree that a major new theorem is usually a stimulus to a field rather than a sign of its impending demise. However, because of the unprecedented thirty-year team effort required for the classification theorem, group theory came to be regarded as a notable exception. The explanation for this is easily found: those practitioners concentrating on the classification theorem—and this included a substantial portion of simple group theorists—were so fixated on this single objective, their energies so bound up with it, that they were incapable of seeing beyond the classification itself or considering any other aspect of the subject. There was even a fear that all the marvelous techniques developed in the process of dissecting and analyzing simple groups were about to become obsolete. This feeling was intense enough to push a number of group theorists into other areas of mathematics altogether.

However, the first "post-classification" conference—a two-day special session on finite simple groups at the annual meeting of the American Mathematical Society in early 1981—quickly dispelled this gloomy prediction. Indeed group theory is "alive and well": although the focus has shifted, its vitality continues unimpaired. By the time of the meeting, "revisionism" had begun in earnest: new, simplified approaches were presented to both the analysis of nonconnected groups and so-called "standard form" problems. Likewise the geometry of the sporadic groups had been further explored since the classification. In addition, important new properties of the known simple groups were described (whose proofs depended upon the full classification theorem); and there were also several striking results in the new field of "amalgams", a rich blend of graph and local theory having significance for both finite and infinite groups.

Besides these, there are other major areas of finite group theory not even touched on within the special session: determination of the ordinary and modular representations and characters of the known simple groups, representation theory of finite groups in general, determination of the maximal subgroups of the known simple groups, the theory of solvable groups, etc., each of which abounds with deep unsolved problems. In particular, the already-mentioned connection between the group $F_1$ and classical areas of mathematics holds forth the promise of future exciting group-theoretic developments. And this does not take into account future applications to other areas of mathematics.

Thus the obituary for finite group theory has been totally premature. Nevertheless the completion of the classification was accompanied by a sense of loss as well as exhilaration, for there was a clear realization that the "team" would now begin to disperse. Group theory is simply too broad a field to sustain the degree of cohesiveness inspired by the pursuit of the finite simple groups.

# 2

# The Known Simple Groups

In this chapter we give a brief account of the known simple groups. These include the trivial groups of prime order, the alternating groups of degree at least 5, the groups of Lie type, and the 26 sporadic groups. The groups of Lie type are analogues over finite fields of the complex Lie groups, and they constitute the broadest category of known simple groups. The finite case involves a few more "types" than in the complex case; in particular, the family of Suzuki groups and the two families of Ree have no direct complex analogues. The groups of Lie type will be described in Section 2.1, the balance of the chapter being devoted to the sporadic groups.

## 2.1. The Groups of Lie Type

There are four families of simple complex Lie groups: $A_n(\mathbb{C})$, $B_n(\mathbb{C})$, $C_n(\mathbb{C})$, $D_n(\mathbb{C})$, corresponding, respectively, to the linear groups $SL(n+1,\mathbb{C})$, the orthogonal groups $SO(2n+1,\mathbb{C})$, the symplectic groups $Sp(2n,\mathbb{C})$, and the orthogonal groups $SO(2n,\mathbb{C})$. In addition, there are five exceptional Lie groups $G_2(\mathbb{C})$, $F_4(\mathbb{C})$, $E_6(\mathbb{C})$, $E_7(\mathbb{C})$, $E_8(\mathbb{C})$. As Elie Cartan showed in the last century, these arise as automorphism groups of the corresponding simple Lie algebras. Finite analogues of many of these groups were known long before Chevalley—certainly the analogues of the classical groups; but also Leonard Dickson had constructed the analogue of $G_2(\mathbb{C})$ and $E_6(\mathbb{C})$ early in the century. However, Chevalley gave the first systematic treatment of the whole subject [66].

He first proved that every complex Lie algebra $L$ has an *integral* basis—i.e., with respect to this basis the coefficients of the Lie product of any two elements of $L$ are all *integers*. (It was known well before Chevalley that a basis existed with respect to which the coefficients were rational.) For any field $K$ there is a natural ring homomorphism of the ordinary integers $\mathbb{Z}$ into the prime field of $K$. Thus, with the aid of this integral basis, Chevalley was able to view the exponentiation formulas describing the automorphisms of $L$ that generate the corresponding Lie group as defined relative to the given field $K$ rather than to the complex numbers $\mathbb{C}$. By this method he was led to define, for each choice of $L$ and $K$, a specific group by means of an appropriate set of generators (the so-called "adjoint" group associated with $L$ and $K$). In particular, these groups are finite when $K$ is finite.

We shall outline Chevalley's procedure, first recalling some basic facts and notation about finite-dimensional complex Lie algebras (see [186, 272] for the details). By definition, such an algebra is a finite-dimensional complex vector space, equipped with a (Lie) product (denoted by $[xy]$) satisfying

$$[[xy]z]+[[zx]y]+[[yz]x]=0, \qquad [(cx)y]=c[xy]$$

for $x, y, z \in L$ and $c \in \mathbb{C}$. $L$ acts on itself by right Lie multiplication, and for $x \in L$ one writes $\mathrm{ad}(x)$ for the linear transformation of $L$ defined by $a \mapsto ax$ for $a \in L$. The mapping $x \mapsto \mathrm{ad}(x)$ gives an (algebra) *representation* of $L$ as an algebra $\mathrm{ad}(L)$ of linear transformations on $L$ and is called the *adjoint* representation of $L$.

The derived algebra $L'=[LL]$ is the subalgebra generated by the sums of all products $[xy]$, $x, y \in L$. $L$ is *abelian* if $L'=0$. Inductively one defines $L^k=[L^{k-1}L]$ for all $k \geqslant 2$ in similar fashion. $L$ is nilpotent if $L^k=0$ for some $k$.

A subalgebra $H$ of $L$ is called a *Cartan* subalgebra if $H$ is nilpotent and its own normalizer (i.e., if for $x \in L$, $[hx] \in H$ for all $h \in H$, then $x \in H$). Every finite-dimensional algebra $L$ possesses a Cartan subalgebra.

For $x, y \in L$, set

$$(x, y)=\mathrm{Trace}(\mathrm{ad}(x)\mathrm{ad}(y)).$$

Then $(\ ,\ )$ is a symmetric bilinear form on $L$ with values in $\mathbb{C}$; it is called the *Killing form* on $L$. By definition, $L$ is *semisimple* if its Killing form is nondegenerate [nonsingular, i.e., $(x, y)=0$ for all $y \in L$ implies $x=0$].

We are interested here only in the semisimple case. In this case, $L \cong \mathrm{ad}(L)$ [i.e., $\mathrm{ad}(x)=0$ if and only if $x=0$]. Furthermore, $L$ is then the (algebra) direct sum of *simple* subalgebras (i.e., algebras possessing no nontrivial proper two-sided ideals), so that the semisimple theory is reducible to the simple case.

Under our continuing assumption of semisimplicity, any Cartan subalgebra $H$ of $L$ is necessarily abelian. Furthermore, the restriction of the adjoint representation of $L$ to $H$ gives a fundamental (vector space) decomposition of $L$ as the direct sum of $H$-invariant subspaces. Its description depends on the basic notions of "root" and "root space." We can define a *root* of $H$ as a complex-valued function $\alpha$ on $H$ with the property that for some $x \neq 0$ in $L$ and all $h \in H$

$$x(h-\alpha(h)1)=0.$$

For a given $\alpha$, the set of such $x$ form an $H$-invariant subspace of $L$, called the *root space* associated with $\alpha$ and denoted by $L_\alpha$.

Since $H$ is a Cartan subalgebra, it is immediate that the root space $L_0$ of the 0-function on $H$ is precisely the subspace $H$ itself.

We let $\Sigma$ denote the set of roots of $H$. Then the main result is as follows:

PROPOSITION 2.1. (i) $L=H\bigoplus_{0\neq\alpha\in\Sigma} L_\alpha$.
(ii) *Each* $L_\alpha$ *is one-dimensional over* $\mathbb{C}$.

The roots are actually *linear* functions on $H$ and so are elements of the *dual* space $H^*$ of $H$. We adopt the convention that $L_\alpha=0$ for $\alpha\in H^*$ and $\alpha$ not a root. We then have the basic relations

PROPOSITION 2.2. $[L_\alpha L_\beta]\subseteq L_{\alpha+\beta}$ *for all* $\alpha$, $\beta\in\Sigma$.

Thus Propositions 2.1 and 2.2 provide a good initial description of the Lie multiplication in $L$.

The dimension $l$ of $H$ over $\mathbb{C}$ is called the *rank* of $L$. Thus $\dim_{\mathbb{C}} H^*=l$. The roots of $H$ span $H^*$, so that $\Sigma$ contains a basis for $H^*$.

The nondegeneracy of the Killing form implies that for each $0\neq\alpha\in\Sigma$, there is a uniquely determined element $t_\alpha\in H$ such that for each $h\in H$

$$(h,t_\alpha)=\alpha(h);$$

in addition, the map: $\alpha\mapsto t_\alpha$ of $H^*$ to $H$ is one-to-one and onto.

We use this correspondence to define a bilinear form [again denoted by ( , )] on $H^*$ by setting

$$(\alpha,\beta)=(t_\alpha,t_\beta)$$

for all $\alpha$, $\beta\in H^*$.

The roots satisfy certain crucial rationality conditions. Indeed, if $\mathbb{Q}$ denotes the field of rational numbers, we have

PROPOSITION 2.3. (i) $(\alpha, \beta) \in \mathbb{Q}$ *for all* $\alpha, \beta \in \Sigma$.

(ii) *If* $\alpha_1, \alpha_2, \ldots, \alpha_l \in \Sigma$ *form a basis for* $H^*$, *then every* $\alpha \in \Sigma$ *is a* $\mathbb{Q}$-*linear combination of* $\alpha_1, \alpha_2, \ldots, \alpha_l$.

In view of this, we can work with the $\mathbb{Q}$-span of $\Sigma$ in $H^*$. For simplicity, we set $E = H^*_{\mathbb{Q}}$. Then $E$ is an $l$-dimensional vector space over $\mathbb{Q}$, $\Sigma \subseteq E$, and $\Sigma$ spans $E$. Furthermore, one shows that the restriction of $(\ ,\ )$ to $E$ is a symmetric nondegenerate positive-definite bilinear form on $E$, so that $E$ is, in fact, an $l$-dimensional rational *Euclidean* vector space with respect to $(\ ,\ )$.

The roots of $H$ have the following further important properties:

PROPOSITION 2.4. (i) *For* $\alpha \in \Sigma$, $-\alpha \in \Sigma$ *and if* $\alpha \neq 0$, *then* $k\alpha \notin \Sigma$ *for any integer* $k \neq 0$, $+1$, *or* $-1$.

(ii) *For* $\alpha, \beta \in \Sigma$, *with* $\beta \neq 0$, $2(\alpha, \beta)/(\beta, \beta)$ *is an integer.*

(iii) *For each* $0 \neq \alpha \in \Sigma$, *the reflection* $w_\alpha$ *of* $E$ *in the hyperplane orthogonal to* $\alpha$ [*orthogonality defined with respect to* $(\ ,\ )$] *leaves* $\Sigma$ *invariant.*

The $w_\alpha$ are defined analytically by the condition

$$w_\alpha e = e - (2(e, \alpha)/(\alpha, \alpha))\alpha$$

for all $e \in E$.

The group $W$ generated by the $w_\alpha$ is called the *Weyl* group of $L$. $W$ is thus a group of *symmetries* of $\Sigma$.

It is natural to use Proposition 2.4(ii) to normalize our inner product. Thus for $e, e' \in E$ with $e' \neq 0$, we set

$$\langle e, e' \rangle = 2(e, e')/(e', e').$$

Then $\langle\ ,\ \rangle$ is a positive-definite (but not symmetric) bilinear form on $E$, and it has the key property that $\langle \alpha, \beta \rangle \in \mathbb{Z}$ for all $\alpha, \beta \in \Sigma$.

We need one final notion before we can state the Chevalley integral basis theorem. Let $\alpha_1, \alpha_2, \ldots, \alpha_l$ be any set of roots that form a basis for $E$. By Proposition 2.3, every $\alpha \in \Sigma$ is a $\mathbb{Q}$-linear combination of $\alpha_1, \alpha_2, \ldots, \alpha_l$.

DEFINITION 2.5. A root $\alpha \in \Sigma - \{0\}$ is called *positive* with respect to the ordered set $\alpha_1, \alpha_2, \ldots, \alpha_l$ if the first nonzero coefficient of $\alpha$ as a $\mathbb{Q}$-linear combination of $\alpha_1, \alpha_2, \ldots, \alpha_l$ is positive; otherwise $\alpha$ is called *negative*. Furthermore, for $\alpha$, $\beta \in \Sigma$, write $\alpha > \beta$ (in the given ordering) if $\alpha - \beta$ is a positive root. Finally, call a positive root *simple* (in the given ordering) if $\alpha$ is not a sum of two positive roots.

Note that the map $\alpha_i \mapsto -\alpha_i$, $1 \leqslant i \leqslant l$, is an element of the Weyl group $W$ of $L$ (of order 2), so the positive and negative roots are interchanged by an involution of $W$.

The following result describes the basic properties of simple roots.

PROPOSITION 2.6. *Let $\alpha_1, \alpha_2, \ldots, \alpha_l$ be roots of $\Sigma$ that form a basis of $E$ and let $\pi$ be the subset of simple roots of $\Sigma$ with respect to the ordering in $E$ determined by the given basis. Then we have*

(i) *$\pi$ is a basis of $E$.*
(ii) *If $\alpha$, $\beta \in \pi$ with $\alpha \neq \beta$, then $\langle \alpha, \beta \rangle \leqslant 0$ and $\alpha - \beta$ is not a root.*
(iii) *If $\alpha$ is a positive root, then $\alpha$ is a linear combination of the roots in $\pi$ with nonnegative integer coefficients.*
(iv) *If $\alpha$ is a positive root not in $\pi$, then $\alpha - \beta$ is a positive root for some $\beta \in \pi$.*

In particular, if one considers the ordering on $E$ determined by the basis $\pi$, the proposition implies that the elements of $\pi$ are precisely the simple roots with respect to that ordering.

Given the various properties enjoyed by simple root systems, it is natural to restrict one's attention to such root bases of $E$.

We can now state Chevalley's integral basis theorem.

THEOREM 2.7. *Let $\alpha_1, \alpha_2, \ldots, \alpha_l$ be a simple system of roots. Also set $h_\alpha = 2t_\alpha/(\alpha, \alpha)$ for all $\alpha \in \Sigma - \{0\}$. Then there exist suitable generators $x_\alpha$ of $L_\alpha$, $\alpha \in \Sigma - \{0\}$, such that*:

(i) $[h_\alpha x_\beta] = \langle \beta, \alpha \rangle x_\beta$ *for all* $\alpha$, $\beta \in \Sigma$, $\beta \neq 0$.
(ii) $[x_\alpha x_{-\alpha}] = h_\alpha$ *is an integral linear combination of the* $h_{\alpha_i}$*'s*, $1 \leqslant i \leqslant 1$.
(iii) $$[x_\alpha x_b] = \begin{cases} 0 \ \textit{if}\ \alpha + \beta \notin \Sigma. \\ (r+1)x_{\alpha+\beta} \ \textit{if}\ \alpha+\beta \in \Sigma, \textit{where } r \textit{ is} \\ \textit{an integer uniquely determined by} \\ \textit{the condition } \beta - r\alpha \in \Sigma,\ \beta - (r+1)\alpha \notin \Sigma. \end{cases}$$

Thus the $h_{\alpha_i}$ and the $x_\alpha$ form a basis for $L$ and the (Lie) multiplication of any two basis elements is an *integral* linear combination of the given basis elements.

The structure of $L$ is completely determined by the *integral* matrix $(\langle\alpha_i, \alpha_j\rangle)$. It is called a *Cartan* matrix of $L$.

Denoting its entries by $a_{ij}$, one associates with the simple roots $\alpha_i$ (equivalently with the corresponding Cartan matrix) a *Dynkin diagram*, which consists of $l$ nodes, with two nodes connected by $a_{ij}a_{ji}$ edges [since $(\alpha_i, \alpha_j)$ is symmetric, $a_{ij}a_{ji}$ is nonnegative]. In addition, one assigns the "weight" $(\alpha_i, \alpha_i)$ to the $i$th node.

The simple root system $\pi = \{\alpha_1, \alpha_2, \ldots, \alpha_l\}$ is called *indecomposable* if it cannot be written as the sum of two disjoint nonempty subsets $\pi_1, \pi_2$ such that $(\alpha_i, \alpha_j) = 0$ for all $\alpha_i \in \pi_1$, $\alpha_j \in \pi_2$. This is equivalent to the *connectedness* of the associated Dynkin diagram. The important fact here is that $\pi$ is indecomposable if and only if $L$ is simple. Thus, in general, $\pi$ decomposes into indecomposable subsystems $\pi_1, \pi_2, \ldots, \pi_m$ which are simple root systems of the simple factors $L_1, L_2, \ldots, L_m$ of the decomposition of $L$.

Thus the classification of simple (and semisimple) Lie algebras is completely determined from the classification of connected Dynkin diagrams. The result is as follows:

THEOREM 2.8. *A finite-dimensional simple complex Lie algebra has a Dynkin diagram of one of the forms shown in Figure 2.1; and every such diagram occurs for some simple Lie algebra. Here either all nodes have the same weight or the diagram is partitioned into two connected subsets by the symbol* $>$, *with the nodes on each side having the same weight and the direction of the vertex designating the side with the smaller weight.*

Each of the above simple Lie algebras can be constructed as a factor algebra of a certain "free" complex Lie algebra generated by $3l$ elements by an ideal whose definition depends upon the corresponding set of integers $\langle\alpha_i, \alpha_j\rangle$ (cf. [186, Chapter 7]).

The simple Lie groups are generated by certain "one-parameter" families of automorphisms of their Lie algebras. To describe these, we first define a *derivation* of $L$ as a linear map $D$ of $L$ into $L$ such that for all $x, y \in L$

$$([xy])D = [(xD)y] + [x(yD)].$$

Thus derivations satisfy the basic rules of differentiation for sums and products. Motivation for the concept comes from the fact that for $x \in L$, the

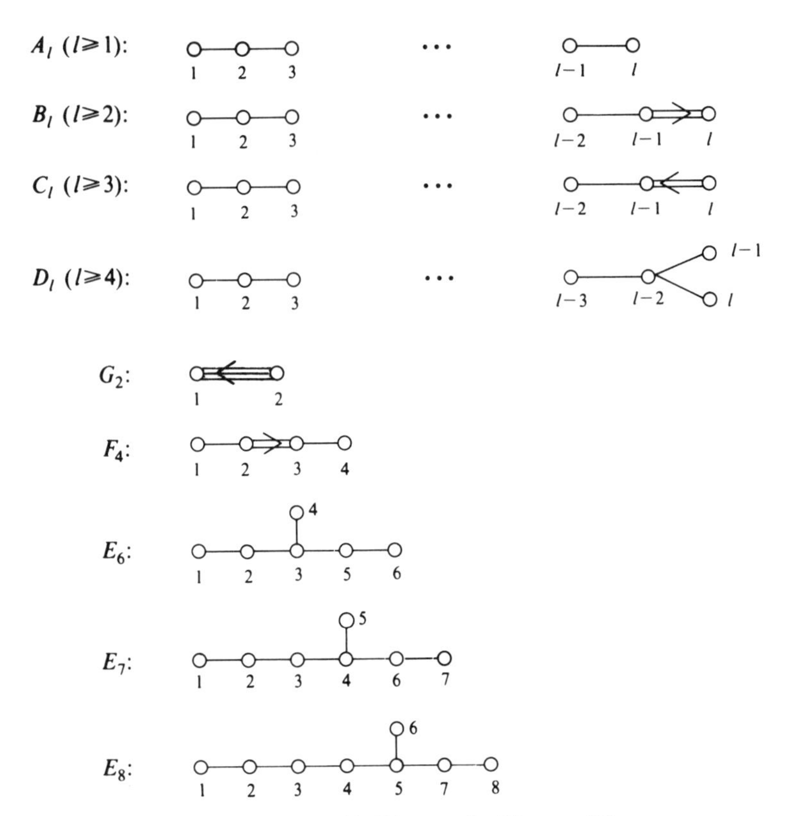


**Figure 2.1.** Dynkin Diagrams for Theorem 2.8.

map $a \mapsto ax$ for $a \in L$ is a derivation of the adjoint algebra $\mathrm{ad}(L)$ of $L$. Since $L$ is semisimple, $L \cong \mathrm{ad}(L)$, so $\mathrm{ad}(x)$ can be viewed as a derivation of $L$.

A derivation $D$ is said to be *nilpotent* if $D^n = 0$ for some positive $n$. If $D$ is nilpotent, the *exponential* of $D$ is defined to be

$$\exp(D) = 1 + D + D^2/2! + D^3/3! + \cdots.$$

A purely formal algebraic calculation gives the following basic fact.

PROPOSITION 2.9. *If $D$ is a nilpotent derivation of $L$, then* $\exp(D)$ *is an automorphism of $L$.*

We know from the multiplication formulas for $L$ that for any root $\alpha \in \Sigma$, $2\alpha$ is not a root and for any root $\beta \neq 0$, $\beta + i\alpha$ is not a root for all sufficiently large values of $i$. It follows that $(\mathrm{ad}(x_\alpha))^n$ is the zero linear transformation of $L$ for some positive $n$, so, in fact, $\mathrm{ad}(x_\alpha)$ is a nilpotent derivation of $L$ for each $\alpha$. Hence so is $\mathrm{ad}(tx_\alpha)$ for each $t \in \mathbb{C}$. Set

$$X_\alpha(t) = \exp(\mathrm{ad}(tx_\alpha))$$

for each nonzero root $\alpha$ and each $t \in \mathbb{C}$.

For a fixed nonzero $\alpha \in \Sigma$, the set of $X_\alpha(t)$ form an abelian group of automorphisms of $L$, isomorphic to $\mathbb{C}$. These are the desired one-parameter families of automorphisms. The fundamental fact here is that the associated Lie group of $L$ is precisely the group

$$\langle X_\alpha(t) | 0 \neq \alpha \in \Sigma, t \in \mathbb{C} \rangle.$$

The existence of the Chevalley integral basis for $L$ permits one to free the above construction from the specific field $\mathbb{C}$ of complex numbers, carrying it over to an arbitrary field $K$ by means of a suitable tensoring process. To describe the generalized construction exactly requires some preliminary formalism.

Let $A$ be any associative algebra over $\mathbb{C}$. If for any $x, y \in A$, we let $[x, y] = xy - yx$, then with respect to this product, $A$ is a Lie algebra, which we denote by $A_{\mathfrak{L}}$.

Now if $L$ is any Lie algebra (over $\mathbb{C}$), a *universal enveloping* algebra $A$ of $L$ is an associative algebra $A$ together with a Lie algebra homomorphism $\phi: L \mapsto A_{\mathfrak{L}}$ such that if $B$ is any associative algebra and $\theta: L \mapsto B$ is a Lie algebra homomorphism, then there exists a unique associative algebra homomorphism $\psi: A \mapsto B$ such that $\theta = \psi \circ \phi$.

Universal enveloping algebras always exist and are unique up to isomorphism. The Birkhoff–Witt theorem [186, Chapter 5] gives a clear description of them, namely, the homomorphism $\phi$ is one-to-one, so that $L$ can be identified with its image in its universal enveloping algebra $A$. Furthermore, with this identification, if $y_1, y_2, \ldots, y_m$ are a linear basis for $L$ over $\mathbb{C}$, then the set of monomials $y_1^{k_1} y_2^{k_2} \cdots y_m^{k_m}$ form a $\mathbb{C}$-basis for $A$.

Now consider our basis $\{h_{\alpha_i}, x_\alpha\}$ of $L$. For simplicity, put $h_i = h_{\alpha_i}$, $1 \leq i \leq l$. Let $A_{\mathbb{Z}}$ denote the subalgebra over $\mathbb{Z}$ generated by the elements $x_\alpha^j / j!$. One next establishes the following key description of $A_{\mathbb{Z}}$. For each choice of positive integers $n_i, j_\alpha$, $1 \leq i \leq l$, $\alpha \in \Sigma$, form the product in $A$ of all elements $\binom{h_i}{n_i} = h_i(h_i - 1) \cdots (h_i - n_i + 1)/1 \cdot 2 \cdots n_i$ and all elements $x_\alpha^j / j\,!$, the products taken in some fixed order (note that these expressions are uniquely determined elements of $A$ inasmuch as $A$ has an associative product and scalar multiplication by $\mathbb{C}$ is well defined). Then the resulting collection of products is a $\mathbb{Z}$-basis for $A_{\mathbb{Z}}$.

Next an abelian subgroup $M$ of $L$ is called a *lattice* in $L$ provided there are elements $y_1, y_2, \ldots, y_m$ in $M$ which form a $\mathbb{C}$-basis for $L$ and a $\mathbb{Z}$-basis for $M$ (i.e., every element of $M$ is an integral linear combination of $y_1, y_2, \ldots, y_m$).

Now let $H_{\mathbb{Z}}$ denote the set of $h \in H$ such that $\alpha(h)$ is an integer for all $\alpha \in \Sigma$ and let $L_{\mathbb{Z}}$ be the $\mathbb{Z}$-span of $H_{\mathbb{Z}}$ and the set of $x_\alpha$, $\alpha \in \Sigma$. The main result is the following:

THEOREM 2.10. (i) *$L_{\mathbb{Z}}$ is a lattice in $L$.*
(ii) *$A_{\mathbb{Z}}$ leaves $L_{\mathbb{Z}}$ invariant.*

We are at last in a position to define the Chevalley groups. First, we form the tensor product $L^K = L_{\mathbb{Z}} \otimes_{\mathbb{Z}} K$ for any field $K$. Then $L^K$ is a vector space over $K$ and the action of $A_{\mathbb{Z}}$ on the lattice $L_{\mathbb{Z}}$ extends to $L^K$.

Just as in the complex case, we now define for each $\alpha$ and each $t \in K$,

$$\exp(tx_\alpha) = \sum_{j=0}^{\infty} t^j x_\alpha^j / j!. \tag{2.1}$$

The right side of the expression is to be interpreted as follows. Each $x_\alpha^j / j!$, being an element of $A_{\mathbb{Z}}$, acts on $L_{\mathbb{Z}}$; it follows that if $\lambda$ is an indeterminate, $\lambda^j x_\alpha^j / j!$ acts on $L_{\mathbb{Z}} \otimes_{\mathbb{Z}} \mathbb{Z}[\lambda]$, where $\mathbb{Z}[\lambda]$ denotes the polynomial ring in $\lambda$ over $\mathbb{Z}$. Since $x_\alpha^j$ acts as 0 for $j$ sufficiently large, it follows that $\exp(\lambda x_\alpha)$ acts on $L_{\mathbb{Z}} \otimes_{\mathbb{Z}} \mathbb{Z}[\lambda]$ and hence on $L_{\mathbb{Z}} \otimes_{\mathbb{Z}} \mathbb{Z}[\lambda] \otimes_{\mathbb{Z}} K$. The mapping $\lambda \mapsto t$ determines a homomorphism of this last space onto $L^K = L_{\mathbb{Z}} \otimes_{\mathbb{Z}} K$ and thus induces an action of $\exp(tx_\alpha)$ on $L^K$.

By analogy with the complex case, we set

$$X_\alpha(t) = \exp(tx_\alpha) \qquad \text{for } \alpha \in \Sigma \text{ and } t \in K.$$

Then we have

DEFINITION 2.11. The group $\langle X_\alpha(t) | \alpha \in \Sigma,\ t \in K \rangle$ is called the (adjoint) *Chevalley group* over $K$ associated with the Lie algebra $L$.

It is easily checked from (2.1) that

$$X_\alpha(t+u) = X_\alpha(t) X_\alpha(u)$$

for a fixed $\alpha$ and $t, u \in K$, so that the set $X_\alpha(t)$, $t \in K$, forms an abelian group $\chi_\alpha$ isomorphic to the additive group of $K$. $\chi_\alpha$ is called a *root* subgroup of the given Chevalley group, which is thus generated by its root subgroups. (We shall have more to say about relations that hold between distinct root subgroups below and in Section 3.1.)

We remark that the Chevalley groups we have constructed are called "adjoint" because they have been constructed from the *adjoint* representation of $L$, with which we have worked throughout the discussion. However, the entire process can be extended to other faithful representations of $L$ and lead to other forms of the Chevalley groups. More precisely, adjoint Chevalley groups always have trivial center, and the other forms turn out to be covering groups of the adjoint version. Furthermore, there exists a representation that leads to a *universal* Chevalley group, the other versions being homomorphic images of universal groups. For example, in the case of $A_l$, the Lie algebra of $(l+1)\times(l+1)$ complex matrices of trace 0 with Lie product $[XY]=XY-YX$ for $X, Y\in A_l$, the adjoint Chevalley group is $PSL(l+1, K)$ while the universal group is $SL(l+1, K)$, the latter obtained from the natural $(l+1)$-dimensional representation of $A_l$. (This is fully discussed in [272].)

We note that in the complex case, the universal Chevalley groups are precisely the *simply connected* Lie groups.

If we take the field $K$ to be finite [whence $K=GF(q)$ for some prime power $q$], we obtain the *finite* (adjoint) Chevalley groups $A_l(q)$, $B_l(q)$, $C_l(q)$, $D_l(q)$, $G_2(q)$, $F_4(q)$, $E_6(q)$, $E_7(q)$, and $E_8(q)$. Likewise there exist universal versions of each of these groups.

The discussion has been so formal that it will be helpful to consider an example. Let then $G^*=SL(n,q)$ (where $n=l+1$). Let $H$ be the Cartan subgroup consisting of diagonal matrices of determinant 1. We let $h_{ij}(t)$ be the diagonal matrix with $0\neq t\in GF(p^m)$ in the $ii$th position and $t^{-1}$ in the $jj$th position, and 1's elsewhere. Also let $U_{ij}$ be the elementary abelian subgroup of $SL(n, p^m)$ of order $p^m$ consisting of the matrices with 1's along the diagonal and 0's elsewhere except in the $ij$th position, $i\neq j$, $1\leqslant i, j\leqslant n$. Then each $U_{ij}$ is irreducible as an $H$-module, $H/C_H(U_{ij})$ is cyclic of order $p^m-1$ [and is covered by $\langle h_{ij}(t)|t\in GF(p^m)\rangle$] and $H/C_H(U_{ij})$ transitively permutes the elements of $U_{ij}^{\#}$. We can identify $H/C_H(U_{ij})$ with the multiplicative group of $GF(p^m)$ and $U_{ij}$ with its additive group. We thus write $U_{ij}(t)$, $t\in GF(p^m)$, for the typical element of $U_{ij}$.

It is easy to check that the $U_{ij}(t)$ generate $G^*$. [As we shall see, the $U_{ij}(t)$ are precisely our "root elements" $X_\alpha(t)$.] Note also that if we set

$$U=\langle U_{ij}(t)|i<j, t\in GF(q)\rangle \qquad \text{and} \qquad V=\langle U_{ij}(t)|i>j, t\in GF(q)\rangle,$$

then $U$ and $V$ are each $H$-invariant Sylow $p$-subgroups of $G^*$, consisting, respectively, of lower and upper triangular matrices with 1's on the diagonal. Thus there is a natural partition of the roots into two subsets—"positive" and "negative." This is a general phenomenon in the groups of Lie type.

Multiplication of the $U_{ij}(t)$ is easily checked to be given by the following rule:

$$[U_{ij}(t), U_{i'j'}(t)] = \delta_{i,j'} U_{i'j}(-tt') + \delta_{j,i'} U_{ij'}(tt') \qquad (2.2)$$

(where $\delta_{i,j}$ is the Kronecker $\delta$), whenever $i \neq j$, $i' \neq j'$ and $i \neq j'$ or $i' \neq j$ (or both).

Furthermore, we can recover the $h_{ij}(t)$ from the $U_{ij}(t)$:

$$h_{ij}(t) = U_{ij}(t) U_{ji}(-t^{-1}) U_{ij}(t) U_{ij}(-1) U_{ji}(1) U_{ij}(-1). \qquad (2.3)$$

Thus we have a good description of the group $SL(n,q)$ in terms of its root subgroups $U_{ij}(t)$. It remains only to reinterpret the $U_{ij}(t)$ in terms of the $X_\alpha(t)$'s.

First of all, the associated Lie algebra of $G^*$ is the algebra $A_l$ of $n \times n$ matrices over $GF(q)$ of trace 0. The subalgebra $H$ of all such diagonal matrices is a Cartan subalgebra. Let $I_{ij}$ be the matrix with 1 in the $ij$th position and 0 in all other positions. Then the $n^2-1$ elements $I_{ij}$, $i \neq j$, $1 \leqslant i$, $j \leqslant n$, and $I_{ii} - I_{11}$, $2 \leqslant i \leqslant n$, clearly form a vector space basis for $L$, with the last $n-1$ elements forming a basis for $H$. Furthermore, the 1-dimensional space $L_{ij}$ spanned by $I_{ij}$ for $i \neq j$ are $H$-variant. Thus we have an $H$-invariant decomposition of $L$ as

$$L = H \oplus \sum_{i \neq j} L_{ij}.$$

On $L_{ij}$, the element $h = \mathrm{diag}(h_1, h_2, \ldots, h_n)$ of $H$ acts as

$$hI_{ij} = (h_i - h_j) I_{ij},$$

so $h_i - h_j$ is a characteristic root of $h$ with associated characteristic vector $I_{ij}$. Thus, for each $i$, $j$ with $i \neq j$, there corresponds a root $\alpha_{ij}$ of $H$, given by

$$\alpha_{ij}(h) = h_i - h_j \qquad \text{for } h = \mathrm{diag}(h_1, h_2, \ldots, h_n) \in H.$$

$L_{ij}$ is the root space of $\alpha_{ij}$, and we can take the element $I_{ij}$ as basis element $x_{ij}$ of $L_{ij}$.

Finally, by definition, $X_{\alpha_{ij}}(t)=\exp(tx_{ij})=\exp(tI_{ij})$ for $t\in GF(q)$, $i\neq j$. But for each $i\neq j$, we see that $I_{ij}^2=0$, so $X_{\alpha_{ij}}(t)=I+tI_{ij}$. Furthermore, by its definition, clearly $U_{ij}(t)=I+tI_{ij}$. Hence in fact

$$X_{\alpha_{ij}}(t)=U_{ij}(t).$$

Commutator formulas for the $X_\alpha(t)$ similar to (2.2) and reconstruction of the "Cartan" subgroup from the $X_\alpha(t)$'s as in (2.3) hold for all Chevalley groups.

THEOREM 2.12. *If $G^*$ is a Chevalley group over $GF(q)$ (adjoint or universal), then*

(i) $[X_\alpha(t), X_\beta(u)]=\Pi X_{i\alpha+j\beta}(c_{ij}t^iu^j)$ *for* $\alpha,\beta\in\Sigma-\{0\}$ *and* $t,u\in GF(q)$, *where* $i$, $j$ *is restricted to positive integral values for which* $i\alpha+j\beta$ *is a root and the* $c_{ij}$ *are suitable integers which depend on* $\alpha,\beta$ *but not on* $t$ *and* $u$.
(ii) *If we set* $h_\alpha(t)=X_\alpha(t)X_{-\alpha}(-t^{-1})X_\alpha(t)X_\alpha(-1)X_{-\alpha}(1)X_\alpha(-1)$, *then* $h_\alpha(tu)=h_\alpha(t)h_\alpha(u)$ *for* $\alpha\in\Sigma-\{0\}$ *and* $t$, $u\in GF(q)-\{0\}$.

Equation (i) is known as the *Chevalley commutator* formula. (ii) shows that for fixed $\alpha$, the set $h_\alpha(t)$ is a cyclic group $H_\alpha$ isomorphic to a homomorphic image of the multiplicative group of $GF(q)$. (For universal groups, this is an actual isomorphism.) Furthermore, it can be shown that the subgroup $H=\langle H_\alpha|\alpha\in\Sigma-\{0\}\rangle$ is abelian and normalizes each root subgroup $\chi_\beta$. $H$ is called a *Cartan* subgroup of $G^*$.

Just as the complex unitary groups $GU(n,\mathbb{C})$ are obtained from $GL(n,\mathbb{C})$, so likewise there exist finite variants of the Chevalley groups. The general theory was worked out by Robert Steinberg [269], who not only constructed the groups but also showed that they had an internal structure similar to that described above for the Chevalley groups. To define these "twisted" groups, we first recall the situation in the complex case. The group $GU(n,\mathbb{C})$ consists of all nonsingular complex matrices $U\in GL(n,\mathbb{C})$ such that

$$U=\left((\overline{U})^t\right)^{-1},$$

where $(\overline{U})^t$ denotes the transpose of the complex conjugate $\overline{U}$ of $U$. One can describe the unitary group more algebraically as follows. For each element

$X \in GL(n,\mathbb{C})$, set

$$\alpha(X) = \left((\bar{X})^t\right)^{-1}.$$

Then $\alpha$ is an automorphism of $GL(n,\mathbb{C})$ of period 2; and the unitary group is precisely the set of elements of $GL(n,\mathbb{C})$ left fixed by $\alpha$. In the Lie notation, the unitary group is denoted by ${}^2A_l(\mathbb{R})$ $(l=n-1)$. It is the compact real form of the linear group.

Similarly, conjugation of the orthogonal group $SO(2n,\mathbb{C})$, by an orthogonal transformation of determinant $-1$ and order 2 induces an outer automorphism of period 2. In the Lie theory, such an automorphism corresponds to a so-called "graph" automorphism of $D_n(\mathbb{C})$ (i.e., automorphisms induced from a symmetry of the associated Dynkin diagram). Moreover, the exceptional group $E_6(\mathbb{C})$ also has a graph automorphism of period 2. Hence, if we let $\alpha$ again denote the product of complex conjugation with the corresponding graph automorphism, we can similarly define the compact real form ${}^2D_n(\mathbb{R})$ of the orthogonal group $D_n(\mathbb{C})$ and the compact real form ${}^2E_6(\mathbb{R})$ of $E_6(\mathbb{C})$.

Steinberg showed that this entire process carries over to the Chevalley groups over arbitrary fields. In particular, he obtained finite analogues ${}^2A_l(q), {}^2D_l(q), {}^2E_6(q)$ of the corresponding compact real Lie groups, where in place of complex conjugation (which is an automorphism of $\mathbb{C}$ of period 2), one uses the Frobenius automorphism $\phi$ of the field $GF(q^2)$ of period 2 defined by

$$\phi(x) = x^q \qquad \text{for } x \in GF(q).$$

In particular, one can define in this way the finite general unitary group $GU(n,q)$ and its subgroup $SU(n,q)$ of elements of determinant 1. In the Lie notation, the adjoint group ${}^2A_l(q)$ has the form:

$${}^2A_l(q) = SU(l+1,q)/(\text{mod scalars}) = U_{l+1}(q) = PSU(l+1,q).$$

A word about the orthogonal groups. Over $\mathbb{C}$, all nondegenerate symmetric bilinear forms are equivalent, and so for each $m$ there is a uniquely determined group $SO(m,\mathbb{C})$. However, over $GF(q)$, there are always *two* such inequivalent forms, which determine two families of orthogonal groups (cf. Section 4.14). The corresponding groups are distinct if $m$ is even and are denoted by $SO^+(m,q)$ and $SO^-(m,q)$, respectively. On the other hand, if $m$ is odd, they are the same groups and are denoted by $SO(m,q)$ [sometimes, for convenience, by $SO^+(m,q)$]. These groups are, in general, nonsimple. One denotes their derived groups by $\Omega_m^+(q)$ and $\Omega_m^-(q)$, respectively, and one has

$$|SO^{\pm}(m,q):\Omega_m^{\pm}(q)|=1,2,\text{ or }4 \quad\text{and}\quad |Z(SO^{\pm}(m,q))|=1\text{ or }2.$$

Finally, the factor groups $\Omega_m^{\pm}(q)/Z(\Omega_m^{\pm}(q))$ are denoted correspondingly by $P\Omega_m^{\pm}(q)$ and $P\Omega_m^-(q)$. For $m$ odd, one also writes simply $\Omega_m(q)$ and $P\Omega_m(q)$. In the Lie notation, one has the following isomorphisms:

$$P\Omega_{2n+1}(q)=B_n(q), \qquad P\Omega_{2n}^+(q)=D_n(q), \qquad\text{and}\qquad P\Omega_{2n}^-(q)={}^2D_n(q).$$

The notation $O_m^{\pm}(q)$ for $SO^{\pm}(m,q)/Z(SO^{\pm}(m,q))$ is also standard [likewise $O_m(q)$ for $m$ odd].

The classical group $D_4(\mathbb{C})$ is the only complex Lie group having a graph automorphism of period 3 [the fixed subgroup of which is the Lie group $G_2(\mathbb{C})$]. Since $\mathbb{C}$ has no automorphism of period 3, it is not possible to use this automorphism to construct a twisted group as in the other cases. On the other hand, in the field $GF(q^3)$ the Frobenius map $\phi(x)=x^q$ for $x\in GF(q^3)$ is an automorphism of period 3, and taking its product with the graph automorphism of period 3, we can construct a twisted group just as before, called the *triality* twisted $D_4(q)$ and denoted by ${}^3D_4(q)$.

These twisted groups are referred to as the *Steinberg variations* of the Chevalley groups.

Although Suzuki did not discover his family of simple groups $Sz(2^n)$, $n$ odd, $n>1$, from the Lie theory, but rather in the process of determining all simple groups in which the centralizer of every involution is a 2-group [276, 278], Ree observed that they could equally well be constructed by a suitable twisting of the orthogonal groups $B_2(2^n)$. Indeed, it was known earlier that because of certain degeneracies in the multiplication coefficients, the three families $B_2(2^n)$, $G_2(3^n)$, and $F_4(2^n)$ possess an "extra" automorphism, not accounted for by the general theory. From Suzuki's defining relations for his group, Ree was led to play a variation of the Steinberg twisting game with each of these automorphisms [239, 240], taking as $\alpha$ the product of the "extra" automorphism together with a suitable "field" automorphism induced from the Galois group of $GF(p^n)$ (the requirement that $n$ be odd is needed to insure that $\alpha$ can be chosen of period 2). He obtained in this way three families of groups, the first of which was just the groups $Sz(2^n)$. These families are denoted by ${}^2B_2(2^n)$, ${}^2G_2(3^n)$, and ${}^2F_4(2^n)$, $n$ odd. Suzuki's and Ree's work will be described in more detail in Section 3.2.

The (finite) groups of *Lie type* consist of the Chevalley groups together with the Steinberg and Suzuki–Ree variations together with all their covering groups. As we have stipulated, we denote by $Chev(p)$ the subset of these that are defined over $GF(p^m)$ for a given prime $p$. The combined analysis of Chevalley, Steinberg, and Suzuki–Ree (plus Tits [306], who proved the simplicity of the group $({}^2F_4(2)')$ yields the following results.

THEOREM 2.13. *If $G^*$ is a group of Lie type with $Z(G^*)=1$, then either $G^*$ is simple or one of the following holds:*

(i) $G^*=A_1(2)$, $A_1(3)$, ${}^2A_2(2)$, *or* ${}^2B_2(2)$, *and* $G^*$ *is solvable.*
(ii) $G^*=B_2(2)\cong\Sigma_6$.
(iii) $G^*=G_2(2)$, $|G^*:(G^*)'|=2$, *and* $(G^*)'\cong{}^2A_2(3)$.
(iv) $G^*={}^2G_2(3)$, $|G^*:(G^*)'|=3$, *and* $(G^*)'\cong A_1(8)$.
(v) $G^*={}^2F_4(2)$, $|G^*:(G^*)'|=2$, *and* $(G^*)'$ *is simple.*

Implicit in this statement is the exclusion of the families $B_1$, $C_1$, $C_2$, $D_2$, $D_3$, and ${}^2D_3$ because of the following isomorphisms ($D_1$ does not exist in the Lie theory):

$$C_1\cong B_1\cong A_1,\qquad C_2\cong B_2,$$

$$D_2\cong A_1\times A_1,\quad D_3\cong A_3,\quad {}^2D_2(q)\cong A_1(q^2),\quad {}^2D_3\cong{}^2A_3. \tag{2.4}$$

Furthermore, one has

$$B_n(q)\cong C_n(q)\qquad \text{for } q\cong 2^m. \tag{2.5}$$

There are some other isomorphisms among the simple groups of Lie type of low order, namely,

$$A_1(4)\cong A_1(5),\quad A_1(7)\cong A_2(2),\quad \text{and}\quad B_2(3)\cong{}^2A_3(2). \tag{2.6}$$

In addition, three groups of Lie type are isomorphic to alternating groups:

$$A_1(4)\cong A_5,\quad A_1(9)\cong A_6,\quad \text{and}\quad A_3(2)\cong A_8. \tag{2.7}$$

The Chevalley groups (as well as the other groups of Lie type) have a uniform description in terms of their so-called *Bruhat* decomposition. This is true for Chevalley groups over arbitrary fields, but we limit our statement to the finite case. The Weyl group $W$ of $L$ lifts in a natural way to the corresponding Chevalley group $G^*$, normalizing the Cartan subgroup $H$. Actually if $N = N_{G^*}(H)$, then $N/H \cong W$, but it may or may not be the case that $N$ splits over $W$. However, it does hold for $G^* \cong SL(n,q)$, where $W$ is the subgroup of permutation matrices (whence $W \cong \Sigma_n$) and $N$ the subgroup of monomial matrices.

If we let $P$ be the subgroup of $G^*$ generated by the set of $\chi_\alpha$ with $\alpha$ a positive root, then as in the $SL(n,q)$ case, $P$ is a Sylow $p$-subgroup of $G^*$ and is invariant under $H$. The subgroup $B = HP$ is called a *Borel* subgroup of $G^*$. The Bruhat decomposition of $G^*$ provides a beautiful description of the multiplication in $G^*$ in terms of the subgroups $B$ and $N$.

To make this precise, we need a preliminary definition, based on certain characteristic properties of the Weyl group $W$.

DEFINITION 2.14. A group $W$ will be said to be a *Coxeter* group (after H. S. M. Coxeter) or a group *generated by reflections* provided:

(1) $W$ is generated by distinct involutions $w_i$, $1 \leqslant i \leqslant m$.
(2) If $w_i w_j$ has order $k_{ij}$, then the relations

$$(w_i w_j)^{k_{ij}} = 1, \qquad 1 \leqslant i, j \leqslant m,$$

constitute a complete set of defining relations for $W$.

For example, the symmetric group $\Sigma_{m+1}$ is a Coxeter group with respect to the transpositions $w_i = (i, i+1)$, $1 \leqslant i \leqslant m$.

The integer $m$ is called the *rank* of $W$. Note that if $m=1$, then $|W| = 2$, while if $m = 2$, then $W$ is a dihedral group of order $2k_{12}$. Moreover, the involutions $w_i$, $1 \leqslant i \leqslant m$, are called a *defining set* for $W$.

THEOREM 2.15. *Let $G^* \in Chev(p)$, let $B$ be a Borel subgroup of $G^*$ $[B = N_{G^*}(P), P \in Syl_p(G^*)]$, and let $H \leqslant B$ be a Cartan subgroup of $G^*$ [$H$ is a complement to $P$ in $B$]. Then $G^*$ possesses a subgroup $N$ with the following properties*:

(i) *$B \cap N = H \lhd N$ and $W = N/H$ is a Coxeter group.*
(ii) *$G^* = BNB$.*
(iii) *Let $w_i$, $1 \leqslant i \leqslant m$, be a defining set for $W$ and let $v_i$ be a representative of $w_i$ in $N$. Then for each $v \in N$ and every $i$, $1 \leqslant i \leqslant m$, we have*

$$BvBv_iB \leqslant (BvB) \cup (Bvv_iB).$$

(iv) *$B^{v_i} \neq B$, $1 \leqslant i \leqslant m$.*

The integer $m$ is called the *Lie rank* of $G^*$ and $W$ is called the *Weyl* group of $G^*$ (the same value of $m$ and group $W$ as in the Lie algebra associated with $G^*$). The double coset multiplication formulas of (iii) give very strong restrictions on the structure of $G^*$. In fact, it is not too difficult to prove that (ii) and the second assertion of (i) are consequences of (iii) and the fact that $G^* = \langle B, N \rangle$.

As an immediate corollary of the theorem, one has

COROLLARY 2.16. *If $G^* \in Chev(p)$ is of Lie rank 1, then $G^*$ is a doubly transitive group in which the stabilizer of a point is a Borel subgroup of $G^*$.*

This doubly transitive permutation representation will be described in more detail in the next chapter.

The Bruhat description of a group $G^*$ of Lie type is very important for understanding its structure. In particular, one can use it to calculate the order of $G^*$ (the list of orders is given in Section 2.11). In addition, based on the Bruhat decomposition alone, one can give a direct definition (which we omit) of the notion of root and root subgroup without recourse to the underlying associated Lie algebra (cf. Curtis [76]).

Furthermore, the subgroups of $G^*$ containing $B$ (together with their $G^*$-conjugates) are of special significance. Such subgroups are called *parabolic* subgroups. Parabolics have the following basic structure:

PROPOSITION 2.17. *Let the notation be as in Theorem 2.15. If $Y$ is a parabolic subgroup of $G^*$ with $B < Y$, then we have*

(i) *$O_p(Y) \leq P$ and $O_p(Y)$ is generated by a suitable proper subset of root subgroups of $P$.*
(ii) *If $\bar{Y} = Y/O_p(Y)$, then $\bar{Y} = \bar{H}\bar{Y}_1\bar{Y}_2 \cdots \bar{Y}_r$, where each $\bar{Y}_1 \lhd \bar{Y}$ and $\bar{Y}_i \in Chev(p)$, $1 \leq i \leq r$.*
(iii) *If we set $N_Y = N_Y(H)$, then $\bar{Y}$ has a Bruhat decomposition with respect to its Borel subgroup $\bar{B} = \bar{H}\bar{P}$ and $\bar{N}_Y$.*

The subgroups $\bar{Y}_1, \bar{Y}_2, \ldots, \bar{Y}_r$ in (ii) are known as the *Levi* factors of the parabolic $Y$. There is a very pretty way of describing the parabolics of $G^*$ and their Levi factors in terms of the Dynkin diagram $D$ of $G^*$. Indeed, there exists a one-to-one correspondence between parabolics $Y$ of $G^*$ containing $B$ and subsets $\mathfrak{S}$ of $D$ (with $Y = B$ corresponding to the empty subset). Moreover, if $\mathfrak{S} \neq \varnothing$ and $\mathfrak{S}$ is expressed as the disjoint union of connected subsets $D_1, D_2, \ldots, D_r$, then

(1) Each $D_i$ is the Dynkin diagram of a simple group of Lie type.
(2) $Y$ has exactly $r$ distinct Levi factors $\overline{Y}_1, \overline{Y}_2, \ldots, \overline{Y}_r$. (2.8)
(3) In a suitable ordering of the $\overline{Y}_i$, $D_i$ is the Dynkin diagram of $\overline{Y}_i$, $1 \leq i \leq r$.

Of particular interest are the *maximal* parabolics, those in which $\mathbb{S} = D - \{\text{one node}\}$, and the *minimal* parabolics, those in which $\mathbb{S}$ is a single node. A minimal parabolic $Y$ has a single Levi factor $\overline{Y}_1$ which is necessarily of Lie rank 1. In the case that $G^*$ is a Chevalley group over $GF(p^n)$, one always has $\overline{Y}_1/Z(\overline{Y}_1) \cong L_2(p^n)$. However, in the twisted groups, other rank 1 groups occur as well. Clearly the number of minimal parabolics containing $B$ is exactly $l$, the number of nodes of $D$. Furthermore, in view of the Bruhat decomposition of $G^*$, one obtains at once:

PROPOSITION 2.18. *Let the notation be as in Theorem* 2.15. *If $G^*$ has Lie rank at least* 2, *then $G^*$ is generated by its minimal parabolic subgroups containing $B$.*

There exists a rich geometry related to the parabolics of $G^*$. Indeed, Tits has completely characterized the groups of Lie type (of Lie rank at least 3) in terms of these geometries. This will be described in Section 3.1.

## 2.2. The Mathieu Groups

In the course of searching for highly transitive permutation groups, Mathieu (about 1860) discovered, two such *quintuply* transitive groups, of respective degrees 12 and 24 and order $8 \cdot 9 \cdot 10 \cdot 11 \cdot 12$ and $3 \cdot 16 \cdot 20 \cdot 21 \cdot 22 \cdot 23 \cdot 24$ [210–212]. They are denoted by $M_{12}$ and $M_{24}$; likewise $M_{11}$ denotes the one-point stabilizer in $M_{12}$, $M_{23}$ the one-point stabilizer in $M_{24}$, and $M_{22}$ the one-point stabilizer in $M_{23}$. (Thus $|M_{11}| = 8 \cdot 9 \cdot 10 \cdot 11$, $|M_{23}| = 3 \cdot 16 \cdot 20 \cdot 21 \cdot 22 \cdot 23$, and $|M_{22}| = 3 \cdot 16 \cdot 20 \cdot 21 \cdot 22$.) Each of these five groups is simple, and together they represent the first five sporadic simple groups. Remarkably, apart from the alternating and symmetric groups themselves, they include the only known quadruply and quintuply transitive permutation groups. Equally surprising is the fact that it took over a hundred years for the sixth sporadic simple group to be discovered.

Since their discovery, many descriptions of the Mathieu groups have been given. We shall present three: the first as permutation groups, the second as groups of transformations of projective lines, and the third as groups of automorphisms of so-called Steiner triple systems (cf. [321]).

Let $A$, $B, C$ be the following three permutations:

$$A=(1,2,3,4,5,6,7,8,9,10,11),$$

$$B=(5,6,4,10)(11,8,3,7),$$

$$C=(1,12)(2,11)(3,6)(4,8)(5,9)(7,10).$$

THEOREM 2.19. *$M_{11}=\langle A, B\rangle$ and $M_{12}=\langle A, B, C\rangle$.*

Let $D$, $E$, $F$ denote the following three permutations.

$$D=(1,2,3,4,5,6,7,8,9,10,11,12,13,14,15,16,17,18,19,20,21,22,23),$$

$$E=(3,17,10,7,9)(5,4,13,14,19)(11,12,23,8,18)(21,16,15,20,22),$$

$$F=(1,24)(2,23)(3,12)(4,16)(5,18)(6,10)(7,20)(8,14)(9,21)(11,17)(13,22)(19,15).$$

THEOREM 2.20. *$M_{23}=\langle D, E\rangle$ and $M_{24}=\langle D, E, F\rangle$.*

Finally let $G$, $H$, $I$ denote the following three permutations:

$$G=(1,2,3,4,5,6,7,8,9,10,11)(12,13,14,15,16,17,18,19,20,21,22),$$

$$H=(1,4,5,9,3)(2,8,10,7,6)(12,15,16,20,14)(13,19,21,18,17),$$

$$I=(11,22)(1,21)(2,10,8,6)(12,14,16,20)(4,17,3,13)(5,19,9,18).$$

THEOREM 2.21. *$M_{22}=\langle G, H, I\rangle$.*

Consider next the projective line $\mathcal{P}_1(q)$ coordinatized by the finite field $GF(q)$, with the usual convention that $a/0=\infty$ if $a\neq 0$, $a\in GF(q)$. There are $q+1$ points on $\mathcal{P}_1(q)$, each represented by $a/b$ with $a$, $b\in GF(q)$ and $a$ or $b\neq 0$. The projective linear group $PGL(2, q)$, or "fractional" linear group as it is often called, can be viewed as a permutation group of these $q+1$ points. As is easily checked, it is triply transitive on this set.

The Mathieu groups $M_{12}$ and $M_{24}$ can be defined as extensions of the groups $L_2(11)$ and $L_2(23)$ [of index 2 in $PGL(2,11)$, $PGL(2,23)$, respectively] by certain polynomial transformations of $\mathcal{P}_1(11)$ and $\mathcal{P}_1(23)$, respectively.

THEOREM 2.22. *If $f$ denotes the transformation of $\mathcal{P}_1(11)$ given by*

$$f\colon x'=4x^2-3x^7,$$

*then $M_{12}=\langle L_2(11), f\rangle$.*

THEOREM 2.23. *If $f$ denotes the transformation of $\mathcal{P}_1(23)$ given by*

$$f\colon x'=-3x^{15}+4x^4,$$

*then $M_{24}=\langle L_2(23), f\rangle$.*

Using a nonpolynomial transformation of $\mathcal{P}_1(23)$, Conway has given the following neat description of $M_{24}$ [69].

THEOREM 2.24. *Let $f$ denote the transformation of $\mathcal{P}_1(23)$ given by*

$$f\colon x'=9^{\varepsilon}x^3,$$

*where $\varepsilon=-1$ if $x$ is a square in $GF(23)$ and $\varepsilon=+1$ if $x$ is a nonsquare in $GF(23)$. Then $M_{24}=\langle L_2(23), f\rangle$.*

Finally we consider Steiner systems for the Mathieu groups.

DEFINITION 2.25. Let $\Omega$ be a set of $n$ elements. A *Steiner triple system* $S(k, m, n)$ on $\Omega$ is defined to be a set of $\binom{n}{k}/\binom{m}{k}$ subsets of $\Omega$, each of size $m$, with the property that every set of distinct $k$ elements of $\Omega$ lies in one and only one of these subsets.

By definition, the automorphism group of $S(k, m, n)$ is the subgroup of $\Sigma(\Omega)$ whose elements transform the subsets of the Steiner system into themselves.

If one considers the subset of $S(k, m, n)$ containing a given element $a$ of $\Omega$, one obtains another Steiner system $S(k-1, m-1, n-1)$ on $\Omega-\{a\}$. In this way the Steiner systems for $M_{11}$, $M_{22}$, $M_{23}$ can be obtained from those of $M_{12}$ and $M_{24}$.

THEOREM 2.26. *There exist unique Steiner triple systems $S(5,6,12)$ and $S(5,8,24)$ such that*

$$\operatorname{Aut}(S(5,6,12)) = M_{12} \quad \text{and} \quad \operatorname{Aut}(S(5,8,24)) = M_{24}.$$

Thus as a corollary one has

COROLLARY 2.27. *There exist Steiner triple systems* $S(4,5,11)$, $S(4,7,23)$, *and* $S(3,6,22)$ *such that*

$$\mathrm{Aut}(S(4,5,11)) = M_{11},$$

$$\mathrm{Aut}(S(4,7,23)) = M_{23},$$

$$\mathrm{Aut}(S(3,6,22)) = \mathrm{Aut}(M_{22}).$$

Note that $M_{22}$ is of index 2 in $\mathrm{Aut}(M_{22})$, while $M_{23}$ and $M_{24}$ are their own automorphism groups. One also has a Steiner system $S(2,5,21)$ and $\mathrm{Aut}(S(2,5,21)) \cong \mathrm{Aut}(L_3(4))$. However, the subgroup $M_{21}$ of $M_{22}$ which acts on $S(2,5,21)$ is just $L_3(4)$ ($M_{21}$ is the stabilizer of a point in $M_{22}$).

## 2.3. Janko's First Group

The family of Ree groups ${}^2G_2(3^{2n+1})$, $n>1$, is a very interesting one. If $G^*$ is one of its members, then $G^*$ has the following properties:

(a) A Sylow 2-subgroup of $G^*$ is elementary abelian of order 8.
(b) If $t^* \in \mathcal{I}(G^*)$, then $C_{G^*}(t^*) \cong Z_2 \times L_2(3^{2n+1})$.
(c) $|G^*| = (3^n-1)3^{3n}(3^{3n}+1)$ and a Borel subgroup $B^*$ of $G^*$ has order $(3^n-1)3^{3n}$.
(d) The permutation representation of $G^*$ on the cosets of $B^*$ is doubly transitive and some three-point stabilizer has order 2.

In attempting to characterize the Ree groups by internal properties, it is natural to ask to what extent conditions (a), (b) imply (c), (d) and also whether an arbitrary group satisfying conditions (c), (d) must be isomorphic to ${}^2G_2(3^n)$. The second question will be discussed in the next chapter; here we focus on the first.

Regarded as a general question, the restrictions on the order of a Sylow 2-subgroup of $G$ and on the order of the characteristic power $q$ in $L_2(q)$ are clearly artificial. First, H. N. Ward [316] and then Janko–Thompson [192] considered this general problem; their combined efforts yield the following result.

THEOREM 2.28. *If G is a simple group with abelian Sylow 2-subgroups and the centralizer of some involution of G is isomorphic to* $Z_2 \times L_2(q)$, $q \geqslant 3$, *then one of the following holds*:

(i) $q=3^n$, *n odd*, $n>1$, $|G|=(3^n-1)3^{3n}(3^{3n}+1)$, *and G is a doubly transitive permutation group of degree* $3^{3n}+1$ *in which some* 3*-point stabilizer has order* 2.

(ii) $q=4$ *or* 5. [*Note that* $L_2(4)\cong L_2(5)$.]

We describe some of the key points of the proof. Let $S\in Syl_2(G)$ and let $t\in\mathcal{I}(S)$ be such that $C_t=\langle t\rangle\times L$, where $L\cong L_2(q)$. Since $S$ is abelian, $S\leqslant C_t$ and so $S=\langle t\rangle\times R$, where $R=S\cap L$ is also abelian. If $q$ is odd, $L$ has dihedral Sylow 2-subgroups, so as $R$ is abelian and $|L|=\frac{1}{2}q(q^2-1)$, the only possibility is $q\equiv 3,5 \pmod 8$ with $R$ a four group. Hence in this case $S$ is elementary of order 8. On the other hand, if $q=2^m$, $R$ is elementary of order $2^m$, whence $S$ is elementary of order $2^{m+1}$.

Since $G$ is simple and $S$ is abelian, an easy transfer argument, using the focal subgroup theorem, (Theorem 1.19) implies that all involutions of $S$ are conjugate in $N=N_G(S)$. Thus $N$ acts transitively on $S^\#$ and $N/S$ is of odd order. We use this to reduce to the case that $q$ is odd. Since $L_2(4)\cong L_2(5)$ (and $q\geqslant 3$ by hypothesis), it will suffice to derive a contradiction from the assumption $q=2^m$, $m\geqslant 3$. But the Borel subgroup $B=N_L(R)$ of $L\cong L_2(2^m)$ is a Frobenius group with kernel $R$ and cyclic complement $H$ of order $2^m-1$. Since $H$ centralizes $t$, $H\leqslant N$. Also $R=C_B(R)=C_L(R)$, so $S=C_{C_t}(S)$ $=C_G(S)$. Hence $\bar{N}=N/S\leqslant \mathrm{Aut}(S)$. Thus $\bar{N}$ is a group of automorphisms of $S$ of odd order (hence, in particular, solvable), acting transitively on $S^\#$ (and hence irreducibly on $S$) and containing the cyclic subgroup $\bar{H}$ of order $2^m-1$.

We prove that no such subgroup $\bar{N}$ exists. Indeed, by a well-known arithmetic result, if $m\neq 6$, there is a prime $d$ dividing $2^m-1$, not dividing $2^i-1$ for $i<m$. This implies that any subgroup $\bar{D}$ of order $d$ in $\bar{N}$ acts trivially on any subgroup of $S$ of order $<2^m$. But now, using standard results on the representations of solvable groups (including Clifford's theorem, Theorem 1.5), one concludes easily from this property of $d$ that $\bar{D}$ centralizes $O_{d'}(\bar{N})$ and also that a Sylow $d$-subgroup of $\bar{N}$ is cyclic (cf. Section 4.1). A similar argument applies when $m=6$, taking $d=7$. (However, we note that the condition $m\geqslant 3$ is required here, for if $m=2$, then $d=3$ and $|S|=8$, in which case $\bar{N}$ might be nonabelian of order 21. Indeed, this exceptional configuration occurs in $J_1$.) It follows that $\bar{D}=$ $\Omega_1(O_d(\bar{N}))\lhd\bar{N}$. But then $\bar{N}$ leaves invariant $\langle t\rangle=C_S(\bar{D})$, contrary to the irreducible action of $\bar{N}$ on $S$.

Thus we can assume that $q$ is odd, whence $q\equiv 3,5 \pmod 8$ and $|S|=8$. Hence $q=p^m$, $p$ odd, and the congruence on $q$ forces $m$ to be odd. Suppose

$q=3$, in which case $L \cong L_2(3) \cong A_4$ and $R \lhd L$, whence $S \lhd C_t$ and so $C_t \leqslant N$. Since all involutions of $S$ are $N$-conjugate, one concludes quite easily now that $N_G(T) \leqslant N$ for every $1 \neq T \leqslant S$. Hence according to Definition 4.20, $N$ is "strongly embedded" in $G$. But then by Bender's classification theorem (Theorem 4.24) and the fact that $G$ has abelian Sylow 2-subgroups of order 8, it follows that $G \cong L_2(8)$.* However, this is a contradiction as centralizers of involutions in $L_2(8)$ have order 8, while $C_t$ has order 24. Thus $q \geqslant 5$. Furthermore, we can suppose that $q>5$, since otherwise the second alternative of the theorem holds.

To one familiar with finite group theory, the preceding argument is entirely routine; however, we have presented it in detail because it is typical of a rather common kind of reduction argument in simple group theory. On the other hand, the Janko–Thompson proof which forces $p=3$ in this remaining case is anything but routine. The key step is the establishment of a formula for the order of $G$, in terms of the degree $f$ of a suitable irreducible character of $G$, obtained using the Brauer character theory methods:

$$|G|=q^3(q^3+1)f.$$

Now a delicate local analytic argument, which depends critically on the given group order formula, yields that some element $a$ of $C_t$ of order 3 centralizes an element $b$ of $C_t$ of order $p$. Since $C_t=\langle t\rangle \times L$, $a$ and $b$, in fact, lie in $L$. On the other hand, it follows from the structure of $L_2(p^m)$ that $C_L(b)$ is a $p$-group. Hence we must have $p=3$.

If one analyzes the case $q=5$ in the same way, one does not reach the anticipated contradiction. However, the "first few times around" numerical errors camouflaged the true situation. But eventually Janko's dogged persistence showed that everything fit together beautifully and, in particular, that such a group $G$ must have a uniquely determined character table and order 175,560. Ultimately, Janko proved the following result, and the modern theory of sporadic groups was launched [187].

THEOREM 2.29. *If $G$ is a simple group with abelian Sylow 2-subgroups (of order 8) and the centralizer of some involution of $G$ is isomorphic to $Z_2 \times L_2(5)$, then $G$ is a uniquely determined simple group of order 175,560. Moreover, $G$ is isomorphic to the subgroup of $GL(7,11)$ generated by the following two*

*Actually this special case of Bender's classification theorem is covered by a prior result of Feit, which had considerable influence on the subsequent developments.

*matrices Y and Z [with coefficients in GF(11)] of order 7 and 5, respectively*:

$$Y=\begin{bmatrix} 0 & 1 & 0 & 0 & 0 & 0 & 0 \\ 0 & 0 & 1 & 0 & 0 & 0 & 0 \\ 0 & 0 & 0 & 1 & 0 & 0 & 0 \\ 0 & 0 & 0 & 0 & 1 & 0 & 0 \\ 0 & 0 & 0 & 0 & 0 & 1 & 0 \\ 0 & 0 & 0 & 0 & 0 & 0 & 1 \\ 1 & 0 & 0 & 0 & 0 & 0 & 0 \end{bmatrix},$$

$$Z=\begin{bmatrix} -3 & 2 & -1 & -1 & -3 & -1 & -3 \\ -2 & 1 & 1 & 3 & 1 & 3 & 3 \\ -1 & -1 & -3 & -1 & -3 & -3 & 2 \\ -1 & -3 & -1 & -3 & -3 & 2 & -1 \\ -3 & -1 & -3 & -3 & 2 & -1 & -1 \\ 1 & 3 & 3 & -2 & 1 & 1 & 3 \\ 3 & 3 & -2 & 1 & 1 & 3 & 1 \end{bmatrix}.$$

Once Janko had the order and character table of the unknown group $G$, he analyzed its modular characters for each prime dividing $|G|$ and eventually showed that such a group $G$ had to possess one and only one absolutely irreducible representation of degree 7 over $GF(11)$.

One would imagine it would be easy to decide whether $GL(7,11)$ contains a subgroup $G$ with the prescribed order and properties, and no doubt it could be settled quickly by a computer. However, it is another matter to accomplish this by hand. Janko uses his character-theoretic information to derive further internal properties of $G$. First, $N=N_G(S)=HS$, where $H$ is nonabelian of order 21. Moreover, if $y\in H$ has order 7 and $x\in S^{\#}$, $N_G(H)$ contains an involution $w$ such that $z=xw$ has order 5 and $C_H(w)$ has order 3. Furthermore, $G=\langle y,z\rangle$. Representing $y$ by the matrix $Y$ above, Janko then argues, using the various relations among $y, x, w$ which he has established, that there is only one possibility for the $7\times 7$ matrix over $GF(11)$ representing $z$, namely, the matrix $Z$ above. At this point, uniqueness has been proved: there exists at most one group satisfying the given conditions.

However, to establish *existence* of such a group, it remains to determine the group $\langle Y,Z\rangle$ and to show that it satisfies the given hypotheses. This reduces without much difficulty to showing that $\langle Y,Z\rangle$ has order 175,560. This highly arduous task was carried out by M. A. Ward. Thus $J_1$ uniquely exists!

The group $J_1$ has no doubly transitive or even rank 3 permutation representations (as follows directly from its character table), so there is no "natural" geometry associated with it. The best that has been done is to show that $J_1$ is a subgroup of $G_2(11)$ (which was proved by Coppell), but again the embedding is not a natural one. Thus no pat reason for the existence of this group has been found. The implication of this last remark is that $J_1$ could have been discovered only in the process of treating some general classification problem. (M. Hall has systematically determined all simple groups of order less than one million and would have hit Janko's group at 175,560 had it not already been known to exist. McKay and also several Caltech students were involved in this effort. However, this exhaustive approach is clearly limited to groups of low orders.)

## 2.4. Sporadic Groups from Centralizers of Involutions

With the construction of $J_1$, it was natural to experiment with other prospective candidates for centralizers of involutions in new simple groups. Given the great effort required to construct $J_1$, coupled with the extremely low probability of success, it is remarkable that four further sporadic groups —Janko's groups $J_2$, $J_3$, $J_4$ and Richard Lyons's group $Ly$—have arisen as a result of just such a judicious guess. Dieter Held's group also arose from a centralizer-of-involution problem, but in that case the centralizer was "given" in advance.

In addition, five other groups—Michael O'Nan's group $ON$, the Fischer "$\{3,4\}$-transposition" group $F_2$, the Fischer–Griess group $F_1$ and its two "offspring," the Harada group $F_5$ and the Thompson group $F_3$—each of which arose in a context distinct from centralizers of involutions, have been investigated beginning with the structure of the centralizer of one of their involutions. Thus a total of 11 of the 26 sporadic groups have been constructed from such a centralizer. (Some of these groups have more than one conjugacy class of involutions.)

Table 2.1 lists each of these 11 groups with the corresponding centralizer of an involution. The new notation occurring in the table will be explained directly. The phrase "2-constrained" means that the group in question is 2-constrained; the symbols $\hat{M}_{22}$, $\hat{A}_{11}$, $\hat{\hat{L}}_3(4)$, ${}^2\hat{E}_6(2)$, $\hat{F}_2$, $\widehat{HS}$ denote perfect central extensions of the corresponding "unhatted" groups by $Z_3$, $Z_2$, $Z_4$, $Z_2$, $Z_2$, $Z_2$, respectively; and the symbol $HS$ denotes the Higman–Sims sporadic simple group, which will be discussed in Section 2.6.

We make a few comments. Except in the $L_3(4)$ case, there is only one possibility for the given extension. However, $L_3(4)$ has two nonisomorphic perfect central extensions by $Z_4$, precisely one of which occurs in O'Nan's group. It is interesting to note that in $He$ the centralizer of a non-2-central

**Table 2.1.** Centralizers of Involutions in Some Sporadic Groups

| Group | Centralizer of an involution |
|---|---|
| $J_1$ | $Z_2 \times A_5$ |
| $J_2$ | $A_5/Q_8 * D_8$, 2-constrained |
| $J_3$ | same |
| $J_4$ | $(\hat{M}_{22}/(D_8)^6)\cdot 2$, 2-constrained |
| $He$ | $L_3(2)/D_8 * D_8 * D_8$, 2-constrained |
| $Ly$ | $\hat{A}_{11}$ |
| $ON$ | $(\hat{L}_3(4))\cdot 2$ |
| $F_2$ | $({}^2\hat{E}_6(2))\cdot 2$ |
| $F_1$ | $\hat{F}_2$ |
| $F_5$ | $(\widehat{HS})\cdot 2$ |
| $F_3$ | $A_9/(D_8)^4$, 2-constrained |

involution is of the form $(\hat{L}_3(\hat{4}))\cdot 2$, where the symbol $\hat{L}_3(\hat{4})$ denotes the (unique) perfect central extension of $L_3(4)$ by $Z_2 \times Z_2$. (Covering groups of the known simple groups will be discussed more fully in Section 4.15.)

Apart from the groups $F_1$, $F_2$, and $F_5$, the involution specified in the table is 2-central. The centralizer of a 2-central involution in these three groups is as follows:

$F_1$ $\quad .1/(D_8)^{12}$, 2-constrained;

$F_2$ $\quad .2/(D_8)^{11}$, 2-constrained;

$F_5$ $\quad (A_5 \int Z_2)/(D_8)^4$, 2-constrained.

Here the symbols .1, .2 denote Conway's first and second simple groups, which will be discussed in Section 2.9.

I would like to explain now the precise context in which each of these groups has arisen, with the exception of the already discussed $J_1$ and the group $F_2$, which is best left to a discussion of Fischer's general theory of groups generated by a conjugacy class of "transpositions."

### A. Janko's Groups $J_2$, $J_3$

In looking for propitious centralizers of involutions to examine, Janko decided to try some that closely resembled those in Mathieu groups, yet were distinct from the centralizers of involutions in any then-known simple group. In $M_{12}$ [as well as in $L_4(2) \cong A_8$], there is an involution with centralizer $\Sigma_3/Q_8 * Q_8$, so Janko tried his luck with $A_5/Q_8 * D_8$. What a wise choice! For Janko was able to prove [188]:

THEOREM 2.30. *If G is a simple group in which the centralizer of an involution is isomorphic to* $A_5/Q_8 * D_8$, *then one of the following holds*:

(i) *G has two classes of involutions and*

$$|G| = 2^7 \cdot 3^3 \cdot 5^2 \cdot 7.$$

(ii) *G has one class of involutions and*

$$|G| = 2^7 \cdot 3^5 \cdot 5 \cdot 17 \cdot 19.$$

Furthermore, Janko determined (as was done for most of the groups to be discussed here) the complete local structure of such a group $G$ as well as its character table in each case. However, this result, in contrast with his work on $J_1$, does not tell you whether there *exists* a simple group of either order or even whether there is *at most one* such group. Thus Janko achieved "strong evidence" for two new simple groups but did not settle the "existence and uniqueness" question.

It is best to postpone a discussion of the actual construction of $J_2$ and $J_3$ as well as the other eight groups on the list until later in the chapter.

## B. Held's Group *He*

Held began with the following interesting fact: The groups $L_5(2)$ and $M_{24}$ have 2-central involutions with isomorphic centralizers. It was natural to attempt to characterize these groups by this property; and this was Held's intent when he began the problem. However, in his analysis of the conjugacy classes of involutions he was led to *three* distinct but self-consistent possibilities. He ultimately proved [165]:

THEOREM 2.31. *If G is a simple group in which the centralizer of an involution is isomorphic to* $L_3(2)/D_8 * D_8 * D_8$, *then one of the following holds*:

(i) $G \cong L_5(2)$ *or* $M_{24}$.
(ii) *G has exactly two conjugacy classes of involutions and*

$$|G| = 2^{10} \cdot 3^3 \cdot 5^2 \cdot 7^3 \cdot 17.$$

Again Held determines the full internal structure and a good deal of the character table. Likewise the theorem says nothing about the existence and uniqueness of a group satisfying (ii). In the next chapter we shall describe the general methods by which one can identify a known simple group from its internal structure. Obviously some procedure is necessary for Held to be able to conclude that $G \cong L_5(2)$ or $M_{24}$.

## C. Lyons's Group $Ly$

John McLaughlin's group $Mc$ (to be described in the next section) arises from a primitive rank 3 permutation group problem. However, once $Mc$ was constructed, its internal structure could obviously be examined. It has only one conjugacy class of involutions and the centralizer of an involution is isomorphic to $\hat{A}_8$ ($\hat{A}_n$ denotes the perfect central extension of $A_n$ by $Z_2$). This was an intriguing answer since it immediately suggests studying the class of groups $G$ having $\hat{A}_n$ as centralizer of an involution $t$ for any $n \geqslant 5$. From the structure of a Sylow 2-subgroup $T_n$ of $\hat{A}_n$, it is immediate that such an involution $t$ must be 2-central in $G$. Also by order considerations, $T_n \cong T_{n+1}$ for $n$ even.

Thompson then made the interesting observation that the group $T_{10}$ ($\cong T_{11}$), of order $2^8$, appears on MacWilliams's list of 2-groups of normal 2-rank 2 which are possible candidates for Sylow 2-subgroups of some simple group, as an exceptional case [207]. Here then was the place to begin! And Thompson suggested the problem to his student Lyons, who proved [206]:

THEOREM 2.32. *If $G$ is a simple group in which the centralizer of some involution $z$ is isomorphic to $\hat{A}_n$, $n=10$ or $11$, then $n=11$, $G$ has only one conjugacy class of involutions, and*

$$|G| = 2^8 \cdot 3^7 \cdot 5^6 \cdot 7 \cdot 11 \cdot 31 \cdot 37 \cdot 67.$$

(That $G$ has only one conjugacy class of involutions is immediate from Glauberman's fundamental $Z^*$-theorem (see Theorem 4.95).

Lyons eliminates the case $n=10$ by an analysis of the 3-structure of such a group $G$. Indeed, if $x$ is an element of order 3 in $C_z = C_G(z)$ which corresponds to a 3-cycle in $C_z/\langle z \rangle$ ($\cong A_{10}$), he argues, on the one hand, that a Sylow 3-subgroup of $C_G(x)$ has order $3^3$ and, on the other, that its order $\geqslant 3^4$.

The following result shows that there is no more gold to be found in this direction!

THEOREM 2.33. *If $G$ is a simple group in which the centralizer of some involution $z$ is isomorphic to $\hat{A}_n$, $n \geqslant 5$, then $n=8$ or 11.*

The cases $n=5,6,7$ follow from a theorem of Brauer and Suzuki on groups with quaternion Sylow 2-subgroups [48] (cf. Theorem 4.88). The $Z^*$-theorem generalizes their quaternion result.) The case $n=9$, due to Janko [189], involves an analysis of the 3-structure of $G$ very similar to that of Lyons's $n=10$ case. (In fact, it was Janko's argument in the $\hat{A}_9$ that suggested Lyons's 3-local approach to the $\hat{A}_{10}$ problem.) Eventually Janko argues that a suitable quasi-dihedral subgroup of $C_z$ of order 16 is forced to normalize a 3-subgroup $D$ of $G$ of order $3^8$ with $|C_D(z)|=3^4$, which is incompatible with the structure of $C_z$.

On the other hand, when $n \geqslant 12$, a more uniform argument can be given. This was done by Thompson and also by Ronald Solomon [264] and will be described in the sequel.

Thus, the only possibilities are $n=8$ or 11, as asserted. Of course, there remains the question whether McLaughlin's and Lyons's groups are the unique solutions in these cases. Sims's computer construction of $Ly$, which we shall describe in the next section, shows that this is indeed the case for Lyons's group. The corresponding assertion for McLaughlin's group was obtained by Janko and S. K. Wong [193], and we should like to say a few words about their proof.

Beginning with $C_z \cong \hat{A}_8$, their aim is to reduce to the precise graph-theoretic conditions under which McLaughlin initially constructed $Mc$. It is again immediate from the $Z^*$-theorem that $G$ has only one class of involutions. Janko and Wong first determine very detailed information about the local structure of $G$, similar to Lyons's work in the $n=11$ case. However, in contrast to his situation, they need no character theory to obtain the order of $G$. Rather they proceed as follows: using the local information, they construct a subgroup $G_0$ of $G$ with $G_0 \cong U_4(3)$, which is identified by the same $(B,N)$-pair approach as Kok Wee Phan had earlier used in his characterization of $U_4(3)$ [236]. They then argue that $G_0$ is maximal in $G$ and that for suitable elements $v \in G-G_0$ and $a \in G_0$, $G$ is the union of the three double cosets:

$$G_0, \quad G_0vG_0, \quad \text{and} \quad G_0vavG_0.$$

This immediately yields that

$$|G:G_0|=275 \qquad (\text{whence } |G|=|Mc|),$$

and that $G$ is a rank 3 permutation group on the right cosets of $G_0$, with the orbits of $G_0$ in this representation having lengths

$$1, \quad 112, \quad \text{and} \quad 162.$$

As we shall see in Section 2.6, the existence of a rank 3 permutation representation implies that there is a natural graph associated with $G$. Finally, Janko and Wong prove that this graph is isomorphic to the one used by McLaughlin in the construction of his group. Hence by his results, $G \cong Mc$.

## D. O'Nan's Group *ON*

Some years ago, Alperin and I were investigating simple groups of 2-rank 3. This work led Alperin to analyze the possible 2-constrained groups of the form $L_3(2)/(Z_{2^n} \times Z_{2^n} \times Z_{2^n})$ for some $n$. He proved [230]:

THEOREM 2.34. *For each value $n \geqslant 1$, there are (up to isomorphism) exactly two 2-constrained groups of the form $L_3(2)/(Z_{2^n} \times Z_{2^n} \times Z_{2^n})$. One extension splits, the other does not. Moreover, each nonsplit group has 2-rank* 3.

Denote these groups by $Alp_n^1$ and $Alp_n^2$, respectively. The group $Alp_1^1$ occurs as a 2-local (mod core in the case of $A_{11}$) in $A_8$, $A_9$, $A_{10}$, $A_{11}$, $M_{22}$, $M_{23}$, the group $Alp_1^2$ in $G_2(q)$ and ${}^3D_4(q)$, $q$ odd, and the group $Alp_2^1$ in $HS$. Here then is another natural question: For what values of $n \geqslant 2$ do the groups $Alp_n^1$ or $Alp_n^2$ occur as 2-local subgroups of a simple group $G$? Note that for $n \geqslant 2$, it is easily shown that $O_2(Alp_n^i)$ is the unique subgroup of $Alp_n^i$ of type $Z_{2^n} \times Z_{2^n} \times Z_{2^n}$, and so $Alp_n^i$ must then necessarily contain a Sylow 2-subgroup of $G$. Thus for a given $n \geqslant 2$ and $i=1$ or 2, $G$ has a uniquely determined Sylow 2-subgroup, isomorphic to that of $Alp_n^i$.

I was hoping to interest a doctoral student in this problem; but O'Nan was forced to consider it in connection with his work on doubly transitive permutation groups of odd degree in which the one-point stabilizer has a nontrivial normal subgroup of odd order [228]. That analysis ultimately reduced to determining all groups $G$ of 2-rank $r$, $r$ a positive integer, such that for some elementary abelian 2-subgroup $A$ of $G$ of rank $r$, $N = N_G(A)$ transitively permutes the "flags" of $A$. Here, by definition, a *flag* of $A$ is a nested sequence of subspaces

$$1 < A_1 < A_2 < \cdots < A_r = A.$$

(Thus $A_i$ is of codimension 1 in $A_{i+1}$, $1 \leqslant i \leqslant r-1$.) To say that $N$ acts transitively on the flags of $A$ means that for any two flags, some element of $N$ conjugates the subspaces of the first into the corresponding subspaces of the second.

In dealing with the case $r=3$, one of the possibilities is that $N/O(N) \cong Alp_n^i$. If $n=1$, then $A \in Syl_2(C_G(A))$, and it is a consequence of the classification of nonconnected groups (to be discussed in the sequel) that $G$ is isomorphic to one of the groups listed above (assuming $G$ is simple). Thus O'Nan could restrict himself to the case $n \geqslant 2$, in which case $G$ has a specified Sylow 2-group for each $n$ and $i$.

O'Nan proved [230], using a characterization of $HS$ by its Sylow 2-group previously obtained by Morton Harris and me [137]:

THEOREM 2.35. *If $G$ is a simple group with Sylow 2-subgroup isomorphic to that of $Alp_n^i$, $n \geqslant 2$, $i=1$ or 2, then $n=2$ and one of the following holds*:

(i) *$i=1$ and $G \cong Hs$.*
(ii) *$i=2$, $G$ has only one conjugacy class of involutions, and*

$$|G| = 2^9 \cdot 3^4 \cdot 5^7 \cdot 7^3 \cdot 11 \cdot 19 \cdot 31.$$

## E. Janko's Fourth Group

The reader will certainly have observed that several of the 11 groups listed in Table 2.1 are of $GF(2)$-type. Moreover, as we have already pointed out, such groups play a basic role in the general theory of simple groups. In recent years, it was felt that this was the one "corner" in which some as-yet-undetected sporadic groups might still be hiding. Given his success with $J_2$ and $J_3$, Janko was quite willing to devote considerable time and energy to such a search. At one point early on, it looked as though he might have hit upon an entire family of new simple groups of $GF(2)$-type, but ultimately their possible existence succumbed to a contradiction. After that encounter, Janko focused primarily on narrow-width cases, and his efforts were finally rewarded by the discovery of $J_4$ in 1976 [191].

However, at about this time work on the general structure of groups of $GF(2)$-type began in earnest. Aschbacher made an important contribution to the effort [14, 15], but the prime mover was Timmesfeld, who broke the problem open, reducing it to a number of special cases [303]. At that point Stephen Smith took over and completed the analysis in these residual cases [261–263] (certain cases had been handled independently by Arthur Reifart [241] and others). This fundamental chapter of simple group theory will be fully described in the sequel; let me only say here that with the closing of this frontier, most finite group theorists felt that the search for sporadic groups was finally over.

Janko's main result is as follows:

THEOREM 2.36. *If G is a simple group in which the centralizer of some involution is isomorphic to* $(\hat{M}_{22}/(D_8)^6)\cdot 2$, *then G has exactly two conjugacy classes of involutions and*

$$|G| = 2^{21}\cdot 3^3\cdot 5\cdot 7\cdot 11^3\cdot 23\cdot 29\cdot 31\cdot 37\cdot 43.$$

Here $\hat{M}_{22}$ denotes a nonsplit extension of $M_{22}$ by $Z_3$.

Janko determined that the centralizer of an involution of the second class has to be a nontrivial split extension of an elementary group $E$ of order $2^{11}$ by Aut($M_{22}$). Furthermore, he showed that a maximal 2-local subgroup containing this centralizer is necessarily a split extension of $E$ by $M_{24}$. In particular, any such group $G$ must be of characteristic 2 type.

Simon Norton, together with others from the University of Cambridge, proved the existence and uniqueness of $J_4$ in 1980 as one of the final steps of the classification proof. Its construction will be described in Section 2.7.

### F. The Fischer–Griess Group $F_1$ and Its Simple Subgroups

That the search for sporadic groups was not totally haphazard can be seen from the remarkable simultaneous realization by Fischer in West Germany and Griess in the United States in 1974 that there might be a simple group having a covering group $\hat{F}_2$ of $F_2$ by $Z_2$ as centralizer of an involution. In such a group $G$, a likely candidate for the centralizer of a certain element of order 3 would have the form $\widehat{Suz}/X$, where $\widehat{Suz}$ denotes the cover by $Z_2$ of the Suzuki sporadic group (to be described in Section 2.6) and $X$ is extra-special of order $3^{13}$, the extension being 3-constrained. [The group $\widehat{Suz}$ is known to have a faithful 12-dimensional modular representation over $GF(3)$.] Examining the centralizer in $G$ of the involution corresponding to a generator of $Z(\widehat{Suz})$, one can then argue that it very likely has the form $.1/(D_8)^{12}$. By a prodigious calculation, using the Thompson order formula (Theorem 2.43), Griess determined the exact order of such a group $G$ [149]. (Conway and Thompson, who became interested in the problem after Fischer's initial work, also obtained the same value, but only as a possible lower bound for the order of $G$.) Griess's result is as follows:

THEOREM 2.37. *Let G be a simple group containing involutions x, y such that* $C_x \cong \hat{F}_2$ *and* $C_y \cong .1/(D_8)^{12}$. *Then G has exactly two conjugacy classes of involutions and*

$$|G|=2^{46}\cdot 3^{20}\cdot 5^{9}\cdot 7^{6}\cdot 11^{2}\cdot 13^{3}\cdot 17\cdot 19\cdot 23\cdot 29\cdot 31\cdot 41\cdot 47\cdot 59\cdot 71.$$

A group satisfying these conditions is called a group of *type* $F_1$. S. Smith has shown that the structure of the first centralizer $C_x$ follows from that of the second $C_y$, so that a group of type $F_1$ can in fact be defined by a single centralizer of an involution [261]. (This case already appears as an exceptional one in Timmesfeld's work [303], which will be discussed in the sequel.) Thus we have:

THEOREM 2.38. *If G is a simple group containing an involution y such that* $C_y \cong .1/(D_8)^{12}$, *then G is a group of type* $F_1$.

The task of proving the existence of a group of type $F_1$ seemed very difficult. As with $J_4$, the proof was not obtained by Griess until 1980, likewise constituting one of the final steps of the classification. Its construction will be described in Section 2.9. However, very near the outset, Thompson was able to show that a group of type $F_1$ would necessarily possess certain elements of order 3 and 5 whose centralizers would have direct factors (of index 3 and 5, respectively), which were themselves new simple groups. Moreover, he determined likely candidates for centralizers of involutions in these groups: namely, $A_9/(D_8)^4$ and $(\widehat{HS})\cdot 2$, respectively. The suggestive terminology $F_3$ and $F_5$ for these groups has now become standard. [Because of the structure of the centralizer $C_x$ of the element $x$ (of order 2) in a group of type $F_1$, this also explains why the notation $F_2$ has been adopted for Fischer's $\{3,4\}$-transposition group. Likewise since the entire group is the centralizer of the identity element (of order 1), the notation $F_1$ is consistent with that of $F_2$, $F_3$, and $F_5$.]

Thompson proved the existence and uniqueness of $F_3$ [295], and Harada and Norton proved the existence and uniqueness of $F_5$ [163]. In both cases some computer calculations by Peter Smith at the University of Cambridge were required. However, as Griess's construction of $F_1$ was carried out by hand, it provides alternate proofs of the existence of $F_3$ and $F_5$ (as well as of the group $F_2$) which are free of computer calculations. (Griess's construction of $F_1$ does not address the uniqueness question for $F_2$, $F_3$, and $F_5$, but very likely uniqueness of simple groups of each of these three types can be established without the aid of a computer.)

For the present we content ourselves with the following statements concerning $F_3$ and $F_5$.

THEOREM 2.39. *If $G$ is a simple group in which the centralizer of some involution is isomorphic to $A_9/(D_8)^4$, then $G$ has only one conjugacy class of involutions and*

$$|G|=2^{15}\cdot 3^{10}\cdot 5^{3}\cdot 7^{2}\cdot 13\cdot 19\cdot 31.$$

THEOREM 2.40. *If $G$ is a simple group in which the centralizer of some involution is isomorphic to $(\widehat{HS})\cdot 2$, then $G$ has two conjugacy classes of involutions and*

$$|G|=2^{14}\cdot 3^{6}\cdot 5^{6}\cdot 7\cdot 11\cdot 19.$$

Let me also say something concerning the method of determining the order of one of these groups $G$ from the structure of the centralizer of an involution. The first step is to obtain, by 2-local analysis, the precise set of possibilities for the conjugacy of involutions of $G$. If $G$ has more than one class of involutions, the next step is to determine the exact structure of the centralizers of involutions in the remaining classes. In this case, one is now in a position to apply the so-called "Thompson order formula" directly to compute the order of $G$. This formula, which is established by very elementary counting arguments, can be viewed as a refinement of classical results of Brauer and K. A. Fowler [46] on properties of involutions in groups of even order. Ultimately these rest on the following elementary but fundamental property of involutions.

PROPOSITION 2.41. *If $x$ and $y$ are involutions of the group $G$, then $\langle x, y\rangle$ is a dihedral group of order $2|xy|$.*

Indeed, $yx=y^{-1}x^{-1}=(xy)^{-1}$ as $x$, $y$ are involutions, so $x^{-1}(xy)x=yx=(xy)^{-1}$. Thus $x$ "inverts" $xy$ (i.e., transforms it into its inverse under conjugation), so $\langle x, y\rangle=\langle x, xy\rangle$ is dihedral of the specified order.

For simplicity, we state Thompson's formula only for groups having exactly two conjugacy classes of involutions. We need a definition.

DEFINITION 2.42. Let $G$ be a group with exactly two conjugacy classes of involutions, represented by the involutions $x$ and $y$. For any involution $z$ of $G$, define $a(z)$ to be the number of ordered pairs $(u, v)$ of involutions $u$, $v\in G$ such that $u$ is conjugate to $x$ and $v$ to $y$ in $G$ and such that $(uv)^i=z$ for some integer $i$.

Note that as $\langle u, v\rangle$ is a dihedral group, it follows that $z$ centralizes both $u$ and $v$ for any such pair $(u, v)$. Hence $a(z)$ is determined entirely within $C_z$ and can be computed once the exact fusion pattern of involutions in $G$ is known.

THEOREM 2.43. *If $G$ is a group with exactly two conjugacy classes of involutions, represented by the involutions $x$ and $y$, then*

$$|G| = a(y)|C_x| + a(x)|C_y|.$$

When $G$ has only one conjugacy class of involutions, the procedure is considerably more complicated. One first determines the $p$-local structure of $G$ (including the conjugacy classes of elements of order $p$) for the set of "visible" odd primes—namely, those dividing the order of the given centralizer of an involution. This determination makes use of the 2-local structure of $G$ and involves prior classification theorems. (Such theorems are usually also involved in determining the possible involution conjugacy patterns of $G$ and its 2-local structure.)

With this information, one can now obtain a *congruence* for the order of $G$ with the aid of Sylow's theorem and a result of Frobenius [154, Theorem 9.1.1] concerning the number of solutions of the equation $x^n = 1$ in a group.* However, in practice, several local structures may be possible at this stage. If we take O'Nan's group as an illustration, two local structures may occur, one in which $G$ has Sylow 7-subgroups of order 7 (Case I) and the other of order $7^3$ (Case II) [230]. Correspondingly O'Nan obtains:

$$\text{Case I:}\ |G| = 2^9\cdot 3^4\cdot 5\cdot 7\cdot m, \qquad \text{where } m \equiv 939{,}551 \pmod{2^9\cdot 3^4\cdot 5\cdot 7}.$$

$$\text{Case II:}\ |G| = 2^9\cdot 3^4\cdot 5\cdot 7^3\cdot m, \qquad \text{where } m \equiv 6{,}479 \pmod{2^9\cdot 3^4\cdot 5\cdot 7^3}. \tag{2.9}$$

To determine $m$, one must get at the "invisible" primes—those dividing $m$. One is primarily interested in those represented by "strongly real" elements—i.e., elements inverted by some involution of $G$. These strongly real elements break up into $n$ disjoint sets, corresponding to certain abelian subgroups of $G$, with each of which one can associate a certain number $w_i$ of exceptional characters of $G$, $w_i$ being determined from the normalizer of the corresponding abelian subgroup. Using elementary character theory and counting arguments (including a result of Brauer and Fowler [46]), O'Nan

*In his book [154, Theorem 9.1.1], M. Hall proves the following slight extension of Frobenius's original theorem: If $C$ is a conjugacy class of elements of the group $G$, then for any positive integer $n$, the number of solutions in $G$ of the equation $x^n = c$, $c \in C$, is a multiple of g.c.d.$(|C|n, |G|)$.

was now able to prove in succession

$$m \leqslant 2^9 \cdot 5 \cdot 7 \left( 23 + \sum_{i=1}^{n} w_i \right). \tag{2.10}$$

(23 is the total number of strongly real classes of elements of order 2, 3, 5 and 7.)

$$n \leqslant 5. \tag{2.11}$$

$$w_i \leqslant 45. \tag{2.12}$$

$$m \leqslant 4{,}500{,}000. \tag{2.13}$$

Now in Case I of (2.9), there are only three possible values of $m$ satisfying (2.13), namely, $m = 939{,}551$, $2{,}391{,}071$, or $3{,}842{,}591 = 71 \cdot 54{,}121$, the first two being primes. Sylow's theorem yields a contradiction in the first two cases. In the third case, one has $n \leqslant 2$, giving the sharper estimate $m \leqslant 2{,}400{,}000$, a contradiction.

Hence Case II must hold. This time, using the bound for $m$ and (2.9), the only solution is $m = 6{,}479 = 11 \cdot 19 \cdot 31$ and the order of $G$ is uniquely determined.

For simplicity, we shall say that any group $G$ having one of these centralizers of an involution plus any set of further properties derivable from this assumption (such as order, local structure, character table) is of *type*—. Thus we have groups of type $J_i$, $1 \leqslant i \leqslant 4$, type $He$, type $Ly$, type $ON$, and type $F_i$, $i = 1, 2, 3$, or 5.

To distinguish type $J_2$ from type $J_3$, we must, of course, include the condition on the number of conjugacy classes of involutions.

In this terminology, Theorem 2.29 asserts that there is a unique simple group of type $J_1$.

## 2.5. Computer Construction of Sporadic Groups

We must now address the question of the existence (and uniqueness) of simple groups $G$ of each of the 11 groups of the previous section: $J_1$, $J_2$, $J_3$, $J_4$, $He$, $Ly$, $ON$, $F_1$, $F_2$, $F_3$, and $F_5$. The case of $J_1$ was discussed in Section 2.3 Apart from it and the groups $J_2$ and $F_1$, each of the remaining 8 listed groups was originally constructed with the assistance of a high-speed computer. However, as previously mentioned, in view of Griess's hand construction of $F_1$, the existence of $F_2$, $F_3$, and $F_5$ (as well as $He$, which arises as a direct factor in the centralizer of a suitable element of order 7 in $F_1$) no longer requires computer calculations (and very likely their uniqueness can also be obtained without any such calculations).

Because $J_2$ is a rank 3 permutation group, it has a natural underlying geometry, on the basis of which it can be directly constructed by hand. This will be described in the next section. Of the eight sporadic rank 3 permutation groups, only the Arunas Rudvalis group *Ru* has required some computer calculations.*

On the other hand, the groups $J_3$,† $J_4$, *He*, *Ly*, and *ON* appear to possess no such natural associated geometry, which could provide the basis for their construction. Moreover, because of their size, the calculations needed for their construction are too extensive to be carried by hand (as was done for the smaller group $J_1$). A similar difficulty occurs with *Ru*, even though it has a natural underlying geometry; thus an alternate procedure was used for its construction, which depended upon computer calculations.

From the local analysis, one can obtain rather complete information about the local structure of a group $G$ of one of these types. Then, using character theory—primarily (a) Brauer's theorem, which asserts that every irreducible character of $G$ is a $\mathbb{Z}$-linear combination of characters induced from so-called "elementary" subgroups [130, Theorem 4.7.1], and (b) Brauer's $p$-block theory for groups whose order is divisible by the prime $p$ only to the first power [60]—one is able to calculate a substantial portion, and in some cases all, of the character table of $G$. Thus the real question we are asking is the following: How does one construct a (simple) group $G$ from (a) its local structure and (b) its character table?

The first point to be made is that we may suspect that our unknown group $G$ must also contain certain *nonlocal* subgroups. The groups $J_2$, $J_3$, *He*, *Ly*, *ON*, and $F_5$ each possess such a nonlocal subgroup. Moreover, what is even more important, each of these groups has been constructed as a permutation group on the cosets of that subgroup. Why should we imagine that our group $G$ contains a particular such subgroup, e.g., that a group of type *Ly* must have a subgroup isomorphic to $G_2(5)$? We known that $G_2(5)$ is generated by two of its *local* subgroups $A^*$, $B^*$. Suppose our local information about a group $G$ of type *Ly* tells us that $G$ must contain subgroups, $A$, $B$ isomorphic to $A^*$, $B^*$, respectively. Isn't it then reasonable to conjecture that for a suitable choice of $A$, $B$, it will be the case that $\langle A, B\rangle \cong G_2(5)$?

*In Brauer's terminology, the direct product of a $p$-group and a cyclic group is called an *elementary* group.

† M. Mirman and R. Weiss have recently obtained computer-free constructions of *Ru* and $J_3$, respectively.

Under such circumstances, a very nice procedure, known as the "Brauer trick," often enables one to answer the question in the affirmative. In fact, apart from generator and relation calculations, it is essentially the only known method for proving the existence of nonlocal subgroups. To explain the procedure, suppose first that the desired subgroup $H$ of $G$ exists. Then the complex representation corresponding to the permutation representation on the cosets of $H$ decomposes into irreducible constituents. Let $\chi$ be the character of a nontrivial such irreducible constituent. Then $\chi$ is a constituent of $(1_H)^G$, the character of $G$ induced from the trivial character $1_H$ of $H$. The so-called Frobenius reciprocity theorem [130, Theorem 4.4.5] applies in this situation and yields that $1_H$ must be a constituent of the restriction $\chi|_H$ of $\chi$ to $H$. It is this observation that underlies Brauer's procedure.

We begin with our subgroups $A$ and $B$ of $G$ (for this purpose we can assume $G$ exists). From the character table of $G$, we can determine a likely irreducible character for $\chi$. We let $\mathfrak{R}$ be the representation associated with $\chi$, acting on the (complex) vector space $V$. For each subgroup $Y$ of $G$, we denote by $V_Y$, the subspace of $V$ on which $\mathfrak{R}(Y)$ acts trivially and we set $d_Y = \dim V_Y$. Thus $d_Y$ is the multiplicity of $1_Y$ in $\chi|_Y$. The integer $d_Y$ is given by the following inner product:

$$d_Y = (\chi|_Y, 1_Y) = \frac{1}{|Y|} \sum_{y \in Y} \chi|_Y(y). \tag{2.14}$$

If we have sufficient information about the character table of $G$ and of the subgroup $Y$, we can compute $d_Y$.

To apply the Brauer trick, we must be able to compute $d_A$, $d_B$, and $d_C$, where $C = A \cap B$. Since $C \leqslant A$ and $C \leqslant B$, we have $V_A \leqslant V_C$ and $V_B \leqslant V_C$, so by counting dimensions of subspaces of $V_C$, we have

$$\dim(V_A \cap V_B) + d_C \geqslant d_A + d_B. \tag{2.15}$$

If our calculations should yield that $d_A + d_B > d_C$, we can conclude from (2.15) that $V_A \cap V_B \neq 0$. This is the goal of the Brauer method, for it immediately implies that $H = \langle A, B \rangle$ is a *proper* subgroup $G$. Indeed, as $\mathfrak{R}(H)$ acts trivially on $V_A \cap V_B \neq 0$ and $\mathfrak{R}$ is a nontrivial irreducible representation of $G$, $\mathfrak{R}(H) \neq \mathfrak{R}(G)$, so $H < G$. (Of course, $\langle A, B \rangle$ may be a proper subgroup of $G$ even if $d_A + d_B = d_C$; but in that case we cannot verify this fact by the Brauer trick.)

The final step in the process requires the *identification* of the subgroup $H$. This is usually accomplished by invoking some prior classification theorem. Thus in the case of a group $G$ of type $Ly$, Lyons was able to prove [206], using the subgroups $A$ and $B$ and the information he had already obtained about $G$, that $H$ had only one class of involutions and the centralizer of an involution in $H$ was isomorphic to that in $G_2(5)$ [it has a subgroup of index 2 isomorphic to $SL(2,5)*SL(2,5)$]. Now he could invoke a theorem of Paul Fong and Warren Wong [103] to conclude that $H \cong G_2(5)$.

With the Brauer procedure, the following results have been obtained. [Here $M(24)$ denotes the largest of the Fischer 3-transposition groups $M(22)$, $M(23)$, $M(24)$. They will be described in Sections 2.6 and 2.8.]

PROPOSITION 2.44. *If $G$ is a group of type $J_2$, $He$, $Ly$, $ON$, or $M(24)'$, then correspondingly $G$ contains a subgroup isomorphic to $U_3(3)$, $Sp(4,4)^*$, $G_2(5)$, $L_3(7)^*$, or $He$.*

Here $Sp(4,4)^*$ and $L_3(7)^*$ denote, respectively, the split extension of $Sp(4,4)$ by a field automorphism of order 2 and of $L_3(7)$ by the transpose-inverse automorphism of order 2.

On the other hand, the generator-relation method has yielded the following results.

PROPOSITION 2.45. *If $G$ is a group of type $J_3$ or $F_5$, then correspondingly $G$ contains a subgroup isomorphic to $SL(2,16)^*$ or $A_{12}$.*

Here $SL(2,16)^*$ denotes the split extension of $SL(2,16)$ by a field automorphism of order 2. It was Thompson who proved the existence of this subgroup of $J_3$ (see [171]). G. Higman and McKay [171, 216], who constructed $J_3$ (and also $He$) using character-theoretic analysis and computer calculations, were only able to give "strong evidence" that a group of type $J_3$ must have such a subgroup but did not settle its actual existence. On the other hand, Harada was able to prove the existence of $A_{12}$ in a group of type $F_5$ fairly directly from the internal properties of such a group [163].

Since the Fischer groups $M(22)$, $M(23)$, $M(24)$, and $F_2$ are constructed as permutation groups on the cosets of the centralizer of an involution in the distinguished conjugacy class, the existence of nonlocal subgroups in these cases does not enter into the analysis.

The Higman–McKay construction of $J_3$ was carried out by a more or less direct process of coset enumeration, leading to a presentation by generators and relations. This method is very effective for permutation groups of not too high a degree [e.g., $J_3$ is of degree 6,156 on the cosets of an $SL(2,16)^*$ subgroup]. However, more elaborate computer techniques are required to construct permutation groups of large degree ($ON$ has degree 122,760, $Ly$ has degree approximately 9 million, and $F_2$ has degree approximately 13 billion). From his construction of $Ly$, $ON$, and $F_2$ (the third group jointly with Jeffrey Leon), Sims has gradually developed a general technique for constructing permutation groups of large degree by computer methods [202, 255, 256]. To appreciate the technical difficulties involved, consider the fact that a modern computer can store just one permutation of the set $1, 2, \ldots, N$, when $N$ is in the range of 150,000 to 200,000. Using secondary storage devices, one can hold much more information; however, access to such devices is comparatively slow. Hence, although it is easy enough to multiply two such permutations, the programming for carrying this out can be quite tricky. It is clear that to perform such multiplications many times, as will certainly be required to construct the required group $G$, a number of very effective algorithms have to be developed, which require not only a profound understanding of both the computer and finite group theory, but also a high order of imaginative insight.

In order to begin, Sims must have a precise description of the representation of the proposed one-point stabilizer $H$ on its coset space. For example, O'Nan provided this information for his group by establishing the following result, based on an analysis of the conjugacy classes of elements and the character table of a group of O'Nan type.

PROPOSITION 2.46. *If $G$ is a group of type $ON$ and $H$ is the subgroup of $G$ isomorphic to $L_3(7)^*$ constructed in Proposition* 2.44, *then we have*

(i) $|G:H| = 122{,}760$ *and the representation of $G$ on the cosets of $H$ has rank* 5.
(ii) *The nontrivial orbits $\Delta_i$, $1 \leq i \leq 4$, of $H$ on its coset space have length* 6,384, 5,586, 58,653, *and* 52,136, *respectively.*
(iii) *If $a_i$ is a point of $\Delta_i$ and $H_i$ is the subgroup of $H$ fixing $a_i$, $1 \leq i \leq 4$, then we have*
 (1) *$H_1$ is a semidirect product of $Z_7 \times Z_7$ by $Z_3 \times Z_2 \times Z_2$.*
 (2) *$H_2 \cong Z_2 \times PGL(2,7) = C_H(t)$, where $t \in \mathcal{I}(H) - \mathcal{I}(H')$.*
 (3) *$H_3$ is a Sylow* 2*-subgroup of $H$.*
 (4) *$H_4 \cong Z_3 \times \Sigma_4$.*

This information describes the required action of $H$ on its coset space. [We note that a group $G$ of type $ON$ possesses a second conjugacy class of subgroups $H^* \cong L_3(7)^*$ and if we consider the representation of $G$ on the cosets of $H^*$, we obtain a similar but not identical description for the action of $H^*$.]

Thus, Sims begins with a group $H$, a collection of subgroups $H=H_0, H_2, \ldots, H_n$ of $H$, the coset spaces $\Delta_i = H/H_i$ of $H_i$ in $H$, $0 \leqslant i \leqslant n$, and he considers the set

$$\Omega = \bigcup_{i=0}^{n} \Delta_i. \tag{2.16}$$

It should be emphasized that this formulation of the construction problem is the end product of the "evidence" stage of the analysis; it is obtained as a consequence of an explicit permutation representation which an arbitrary group $G$ of the type under investigation has been shown to possess, deducible from its character table. Moreover, determination of the character table of such a $G$ depends in turn on a knowledge of its conjugacy classes and local structure. In addition, if a one-point stabilizer of the given permutation representation is a nonlocal subgroup of $G$, it, too, must be constructed as part of the evidence stage.

Once all this has been achieved, the existence of a group of the given type is thus reduced to the construction of a *transitive permutation group* $G$ on $\Omega$ having $H$ as its one-point stabilizer with the specified action of $H$ on $\Omega$. (The subgroups $H_i$ will be precisely the set of two-point stabilizers of $G$ on the corresponding orbits of $H$.)

There are four major components to the computer analysis.

A. Algorithms for determining the order of a group generated by a given set of permutations.
B. Techniques for defining permutations on large sets.
C. Techniques for verifying relations satisfied by permutations defined as in B.
D. Ad hoc methods for finding the right permutations to generate the group in question.

Let $X$ be a set of permutations of a set $\Omega$. Under $A$, one is interested in algorithms for determining the order of the group $G$ generated by $X$. Let $a \in \Omega$ and let $\Delta$ be the orbit of $a$ under $G$. The elements of $\Delta$ can be determined easily from a knowledge of $X$. Let $X_a$ be the subset of $X$ consisting of those permutations fixing $a$ and set $H_a = \langle X_a \rangle$. Obviously $H_a \leqslant G_a$, the subgroup of $G$ fixing $a$. The approach is to try to show that either $H_a = G_a$ or else to produce an element $y \in G_a - H_a$. In the latter case, one replaces $X$ by $X \cup \{y\}$ and repeats the process.

There are two procedures for carrying this out.

*The Schreier Method.* For each $b \in \Delta$, choose an element $u(b)$ in $G$ taking $a$ to $b$. For each $b$ in $\Delta$ and $x \in X$, form the element

$$y(b,x)=u(b)x\left(u(b^x)^{-1}\right),$$

where $b^x$ is the image of $b$ under $x$.

PROPOSITION 2.47. *$H_a = G_a$ if and only if every $y(b,x) \in H_a$.*

*The Schreier–Todd–Coxeter Method.* Determine some set of relations satisfied by the elements in $X$ and let $G^*$ be the abstract group generated by $X$ defined by these relations. Let $H^*$ be the subgroup of $G^*$ generated by the subset $X_a$.

PROPOSITION 2.48. *If $|G^*:H^*|=|\Delta|$, then $G_a = H_a$.*

Computation of $|G^*:H^*|$ is carried out by the standard methods of what is called "coset enumeration." If this coset enumeration does not show that $|G^*:H^*|=|\Delta|$, one can find elements $y_1^*, y_2^* \in G^*$ such that $H^*y_1^*$ and $H^*y_2^*$ appear to be distinct cosets of $G^*$, but if $y_1$, $y_2$ denote the corresponding elements of $G$, then the cosets $G_a y_1$ and $G_a y_2$ are equal. Form $y=y_1 y_2^{-1}$, so that $y \in G_a$. If $y$ is not in $H_a$, express $y$ as a product of the elements of $X$ and add this word as a new relation.

To carry this out, one normally assumes that one has first obtained a presentation for $H_a$ in terms of the generating set $X_a$.

Now for B. There are two methods for defining permutations on a large set $\Omega$ in such a way that one can work with them effectively on a computer. The first of these is to begin with a group $H$ and subgroups $H_i$, as above, and then to form the coset space $\Omega = \bigcup \Delta_i$, as in (2.16). If $K$ is any subgroup of $H$, one can determine the orbits of $K$ on $\Omega$ when $|H:K| \leqslant 200{,}000$. The orbits of $K$ on a given $\Delta_i$ are, in fact, in one-to-one correspondence with the orbits of $H_i$ on the coset space of $K$ in $H$. This is the first method.

Suppose next that we have defined a permutation group $K$ on a set $\Omega$ and we know the orbits $\Gamma_1, \Gamma_2, \ldots, \Gamma_m$ of $K$ on $\Omega$. Assume now that we are given points $a_i, b_i \in \Gamma_i$, $1 \leqslant i \leqslant m$, a permutation $\pi$ of the set $\{1,2,\ldots,m\}$, and an automorphism $\sigma$ of $K$ (as an abstract group), satisfying the following condition for each $i$, $1 \leqslant i \leqslant m$:

$$\left(K_{a_i}\right)^{\sigma}=K_{b_{\pi(i)}}. \tag{2.17}$$

Here $K_c$ denotes, as usual, the subgroup of $K$ fixing the point $c \in \Omega$. Thus $\sigma$ transforms the one-point stabilizers of $K$ in a well-defined fashion.

PROPOSITION 2.49. *Under the assumption of* (2.17), *there exists a unique permutation $z$ of $\Omega$ such that*

(i) *$z$ normalizes $K$ and induces by conjugation the given automorphism $\sigma$ of $K$.*
(ii) $(a_i)^z = b_{\pi(i)}$, $1 \leq i \leq m$.

For example, Leon and Sims [202] constructed $F_2$ by beginning with the subgroup $H \cong \mathrm{Aut}(M(22))$, which they believed to be a subgroup of $({}^2\hat{E}_6(2)) \cdot 2$, the centralizer of an involution in a group of type $F_2$. [Their construction proved that $M(22)$ is indeed a subgroup of ${}^2E_6(2)$.] They then determined approximately 40 subgroups $H_i$ of $H$ in roughly the same way as described above for a group of type $ON$. Then they constructed the coset space $\Omega$ from $H$ and its subgroups $H_i$ as in (2.16) above, the order of $\Omega$ being approximately $13 \cdot 10^9$. At this point, they took two subgroups for $K$, namely, $K_1 = H$ and $K_2$ a subgroup of $H$ of index approximately 145,000. Then, by the method just described, they constructed two permutations $z_1, z_2$ on $\Omega$ relative to $K_1, K_2$, respectively with $z_1$ centralizing $K_1$ and $z_2$ centralizing $K_2$. Ultimately they showed that the group $\langle H, z_1, z_2 \rangle$ had the required order.

Now consider C. In order to use the Schreier–Todd–Coxeter method to determine the order of the group generated by a set of permutations defined as in B, one needs to be able to verify relations satisfied by them. Take a group of type $F_2$ as an example. Let $w$ be a word in $z_1$, $z_2$, and the elements of $H$, which we would like to prove is the identity element of $G$. Since $z_1$ and $z_2$ centralize $K_2$, we can determine from the elements of $H$ appearing in the word $w$ a subgroup $K_0$ of $K_2$ which centralizes $w$. Assuming that $|K_2 : K_0|$ is not too large, it is possible to find representatives for each of the orbits of $K_0$ on $\Omega$. Since $w$ centralizes $K_0$, to prove that $w = 1$, it therefore suffices to show that $w$ fixes each of these representatives.

Finally, as to D, how does one go about making the right choices for the ultimate generators of $G$? Since one begins with a subgroup $H$ of $G$, which is assumed to be known, one has a set of generators for $H$ (and as remarked, an actual presentation), so it is the additional generators that we are talking about. In the type $F_2$ case, the group $G$ is by definition generated by a certain conjugacy class of involutions with very special properties, so many relations were known to hold and these largely forced the definition of the permutations $z_1$ and $z_2$. On the other hand, in the construction of *Ly* and *ON* a process of trial and error was required.

In fact, in the case of a group $G$ of type *ON*, in order to simplify the calculations, Sims *assumed* that $G$ possessed an outer automorphism $\alpha$ of period 2 and then constructed the extended group $G\langle\alpha\rangle$. Hence Sims proved the existence and uniqueness of a simple group of type *ON* which admits an automorphism of period 2. Subsequently a student of Sims, Steven Andrilli, in his doctoral thesis [6], extended Sims's analysis to the general case (i.e., without any assumption on the existence of automorphisms) and thus proved the existence and uniqueness of simple groups of type *ON*.

As previously mentioned, the original constructions of $F_3$ and $F_5$ also required some computer calculations. Thompson constructed $F_3$ from a certain 248-dimensional Euclidean lattice associated with the complex Lie group $E_8$ (which possesses a representation of this degree). We prefer to discuss Thompson's analysis in Section 2.9, along with the Conway groups and the Leech lattice.

Although $F_5$ is not a rank 3 permutation group, its original construction was carried out by the same general method which Conway and David Wales used to construct *Ru* (see Section 2.6). Their construction was made from a certain 28-dimensional complex representation, whereas $F_5$ was constructed from a 133-dimensional complex representation. It was Norton who actually carried out this construction, on the basis of the group-theoretic information provided by Harada. The necessary computer calculations for *Ru* were carried out by McKay and C. Landauer and for $F_5$ by P. Smith. This is all we shall say about the construction of $F_5$.

Summarizing the results of this section (but omitting $J_2$, $F_3$, $F_1$, and *Ru*, which are to be discussed later), we have

THEOREM 2.50. *There exists a unique (simple) group of each of the types* $J_3$, *He*, *Ly*, *ON*, *and* $F_2$.

## 2.6. Sporadic Groups and Rank 3 Permutation Groups

It was known that the classical groups have primitive rank 3 representations as permutation groups: The linear groups acting on the set of lines of projective space, the symplectic and unitary groups acting on the set of "absolute points," and the orthogonal groups on the set of "singular points." These representation are all clearly related to their structure as groups of Lie type, since the one-point stabilizers are parabolic subgroups in each case. $E_6$ has a similar such rank 3 representation on the cosets of a suitable parabolic. Likewise any quadruply transitive group (hence the alternating, symmetric, and Mathieu groups) have such a representation, when considered as permutation groups on the set of unordered pairs of distinct letters. In addition, Helmut Wielandt has shown that a primitive permutation group of degree $2p$, $p$ a prime, is either doubly transitive or of rank 3 [319, 320].

Thus these primitive rank 3 groups constitute an important class of permutation groups, which had been studied first by Wielandt and then in the 1960s by Donald Higman [164]. But until Janko's group $J_2$ appeared on the scene [188], no new simple groups had ever been found through any of these investigations. However, with the discovery of $J_2$, there followed in rapid succession the construction of four more primitive rank 3 permutation groups: McLaughlin's group $Mc$, the Higman–Sims group $HS$, the Suzuki (sporadic) group $Suz$, and the Rudvalis group $Ru$. Furthermore, Fischer's first three sporadic groups $M(22)$, $M(23)$, $M(24)$, which arose from his analysis of groups generated by a conjugacy class of "3-transpositions" (to be described in the Section 2.7) were each constructed from a rank 3 permutation representation.

Table 2.2 lists each group, its one-point stabilizer, its degree (the number of cosets of the one-point stabilizer), and the sizes of the two nontrivial orbits of the one-point stabilizer, which are referred to as the *subdegrees*.

**Table 2.2.** The Rank 3 Sporadic Groups

| Group | One-point stabilizer | Degree | Subdegrees | |
|---|---|---|---|---|
| $J_2$ | $U_3(3)$ | 100 | 36 | 63 |
| $HS$ | $M_{22}$ | 100 | 22 | 77 |
| $Mc$ | $U_4(3)$ | 275 | 112 | 162 |
| $Suz$ | $G_2(4)$ | 1,782 | 416 | 1,365 |
| $Ru$ | ${}^2F_4(2)$ | 4,060 | 1,755 | 2,304 |
| $M(22)$ | $\widehat{U_6(2)}$ | 3,510 | 693 | 2,816 |
| $M(23)$ | $\widehat{M(22)}$ | 31,671 | 3,510 | 28,160 |
| $M(24)$ | $Z_2 \times M(23)$ | 306,936 | 31,671 | 275,264 |
| $M(24)'$ | $M(23)$ | 306,936 | 31,671 | 275,264 |

The group $M(24)$ is not simple but has a simple subgroup $M(24)'$ of index 2; and the intersection of the one-point stabilizer with $M(24)'$ is $M(23)$.

The first method of studying rank 3 groups was in terms of an associated combinatorial block design and certain related incidence matrices. Its general theory was developed by D. Higman, who characterized some of the classical groups from this point of view. The initial construction of $J_2$ by M. Hall and Wales was carried out along these lines [155]. However, this point of view has for the most part been superceded by the use of a natural graph associated with any transitive permutation group, an idea first introduced and exploited by Sims [254].

DEFINITION 2.51. Let $G$ be a transitive permutation group on a set $\Omega$. For each $a \in \Omega$, let $G_a$ denote the subgroup of $G$ fixing $a$, and let $\Delta(a)$ be one of the remaining orbits of $G_a$ on $\Omega - \{a\}$, chosen so that for $a$, $b \in \Omega$, if $b = a^g$, then $\Delta(b) = (\Delta(a))^g$. In other words, beginning with a fixed $a \in \Omega$ and fixed orbit $\Delta(a)$, we consider the translates of $(a, \Delta(a))$ under the action of $G$.

We now define a *directed graph* $\Gamma$ as follows. The vertices of $\Gamma$ are the elements of $\Omega$, so that $|\Gamma| = |\Omega|$. By definition, the vertex $a \in \Gamma$ is connected by an edge of $\Gamma$ to the vertex $b \in \Gamma$ if and only if $b \in \Delta(a)$.

It is clear from this definition that the action of $G$ on $\Omega$ induces a transitive action of $G$ on the vertices of $\Gamma$, which transforms edges into edges. We say that $G$ is a (transitive) *group of automorphisms* of $\Gamma$. We define $\mathrm{Aut}(\Gamma)$ to be the group of all edge-preserving permutations of $\Gamma$.

All this is fine if we *have* a transitive permutation group $G$ to begin with. However, our problem is rather to construct such a group $G$, given a prospective one-point stabilizer $H$ of $G$ on $\Omega$ and a permutation representation of $H$ on $\Omega - \{a\}$. Such a group $G$ is called a *transitive extension* of $H$. Usually we impose additional internal restrictions on $G$. We have seen how difficult such construction can be in the previous section. Is it therefore likely that the process can be simplified by turning the problem into the construction of a graph having a *transitive* automorphism group? Well, this is indeed the case when the proposed group has rank 3. In that case, one has $a$, $\Delta(a)$, and a single additional orbit $\Phi(a)$; from the internal information one can "see" which points of $\Delta(a)$ and $\Phi(a)$ and which pairs of points of $\Delta(a), \Phi(a)$, respectively, to connect by an edge in order for the resulting graph to possess an automorphism *moving a* (which is the precise requirement for transitivity). Except for the Rudvalis group, it has been possible to carry out the construction of the appropriate graph and the proof of transitivity of its automorphism group entirely by hand.

For example, Higman and Sims began with the Steiner triple system $S(3,6,22)$ associated with $M_{22}$ and defined the vertices of their graph $\Gamma$ to be

$$\{*\} \cup \Omega \cup \Lambda,$$

where $\{*\}$ is a new symbol, $\Omega$ is the set of 22 points of $S(3,6,22)$, and $\Lambda$ is the set of 77 hexads of $S(3,6,22)$. To define $\Gamma$, connect $\{*\}$ to each point of $\Omega$, connect each point of $\Omega$ to those hexads of $\Lambda$ that contain it, and connect two hexads of $\Lambda$ if they are disjoint.

Higman and Sims proved that $\mathrm{Aut}(\Gamma)$ is a transitive group having a simple subgroup of index 2 and order 44,352,000—a new sporadic group [169]. Their construction was carried out in the course of a single 24-hour period, following a lecture by M. Hall at Oxford on the group $J_2$!

It is certainly not obvious that $\Gamma$ possesses an automorphism moving $\{*\}$. However, there is a more geometric description of the Steiner system $S=S(3,6,22)$ in which the existence of such an automorphism becomes almost transparent.

Indeed, choose a point of $\Omega$, which we denote by $\infty$. Then $M_{21}$ $[\cong L_3(4)]$ is the subgroup of $M_{22}$ fixing $\infty$. The $21=4^2+4+1$ points of $\Omega-\{\infty\}$ can therefore be identified with the projective plane $\mathcal{P}=\mathcal{P}_2(4)$ over $GF(4)$ in such a way that $M_{21}$ acts on $\mathcal{P}$.

Furthermore, $M_{21}$ has two orbits $\Lambda_1$ and $\Lambda_2$ in its action on $\Lambda$, with $|\Lambda_1|=21$ and $|\Lambda_2|=56$. In terms of the projective plane $\mathcal{P}$, one has the following description of $\Lambda_1$ and $\Lambda_2$.

(a) If $B\in\Lambda_1$, then $B=\{\infty\}\cup L$, where $L$ consists of the five points of a line of $\mathcal{P}$. (Note that $\mathcal{P}$ has exactly 21 distinct lines.)

(b) If $B\in\Lambda_2$, then $B$ consists of six points of $\mathcal{P}$, no three collinear.

There are actually $3\cdot 56$ sets of six points of $\mathcal{P}$, no three collinear, the full projective group $PGL(3,4)$ in its natural action of $\mathcal{P}$ permuting this set $\mathcal{O}$ of 168 hexads transitively. However, the "little" projective group $L_3(4)$ has three orbits $\mathcal{O}_1, \mathcal{O}_2, \mathcal{O}_3$ on $\mathcal{O}$ and $\Lambda_2=\mathcal{O}_i$ for some $i$, $1\leqslant i\leqslant 3$.

Now the transpose-inverse map $\alpha$ of $PGL(3,4)$ can be realized as a "polarity" of $\mathcal{P}$, i.e., a one-to-one transformation of points to lines and lines to points that preserves the incidence relation and has period 2. The remarkable fact in this particular case is that $\alpha$ induces a permutation (which we denote by the same letter) of the set $\mathcal{O}$. Indeed, any $C\in\mathcal{O}$ is taken by $\alpha$ into 6 lines of $\mathcal{P}$ (no three copunctal), which intersect in exactly 15 points. The *complementary* set $D$ of 6 points of $\mathcal{P}$ has the property that no three are collinear, so $D\in\mathcal{O}$. One now defines

$$\alpha(C)=D.$$

Since $\alpha$ normalizes the group $L_3(4)$, we see that $\alpha$ transforms the orbits $\mathcal{O}_1, \mathcal{O}_2, \mathcal{O}_3$ among themselves. But $\alpha$ has period 2, so its leaves at least one $\mathcal{O}_i$ invariant. Thus, without loss, we can assume that $\alpha$ leaves $\Lambda_2$ invariant.

Furthermore, this polarity $\alpha$ induces a permutation of the set $\mathcal{P}\cup\Lambda_1=(\Omega-\{\infty\})\cup\Lambda_1$, which interchanges the subsets $\Omega-\{\infty\}$ and $\Lambda_1$ (again denoted by the same letter): namely, if $\alpha$ interchanges the point $a\in\mathcal{P}=\Omega-\{\infty\}$ and the line $L$ of $\mathcal{P}$, define

$$\alpha(a)=\{\infty\}\cup L \quad \text{and} \quad \alpha(\{\infty\}\cup L)=a.$$

Thus $\alpha$ is defined on the 98 points $(\Omega-\{\infty\})\cup\Lambda$ of $\Gamma$. One extends $\alpha$ to a map of $\Gamma$ by defining

$$\alpha(\{*\})=\infty \quad \text{and} \quad \alpha(\infty)=\{*\}.$$

It is now an easy matter to check directly that $\alpha$ preserves the incidence relations of the graph and so is an automorphism of $\Gamma$ which moves the point $\{*\}$.

Suzuki has described a lovely construction of a sequence of graphs, which produce larger rank 3 permutation groups from smaller ones, beginning with $\Sigma_4$ and ending with the Suzuki sporadic group, and on the way picking up $J_2$ [281].

Let $H$ be a transitive permutation group and let $\Gamma_0$ be an associated graph. Construct a graph $\Gamma$ with vertices

$$\{*\}\cup\Gamma_0\cup\Lambda,$$

where $\{*\}$ is a new symbol and $\Lambda$ denotes the set of involutions of $H$. Connect $\{*\}$ to each point of $\Gamma_0$, connect two points of $\Gamma_0$ if they are connected as points in the graph of $H$, connect a point $a$ of $\Gamma_0$ to an involution $b$ in $\Lambda$, if $b\in H_a$, the subgroup of $H$ fixing $a$, and connect two involutions $a, b\in\Lambda$ if $a$ and $b$ do not commute, but there is a third involution $c\in\Lambda$ centralizing each of them.

Beginning with $\Sigma_4$ and its natural graph with four points and no edges $\Gamma_0$, one obtains a graph $\Gamma_1$ of order 14 with $\mathrm{Aut}(\Gamma_1)\cong PGL(2,7)$. If one repeats the process with $H$ and $PGL(2,7)$ and with $\Gamma_1$ as $\Gamma_0$, one obtains a graph $\Gamma_2$ of order 36 with $\mathrm{Aut}(\Gamma_2)\cong G_2(2)$. Continuing the process as long as one can, one obtains the graphs and automorphism groups shown in Table 2.3 [the final construction leading to $\mathrm{Aut}(Suz)$ requires a slight modification of the procedure, with $\Lambda$ consisting suitable four groups rather than involutions of $H$].

Unfortunately this process collapses if one tries to repeat it with $\mathrm{Aut}(Suz)$ as $H$ and $\Gamma_5$ as $\Gamma_0$. We note also that $J_2$ and $Suz$ are of index 2 in their automorphism groups.

McLaughlin's construction of his graph and group was carried out along similar lines [220]; however, Conway and Wales's construction of the Rudvalis group was trickier [71]. Griess had shown that if $Ru$ existed, it must have a nontrivial cover $\widehat{Ru}$ by $Z_2$. Rudvalis gave evidence that $\widehat{Ru}$ would have to possess a 28-dimensional complex representation [244] and then Feit proved that if $Ru$ existed, $\widehat{Ru}$ would indeed have a representation of this degree. Conway and Wales proceeded from this presumed representation, and their argument involved the determination of a set of 4060 quadruples of vectors $(v, iv, -v, -iv)$ in complex 28-space, whose automorphism group is transitive of rank 3 as a permutation group on these 4060 quadruples. Moreover, the stabilizer of a quadruple is an extension of ${}^2F_4(2)$ by $Z_2$. Thus they constructed $\widehat{Ru}$. The group $Ru$ $(=\widehat{Ru}/\text{center})$ can be viewed as a permutation group of the corresponding 4060 one-dimensional subspaces.

As for the Fischer groups, their graph is intimately related to the conjugacy class of involutions that define them and will be described in Section 2.8.

So much then for existence. What about the question of the uniqueness of these groups? Since each is determined from the automorphism group of a graph, the question clearly reduces to the uniqueness of the graph.

**Table 2.3.** The Suzuki Sequence

| Graph | Order | Automorphism group |
|---|---|---|
| $\Gamma_3$ | 100 | $\mathrm{Aut}(J_2)$ |
| $\Gamma_4$ | 416 | $\mathrm{Aut}(G_2(4))$ |
| $\Gamma_5$ | 1782 | $\mathrm{Aut}(Suz)$ |

However, the specified action of the one-point stabilizer $H$ on each orbit is such a strong restriction that it forces in each case only one possibility for the graph of a transitive extension of $H$ in which $H$ has the given action.

Finally, defining groups of type $HS$, $Mc$, $Suz$, and $Ru$ by the conditions of the table, we can summarize the above discussion as follows:

THEOREM 2.52. *There exists a unique (simple) group of type $J_2$, $HS$, $Mc$, $Suz$, and $Ru$.*

## 2.7. Janko's Group $J_4$

As remarked in Section 2.6, the existence and uniqueness of $J_4$ was not achieved until 1980, four years after Janko's initial evidence for such a sporadic group. Norton was the principal architect of the construction [224], but he acknowledges the assistance of several University of Cambridge mathematicians: D. Benson, Conway, R. Parker, and J. Thackray. In addition, Thompson made some important observations that underlie the construction, so to a certain extent this work on $J_4$ can be considered a group effort.

As described earlier, if $G$ is a group of type $J_4$, then by definition, the centralizer $H$ of a 2-central involution $z$ of $G$ has a subgroup of index 2 of the form $\hat{M}_{22}/(D_8)^6$ ($\hat{M}_{22}$ denoting the 3-fold cover of $M_{22}$). Janko also showed that such a group $G$ must have a maximal 2-local subgroup $M$ of the form $EK$, where $E$ is elementary of order $2^{11}$ and $K \cong M_{24}$. One can take $z$ to lie in $E$, in which case $H \cap M = EI$, where $I \cong \hat{\Sigma}_6/E_{2^6}$, and $\hat{\Sigma}_6$ is the 3-fold cover of $\Sigma_6$, with $E \cap I = 1$. Since $M$ is maximal, $G = \langle H, M \rangle$. Ultimately $J_4$ was constructed as a subgroup of $GL(112, 2)$ generated by two groups of the form $H$ and $M$ with intersection of the form $EI$.

We explain briefly how this was done. Using the total information supplied by Janko, Norton first determined (uniquely) the full character table of $G$. The minimal degree of a nontrivial irreducible character was 1,333, and the character values involved irrationalities of the form $\sqrt{-7}$. To attempt a computer construction of $J_4$ as a subgroup of $GL(1{,}333, \mathbb{C})$ did not seem feasible. At this point, Thompson turned to the 2-modular representations of $G$ and on the basis of his calculations conjectured that such a group $G$ must have a $GF(2)$-representation of degree 112. Thus it seemed reasonable to attempt to construct $J_4$ as a subgroup of $GL(V)$, where $V$ is 112-dimensional vector space over $GF(2)$.

Other conjectured properties of this representation of $G$ suggested a direction for carrying out the construction. Indeed, it appeared that in its action on $V$, $M$ fixed a 12-dimensional subspace $W$ and a 1-dimensional subspace $\langle u \rangle$ of $W$. Furthermore, in the wedge product $V \wedge V$, of dimension 6,216 over $GF(2)$ (see Section 1.4 for the definition), it appeared that the $G$-images of the vectors $u \wedge w$, $w \in W$, under $G = \langle H, M \rangle$ generated a subspace $X$ of dimension 4,995. Thus $G$ appeared to be contained in the full stabilizer $G^*$ of $X$ in $GL(V)$, i.e., the set of all elements of $GL(V)$ leaving the subspace $X$ of $V \wedge V$ invariant. Furthermore, it also seemed likely that $G^*$ was actually equal to $G$.

This suggested a procedure for constructing $J_4$:

1. Begin with an appropriate 112-dimensional $GF(2)$-representation $V$ of the group $M$ (in particular, leaving invariant subspaces $\langle u \rangle < W$ of dimension 1 and 12).
2. Use step 1 to determine an $M$-invariant subspace $X$ of $V \wedge V$ of dimension 4,995.

Now define $G^*$ to be the full stabilizer in $GL(V)$ of $X$ and analyze $G^*$. The goal is to prove

3. For an appropriate involution $z$ of $E$ ($\leqslant M$), the group $C_{G^*}(z) \cong H$.

Once step 3 was proved, it followed that $G^*$ was, in fact, of type $J_4$. [In addition, identifying $C_{G^*}(z)$ with $H$, also $G^* = \langle H, M \rangle$ with $H \cap M = EI$, as before.] This proves the existence of a group of type $J_4$. As expected, verification of step 3 required considerable computer calculations.

The preceding argument proves only the existence of a unique simple group of type $J_4$, *which is a subgroup of* $GL(112,2)$. Since we do not know whether an arbitrary group $G$ of type $J_4$ must have a faithful irreducible $GF(2)$-representation of dimension 112, it does not completely settle the uniqueness question. To achieve this, Norton worked instead with the known 1,333-dimensional irreducible complex representation of $G$, following a procedure that Thompson used earlier to prove the uniqueness of a group of type $F_1$ which possesses an irreducible representation of degree 196,883 [297]. That argument depends on the generation of $G$ by two subgroups whose intersection has a small index in one of them. The subgroups $H$ and $M$ above do not meet this requirement, so Norton must shift to a second pair of generating subgroups $K$ and $N$, whose existence was known from Janko's prior work (in fact, we can take $N \leqslant M$):

$$K \cong L_5(2)/E_{2^{10}}, \qquad N \cong (L_4(2)/E_{16})/E_{2^{11}}.$$

These subgroups have the key property

$$|N:K\cap N|=2.$$

(Note that this last condition implies that $K\cap N\lhd N$.)

The idea is the following. Let $V$ now denote the 1,333-dimensional complex space on which $G$ acts as a group of linear transformations. One proves first that the restrictions of $G$ to both $K$ and $N$ are *uniquely determined* (reducible) representations of $K$ and $N$, respectively. Indeed, the representation of $K$ is the sum of 465- and 868-dimensional irreducible representations of $K$, and the corresponding degrees for $N$ are 45, 840, and 448. Thus the actions of $K$ and $N$ on $V$ are *unique up to similarity.*

Fix a basis of $V$. Then with respect to this basis, we can regard the action of $K$ on $V$ to be unique. Hence also the action of $K\cap N$ on $V$ is unique. But $N=(K\cap N)\langle a\rangle$ for a suitable involution $a$ of $N$ and the action of $a$ on $V$ is determined up to similarity. The goal now is to show (by appropriate calculations) that these conditions (together with the known action of $a$ on $K\cap N$) suffice to force a *unique* solution for the action of $a$ on $V$ (with respect to the given basis). Thus does Norton establish the uniqueness of a group of type $J_4$.

Hence we have

THEOREM 2.53. *There exists a unique simple group of type* $J_4$.

## 2.8. Transpositions and the Fischer Groups

From the very outset of his career, Fischer, who had been a student of Reinhold Baer, was interested in the generation of a group by a conjugacy class of its involutions with conditions placed on the order of the product of two elements of the class. (Proposition 2.42 shows that this is equivalent to placing conditions on the subgroup generated by pairs of elements of the class.)

Fischer first focused on what he called "$p$-transpositions."

DEFINITION 2.54. A conjugacy class $D$ of involutions of a group $G$ is said to be a class of *$p$-transpositions*, $p$ an odd prime, provided the product of any two elements of the class $D$ has order 1, 2, or $p$.

Fischer investigated groups which were generated by a class of $p$-transpositions, and his first classification [96] foreshadowed the major achievements to come.

THEOREM 2.55. *Let $G$ be a group generated by a class of $p$-transpositions $D$ and assume that the following conditions hold*:

(a) *Any three pairwise distinct elements of $D$ do not generate a 2-group.*
(b) *If $x \in D$, then $C_D(x) \neq x$.*

*Then $G \cong \Sigma_4$ or $\Sigma_5$. In particular, $p=3$.*

The proof of this result made use of the following earlier result of Fischer's, which he established by transforming the problem into a question about distributive quasigroups [97]. This was a highly original way of dealing with a purely group-theoretic question.

THEOREM 2.56. *Let $G$ be a group generated by a conjugacy class of involutions $D$, and suppose that whenever $x, y \in D$ with $x \neq y$, the order of $xy$ is a power of a fixed odd prime $p$. Then $G'$ is nilpotent.*

Note that the hypotheses imply that distinct elements of $D$ never commute.

Using his result on quasigroups, Fischer similarly proved:

THEOREM 2.57. *Let $G$ be a group generated by a conjugacy class of involutions $D$ such that distinct elements of $D$ do not commute. Assume that for all $x \in D$,*

$$C_x = O_2(C_x) \times O(C_x).$$

*Then $G$ is solvable.*

(Theorems 2.55 and 2.56 can now be derived from the odd-order theorem and Glauberman's $Z^*$-theorem, Theorem 4.95 below.)

This last result was used by Fischer to prove the solvability of a finite group $G$ admitting a fixed-point-free automorphism $\alpha$ of order $2p$, $p$ an odd prime [with some restriction on $C_G(\alpha^p)$] [98].

These results were just "warming up" exercises for Fischer, in preparation for the main event, which consisted of the following magnificent theorem [99, 100]. (See Section 4.14 for a discussion of the geometry of the classical groups. In particular, a *transvection* here is by definition an involutory linear transformation acting trivially on a hyperplane of the underlying vector space.)

THEOREM 2.58. *Let $G$ be a group with $Sol(G)=1$ and $G'=(G')'$, which is generated by a conjugacy class of 3-transpositions. Then one of the following holds*:

(i) *$G \cong \Sigma_n$ for some $n$ and except when $n=6$, $D$ is the set of transpositions.*

(ii) *$G \cong Sp(2n,2)$ and except when $n=2$, $D$ is the set of symplectic transvections.*

(iii) *$G \cong O_{2n}^{\pm}(2)$, and $D$ is the set of transvections leaving the corresponding quadratic form invariant.*

(iv) *$G \cong U_n(2)$ and $D$ is the set of unitary transvections.*

(v) *$G \cong O_n^{\mp}(3)\langle d \rangle$, where $d \in D$ and given $\pi = \pm 1$, $D$ is the set of reflections $x \mapsto x + \pi(x,a)a$, where $a$ is a vector such that the inner product $(a,a)=\pi$.*

(vi) *$G$ is one of three new finite groups, denoted by $M(22)$, $M(23)$, and $M(24)$, and $D$ is uniquely determined for each group.*

The exceptions in (i) and (ii) are due to the fact that $\Sigma_6 \cong Sp(4,2) \cong O_4^-(3)\langle d \rangle$. Note that the hypothesis $G'=(G')'$ implies that $G$ is nonsolvable.

The theorem clearly suggests that the given hypothesis must be closely connected with some natural geometric problem, the various solutions of which lead to the alternatives of the theorem, the groups $M(22)$, $M(23)$, $M(24)$ arising as exceptional cases in roughly the same way as $G_2$, $F_4$, $E_6$, $E_7$, $E_8$ occur in the classification of simple Lie algebras. Thus the method of attack must be to build a geometry from the class of 3-transpositions on which the group $G$ will act as a group of automorphisms and then, from the properties of these geometries, to determine the possibilities for $G$.

I would like to describe some of the central steps involved in the proof. Denote by $E$ a set of pairwise commuting involutions of $D$ of maximum size. Fischer proves

PROPOSITION 2.59. (i) *$N_G(E)$ contains a Sylow 2-subgroup of $G$.*

(ii) *$N_G(E)$ acts doubly transitively on the set $E$.*

(iii) *$N_G(E)/C_G(E)$ is isomorphic to one of the following groups: $\Sigma_n$, $A_n$, $GL(n,2)$, $L_m(4)$, where $m=[n/2]$, $E_{2^n} \cdot GL(n,2)$, $M_{22}$, $M_{23}$, or $M_{24}$.*

It is the last three possibilities in (iii) that give rise to the Fischer groups and explain why he adopted the notation $M(22)$, $M(23)$, $M(24)$ for them.

Proposition 2.60. *For $x \in D$, set $D_x = C_D(x) - \{x\}$. Then $D_x$ is a class of conjugate elements of $\langle D_x \rangle$. In particular, $\langle D_x \rangle \leqslant C_G(x)$ acts transitively on the set $D_x$.*

This remarkable fact allows Fischer to use induction in the analysis. However, there are cases in which $\langle D_x \rangle$ has a nontrivial center or is solvable, so these must be analyzed separately.

Proposition 2.61. (i) *If $O(\langle D_x \rangle) \nleqslant Z(\langle D_x \rangle)$, then $G \cong \Sigma_5$.*

(ii) *If $O_2(\langle D_x \rangle) \nleqslant Z(\langle D_x \rangle)$, then $G \cong Sp(n,2)$ or $U_n(2)$.*
(iii) *If $\langle D_x \rangle$ is solvable, then $G \cong \Sigma_5$, $\Sigma_6$, $U_4(2)$, or $U_5(2)$.*

The following result is fundamental and reveals the geometry that lies at the heart of this class of groups.

Proposition 2.62. *For $x \in D$, set $F_x = D - (D_x \cup \{x\})$. Then we have*

(i) *$\langle D_x \rangle$ acts transitively on $F_x$.*
(ii) *As a permutation of $D$, $G$ has rank 3 with one-point stabilizer $\langle D_x \rangle = C_G(x)$ having orbits $\{x\}$, $D_x$, and $F_x$.*

Thus the geometry underlying $G$ is that of a graph associated with a rank 3 permutation group whose vertices can be identified with the elements of $D$, a given vertex $x \in D$ being connected to those elements of $D - \{x\}$ with which it commutes. As we have pointed out in the previous section, the symmetric, alternating, unitary, symplectic, and orthogonal groups all have natural representations as automorphism groups of such graphs. What Fischer has done is to construct these classical graphs [as well as new ones for $M(22)$, $M(23)$, $M(24)$] from internal properties of $G$ that follow from the existence of the class of generating 3-transpositions. The graphs associated with $D$ are, in general, of a special type, which Fischer calls *triple-graphs*, and the analysis is reduced to the study of these triple-graphs and their so-called *triple-maps*. (We shall not attempt to define these terms here.)

In the previous section, we have also listed the subdegrees and one-point stabilizers of $M(22)$, $M(23)$, $M(24)$.

At this point, Aschbacher established a beautiful extension of Fischer's theorem [9]. To state it, we need the following definition.

Definition 2.63. A conjugacy class $D$ of involutions of a group $G$ is said to be a class of *odd transpositions* if the product of any two noncommuting elements of $D$ has odd order.

THEOREM 2.64. *Let $G$ be a group such that $Sol(G)=1$ and $G'=(G')'$, which is generated by a conjugacy class $D$ of odd transpositions. Then one of the following holds*:

(i) *$D$ is a class of 3-transpositions (whence $G$ is determined from Fischer's theorem).*
(ii) *$G \cong Sp(n,q)$, $U_n(q)$, or $O_n^{\pm}(q)$, where $q=2^m$ for some $m$, and $D$ is a class of transvections.*
(iii) *$G \cong$ an extension of $O_n^{\pm}(5)$ by $Z_2$ and $D$ is a class of orthogonal reflections.*
(iv) *$G \cong Sz(q)$, $q=2^m$, $m$ odd, $m>1$.*
(v) *$G \cong L_2(q) \int \Sigma_n$, $q=2^m$, $m>1$.*

Thus Aschbacher's theorem includes all symplectic, unitary, and orthogonal groups of characteristic 2, whereas Fischer's hypothesis picks up only those over the prime field $GF(2)$.

Fischer showed that his groups were all rank 3 as permutation groups on the set of 3-transpositions. In the course of proving his main theorem, Aschbacher derives a converse of this result.

THEOREM 2.65. *Let $G$ be a group with $Sol(G)=1$ which is generated by a conjugacy class $D$ of involutions. If $G$ has rank 3 as a permutation group on the set $D$, then $D$ is a class of 3-transpositions of $G$.*

As would be expected, Aschbacher's proof of Theorem 2.64 utilizes many of the concepts introduced by Fischer; but if anything, it is even more geometric than the 3-transposition theorem. The action of $G$ on the geometry determined by the class $D$ and, in the case of the situations giving rise to the classical groups, the associated bilinear form preserved by $G$ dominates the analysis. For example, to identify the symplectic groups, Aschbacher invokes a theorem of Peter Dembowski and Otto Wagner which characterizes projective space among "symmetric block designs" [78] to conclude that $G$ is acting on projective $n$-space over $GF(q)$, $n$ odd, $q=2^m$. Since he also knows in this case that $D$ is the set of nontrivial "elations" commuting with a certain "symplectic polarity," he is then able to assert that $G=\langle D\rangle \cong Sp_{n+1}(q)$.

In cases in which the action of $G$ on the set $D$ has rank 3—in particular, in the proof of Theorem 2.64—Aschbacher makes strong use of various arithmetical relations established by D. Higman for arbitrary rank 3 permutation groups.

The next stage in the analysis of groups generated by a class $D$ of involutions allows products of pairs of elements of $D$ to have order divisible by 4. A theorem of Baer and Suzuki [130, Theorem 3.8.2], disposes of a trivial case.

THEOREM 2.66. *If $D$ is a conjugacy class of elements of prime order $p$ in a group $G$ with the property that $\langle x, y\rangle$ is a $p$-group for every pair $x, y \in D$, then* $D \leqslant O_p(G)$.

Thus as a corollary, one has

COROLLARY 2.67. *If a group $G$ is generated by a conjugacy class $D$ of involutions such that the product of any two elements of $D$ has order a power of* 2, *then $G$ is a* 2*-group.*

Hence the first interesting case is incorporated in the following definition.

DEFINITION 2.68. A conjugacy class $D$ of involutions of a group $G$ is said to be a class of $\{3,4\}$*-transpositions if* the product of any two elements of $D$ has order 1, 2, 3, or 4. In order to exclude the case of 3-transpositions, we say that $D$ is *nondegenerate* if some product of elements of $D$ has order 4. Furthermore, we say that $D$ is a class of $\{3,4\}^+$-transpositions provided $(xy)^2 \in D$ whenever $x, y \in D$ and $|xy|=4$.

A natural example is the group $GL(n,2)$, which is generated by its transvections, these transvections forming a conjugacy class of $\{3,4\}^+$-transpositions.

Timmesfeld began his investigations with an analysis of groups generated by a nondegenerate class of $\{3,4\}^+$-transpositions and obtained the following lovely classification theorem [299–301].

THEOREM 2.69. *Let $G$ be a group with $Z(G)=1$ and $O(G)=1$ which is generated by a nondegenerate class of $\{3,4\}^+$-transpositions. Then one of the following holds*:

(i) $G \cong GL(n,2)$, $n \geqslant 3$.
(ii) $G \cong Sp(2n,2)$, $n \geqslant 3$.
(iii) $G \cong O_{2n}^{\pm}(2)$, $n \geqslant 4$.
(iv) $G \cong G_2(2)'$, ${}^3D_4(2)$, $F_4(2)$, ${}^2E_6(2)$, $E_7(2)$, *or* $E_8(2)$.

*Moreover, in each case D is a uniquely determined class of 2-central involutions of G.*

Timmesfeld's analysis is again much in the spirit of Fischer's and Aschbacher's. In the special case that leads to $GL(n,2)$, using internal properties of $G$ that follow from the existence of the class $D$, Timmesfeld constructs a vector space over $GF(2)$ on which $G$ acts in a prescribed fashion. He actually considers a slightly more general case, which leads to the groups $G_2(2)'$ and ${}^3D_4(2)$ as well as $GL(n,2)$, the precise hypothesis being that $D$ possesses a nontrivial proper subset $E$ such that

$$E \cap E^x = \varnothing \quad or \quad E \qquad \text{for all } x \in D.$$

To obtain the groups $Sp(2n,2)$ and $O_{2n}^{\pm}(2)$, Timmesfeld shows under suitable conditions that $G$ is also generated by a class of 3-transpositions, so that the groups can be identified from Fischer's classification theorem. The cases leading to $F_4(2)$, ${}^2E_6(2)$, $E_6(2)$, $E_7(2)$, and $E_8(2)$ depend upon a set of conditions on the class $D$ which suffice to imply that the multiplication table of $G$ is uniquely determined.

The next step in Timmesfeld's program was to weaken his assumption on the class $D$ by replacing "3" by "odd." Thus we have the following definition.

Definition 2.70. A conjugacy class $D$ of involutions of a group $G$ is said to be a class of $\{odd, 4\}^+$-transpositions if the product of any two elements $x$, $y$ of $D$ has order 1, 2, 4, or $k$, where $k$ is odd, with the additional restriction that $(xy)^2 \in D$ whenever $|xy| = 4$. Again $D$ is called *nondegenerate* if some product of elements of $D$ has order 4.

This is a fundamental notion, since every group of Lie type of characteristic 2, apart from the Ree groups ${}^2F_4(2^n)$, is generated by a class $D$ of $\{\text{odd}, 4\}^+$-transpositions, the elements of $D$ being so-called "root involutions." For this reason, Timmesfeld refers to a class of $\{\text{odd}, 4\}^+$-transpositions as a class of *root involutions*.

Timmesfeld's major "root-involution" theorem gives a classification of all groups generated by a nondegenerate class of root involutions [301], the degenerate case being covered by Aschbacher's and Fischer's prior work.

Theorem 2.71. *Let G be a group with* $Z(G)=1$ *and* $O(G)=1$ *which is generated by a nondegenerate class of root involutions. Then one of the following holds*:

(i) $G \in Chev(2)$, *with* $G \not\cong U_n(2^m)$ *or* $Sp(4,2^m)$.
(ii) $G \cong A_6$ *or* $J_2$.

*Moreover, in* (i), *either D is the uniquely determined class of root elements corresponding to the "long roots" of the associated Lie algebra of G or* $G \cong F_4(2^m)$, *in which case D may also be the class of root elements corresponding to the "short roots."*

The proof is again similar to those of the preceding theorems, in that Timmesfeld must build up the internal structure of $G$ from properties of the class of $D$. The "root subgroups" play an important role in this construction. For $x \in D$, set

$$D_x = \{y \in D | C_D(y) = C_D(x)\} \quad \text{and} \quad E_x = D_x \cup \{1\}.$$

Then $E_x$ is called a *root subgroup* of $G$. Clearly $E_x = E_y$ for $y \in D_x$.

Fix $E_x$ and let $\Sigma$ be the conjugacy class of $E_x$ in $G$. Then $\Sigma$ is called a *class of root subgroups* of $G$.

Timmesfeld proves the following results.

Proposition 2.72. *If* $\Sigma$ *is a class of root subgroups of* $G$, *then* $|E_x| = q = 2^n$ *for some* $n$ *for any* $E_x \in \Sigma$. *Moreover, if* $E_x, E_y \in \Sigma$, *then one of the following holds*:

(i) $\langle E_x, E_y \rangle$ *is elementary abelian.*
(ii) $\langle E_x, E_y \rangle$ *is special of order* $q^3$ *with center of order* $q$ *and* $[E_x, E_y] \in \Sigma$.
(iii) $\langle E_x, E_y \rangle \cong L_2(q)$.

This result is analogous to that proved by Thompson for *odd* primes in his study of so-called "quadratic pairs" (see Section 4.7).

The case $q=2$ of the root involution theorem is essentially covered by the $\{3,4\}^+$-transposition theorem, so Timmesfeld can assume $q>2$ at critical points in the argument. The main idea of the proof is to construct a certain graph $\Gamma(\Sigma)$ whose vertices are the element of $\Sigma$ and then to prove that $\Gamma(\Sigma)$ is isomorphic to the corresponding graph $\Gamma(\Sigma^*)$ of a suitable group $G^*$ of Lie type of characteristic 2. To obtain this isomorphism, Timmesfeld proves a "graph extension" theorem, which asserts that a certain collection of "local" isomorphisms of $\Gamma(\Sigma)$ on $\Gamma(\Sigma^*)$ can be extended to an actual isomorphism of $\Gamma(\Sigma)$ on $\Gamma(\Sigma^*)$. He also argues that the group $G^*$ is uniquely determined by its graph and is thus able to conclude that $G \cong G^*$.

While Timmesfeld was investigating $\{3,4\}^+$-transposition and root involution groups, Fischer was looking at the broader class of $\{3,4\}$-transposition groups. Fischer had developed a general method for attempting to form, from a class $E$ of transpositions of a group $H$, a larger group $G$ containing $H$, generated by a class $D$ of transpositions containing $E$, which can be viewed as an analogue of the construction of a transitive extension with a given one-point stabilizer. The passages from $U_6(2)$ to $M(22)$, from $M(22)$ to $M(23)$, and from $Z_2 \times M(23)$ to $M(24)$ are examples in which the extended groups exist.

Now Fischer knew that his group $\mathrm{Aut}(M(22))$ (a split extension of $M(22)$ by $Z_2$) was generated by a class $E$ of $\{3,4\}$-transpositions [the elements of $E$ necessarily induce outer automorphisms of $M(22)$]. There was also strong evidence that $\mathrm{Aut}(M(22))$ occurred as a subgroup of ${}^2\hat{E}_6(2)\cdot 2$. This was a natural situation for Fischer to ask whether the latter group could possibly be imbeddable in a larger group $G$ in such a way that the class $E$ extended to a class of generating $\{3,4\}$-transpositions $D$ of $G$. Fischer was eventually able to derive almost complete information about the internal structure of his presumed group $G$, and, in particular, the information which Leon and Sims required for their construction of $G$ [101]. As remarked earlier, this group has come to be denoted by $F_2$. Thus their existence and uniqueness theorem yields the following additional fact about $F_2$.

THEOREM 2.73. *$F_2$ is generated by a class $D$ of $\{3,4\}$-transposition such that $C_{F_2}(x) \cong {}^2\hat{E}_6(2)\cdot 2$ for all $x \in D$.*

This is by no means the end of the Fischer story. In Section 4.10 we shall describe the general fusion theorems of Timmesfeld which are consequences of his root involution theorem.

## 2.9. The Leech Lattice and the Conway Groups

Conway [68] constructed his three simple groups from the automorphism group of the remarkable 24-dimensional Leech lattice, which had its origins in the study of close sphere packings [199, 200].

DEFINITION 2.74. A *lattice* $\Lambda$ of Euclidean $\mathbb{R}^n$ space is the set of all integral linear combinations of $n$ linearly independent vectors $w_1, w_2, \ldots, w_n$ of $\mathbb{R}^n$ having the property that the inner product $(w_i, w_j)$ is an integer for all

$i, j$, $1 \leqslant i, j \leqslant n$. (This definition is compatible with the notion of lattice given in Section 2.1.)

In particular, $\Lambda$ is an abelian group. If $v_j$, $1 \leqslant i \leqslant n$, is an orthonormal basis of $\mathbb{R}^n$, $\Lambda$ is said to be *integral* (*rational*) if the coordinates of each $w_j$ as a linear combination of the $v_i$'s are all integers (rational numbers). Moreover, $\Lambda$ is said to be *unimodular* if the matrix of the change of basis from the $v_i$'s to the $w_i$'s has determinant 1.

A rational lattice can always be made into an integral lattice by replacing the given basis vectors by suitable scalar multiples of themselves. Of course, if the original lattice is unimodular, this property will be lost when shifting to a corresponding integral lattice. The Leech lattice is, in fact, a rational unimodular lattice. However, in describing its related geometry, it was easier for Conway to work with an associated integral form of the lattice.

As remarked in Section 2.2, a set of so-called "fundamental roots" of a complex semisimple Lie algebra $\mathfrak{L}$ determines a lattice in $\mathbb{R}^n$, where $n$ is the dimension of a Cartan subalgebra of $\mathfrak{L}$, the so-called "root lattice" of $\mathfrak{L}$. These root lattices are, in general, nonrational.

By definition, the automorphism group $\mathrm{Aut}(\Lambda)$ of a lattice $\Lambda$ in $\mathbb{R}^n$ is the subgroup of the rotation group of $\mathbb{R}^n$ which transforms $\Lambda$ into itself. The elements of $\mathrm{Aut}(\Lambda)$ are called *rotations* of $\Lambda$.

To describe the Leech lattice, one begins with the Steiner system $S = S(5,8,24)$ on 24 letters $\Omega$, whose automorphism group is $M_{24}$. The set $\Omega^*$ of all $2^{24}$ subsets of $\Omega$ can be viewed as a vector space of dimension 24 over $GF(2)$ if $A+B$ is defined to be the symmetric difference $(A-B)\cup(B-A)$ for $A, B \in \Omega^*$ (i.e., $A, B \subseteq \Omega$).

Proposition 2.75. *The octads of $S$ span a* 12*-dimensional subspace $\mathcal{C}$ of $\Omega^*$ consisting of $\varnothing$, $\Omega$, the* 759 *octads of $S$, their* 759 *complements in $\Omega$, and* 2576 *sets of cardinality* 12.

(The notation $\mathcal{C}$ has been chosen because of the connection with coding theory—the subspace $\mathcal{C}$ being the "extended binary Goley code".)

We let $\mathcal{C}_n$ be the subset of elements of $\mathcal{C}$ of cardinality $n$, so that

$$\mathcal{C} = \mathcal{C}_0 \cup \mathcal{C}_8 \cup \mathcal{C}_{12} \cup \mathcal{C}_{16} \cup \mathcal{C}_{24},$$

with $\mathcal{C}_8$ consisting of the octads of $S$.

Now let $\{v_i \mid 1 \leqslant i \leqslant 24\}$ be an orthonormal basis of $\mathbb{R}^{24}$, the index set being $\Omega$, which we may identify with $\{1,2,\ldots,24\}$. For any $T \in \Omega^*$ and any $m \in \mathbb{Z}$, define

$$[T, m] = \Big\{ v \mid v = \sum x_i v_i, \text{ where } x_i \in \mathbb{Z},\ \sum x_i = 4m \quad \text{and}$$

$$x_i \equiv m \pmod 4 \text{ if } i \notin T,\ x_i \equiv m+2 \pmod 4 \text{ if } i \in T \Big\}.$$

For each pair, $T, U \in \Omega^*$ and $m, n \in \mathbb{Z}$, one has

$$[T, m] + [U, n] = [T+U, m+n]. \tag{2.18}$$

DEFINITION 2.76. Set $\Lambda = \bigcup [T, m]$, the union over all $T \in \mathcal{C}$, $m \in \mathbb{Z}$. In view of (2.18) and the fact that $\mathcal{C}$ is a subspace, $\Lambda$ is an integral lattice. This is the *Leech* lattice. For reasons that will soon become clear, Conway sets $.0 = \mathrm{Aut}(\Lambda)$.

Conway analyzes $\Lambda$ and .0 and establishes all the following facts. For each $T \in \Omega^*$, define

$$v_T = \sum_{i \in T} v_i.$$

It is immediate from the definitions that the vector $2v_T$ is in $\Lambda$ for each $T \in \mathcal{C}_8$.

PROPOSITION 2.77. (i) *$\Lambda$ is spanned by the* 759 *vectors $2v_T$ with $T \in \mathcal{C}_8$ and any one vector of $\Lambda$, all of whose coordinates are odd integers*, e.g., *the vector $v_\Omega - 4v_{24}$.*

(ii) *The* 759 *vectors $2v_T$, $T \in \mathcal{C}_8$, span the sublattice $[\mathcal{C}, 4\mathbb{Z}]$.*

(iii) *If $v, w \in \Lambda$, then the inner product $(v, w)$ is a multiple of* 8 *and $(v, v)$ is a multiple of* 16.

Here $[\mathcal{C}, 4\mathbb{Z}]$ denotes the set of $[T, m]$ taken over all $T \in \mathcal{C}$, $m \in 4\mathbb{Z}$.

DEFINITION 2.78. If $(v, v) = 16n$ for $v \in \Lambda$, $v$ is said to be of *type n*. The set of all $v \in \Lambda$ of type $n$ is denoted by $\Lambda_n$.

The set $\Lambda_1$ is empty, so that $\Lambda_2$ consists of the vectors of minimum type. This set plays an important role in the analysis.

PROPOSITION 2.79. (i) $\Lambda_2$ *has cardinality* 196,560.

(ii) *If* $\Lambda_2(v)$ *denotes the subset of* $\Lambda_2$ *orthogonal to the vector* $v \in \Lambda_2$, *then* $\Lambda_2(v)$ *has cardinality* 93,150.

Observe next that any permutation $\pi$ of $\Omega$ extends to an orthogonal transformation of $\mathbb{R}^{24}$ (denoted by the same letter) under the definition

$$(v_i)\pi = v_{i\pi} \qquad \text{for all } i,\ 1 \leqslant i \leqslant 24.$$

Clearly $\pi$ will be a rotation of $\Lambda$ provided that it preserves $\mathcal{C}$. Hence the rotations induced from the automorphism group of the Steiner system $S = S(5,8,24)$ determine a subgroup $M \cong M_{24}$ of .0.

Furthermore, to each $T \in \Omega^*$, one can associate an orthogonal reflection $\varepsilon_T$ of $\mathbb{R}^{24}$ by setting

$$(v_i)\varepsilon_T = v_i \quad \text{or} \quad -v_i \qquad \text{according as } i \notin T \quad \text{or} \quad i \in T.$$

Likewise $\varepsilon_T$ is a rotation of $\Lambda$ provided $T \in \mathcal{C}$. Moreover, any two such rotations commute and so as $|\mathcal{C}| = 2^{12}$, they generate an elementary abelian subgroup $E$ of .0 of order $2^{12}$.

The rotations of $M$, arising as they do from elements of Aut($S$), induce an action on the elements of $E$ and so $M$ normalizes $E$. Thus .0 contains a subgroup

$$N = EM,$$

a split extension of $E_{2^{12}}$ by $M_{24}$.

Conway proves

PROPOSITION 2.80. *$N$ contains the subgroup of .0 fixing any one coordinate vector $v_i$.*

The crucial result that must be established is the following:

THEOREM 2.81. *$N$ is a proper subgroup of* .0.

Thus Conway must produce rotations of $\Lambda$ that move the coordinate vectors. Indeed, let $T$ be any element of $\Omega^*$ of cardinality 4 (a tetrad). From properties of the Steiner system $S$, $T$ is contained in exactly 5 octads of $S$, which can therefore be expressed as $T + T_i$, $1 \leqslant i \leqslant 5$, where each $T_i$ is a tetrad. Set $T_0 = T$ and

$$\Phi(T) = \{T_i | 0 \leqslant i \leqslant 5\}.$$

Since any 5 elements of $\Omega$ lie in a unique octad, it follows that the $T_i$ are pairwise disjoint and hence that $\Omega = \cup T_i$, $T_i \in \Phi(T)$.

Then for each $j$, $1 \leqslant j \leqslant 24$, $j$ lies in a unique tetrad $T_i$ of $\Phi(T)$ ($i$ depending on $j$). Define the map $\eta_T$ by setting

$$(v_j)\eta_T = v_j - \tfrac{1}{2} v_{T_i}, \qquad 1 \leqslant j \leqslant 24.$$

Conway shows that for each tetrad $T$, the transformation $\eta_T \varepsilon_T$ is a rotation of $\Lambda$ and so is an element of .0. Thus $N < .0$, as asserted.

With this information available, Conway is now able to argue:

THEOREM 2.82. (i) .0 *is transitive on* $\Lambda_2$.

(ii) *If* $v \in \Lambda_2$ *and* $w \in \Lambda_2(v)$, *then the subgroup* $.0_{vw}$ *of* .0 *fixing* $v$ *and* $w$ *is contained in* $N$ *and is a split extension of* $M_{22}$ *by an elementary group of order* $2^{10}$.

Note that as .0 consists of orthogonal transformations and .0 leaves $\Lambda_2$ invariant, $.0_v$ leaves $\Lambda_2(v)$ invariant for any $v \in \Lambda_2$.

Combined with Proposition 2.79, Conway immediately obtains the following corollary.

THEOREM 2.83. (i) $|.0| = 196{,}560 \cdot 93{,}150 \cdot 2^{10} \cdot |M_{22}|$.

(ii) $N$ *is a maximal subgroup of* .0.

The group .0 has a center of order 2, generated by the reflection $\varepsilon_\Omega$.

For $v \in \Lambda_2$, the pair $\{v, -v\}$ is called a *diameter* and the set of diameters is denoted by $\overline{\Lambda}_2$. Thus $\overline{\Lambda}_2$ has cardinality 98,280. Also set $.1 = .0/\langle \varepsilon_\Omega \rangle$, so that .1 acts on $\overline{\Lambda}_2$. The analysis of the action of .1 on $\overline{\Lambda}_2$ is a consequence of the following property of .0.

PROPOSITION 2.84. .0 *acts transitively on the set of ordered pairs of vectors of* $\Lambda_2$ *with any given scalar product.*

The number of $w \in \Lambda_2$ having scalar product $-32, -16, -8, 0, 8, 16, 32$ with a given $v \in \Lambda_2$ is

$$1,\ 4{,}600,\ 47{,}104,\ 93{,}150,\ 47{,}104,\ 4{,}600,\ 1,$$

respectively, and these are the only possible scalar products.

With this information, Conway is able to prove

THEOREM 2.85. .1 *is a simple group* (*of order* $\frac{1}{2}|.0|$).

Now define .2 to be the subgroup of .0 fixing any one vector of $\Lambda_2$. Conway also proves

THEOREM 2.86. .2 *is a simple group of order* $93{,}150 \cdot 2^{10} \cdot |M_{22}|$.

Conway determines yet a third simple group inside .0.

THEOREM 2.87. (i) .0 *acts transitively on the set* $\Lambda_3$.
(ii) *If* .3 *denotes the subgroup of* .0 *fixing any one vector of* $\Lambda_3$, *then* .3 *is a simple group of order* $2^{11} \cdot 3^7 \cdot 5^3 \cdot 7 \cdot 11 \cdot 23$.

The groups .1, .2, and .3 are sporadic groups, distinct from any of those previously discovered.

As is clearly evident, there is a rich geometry associated with .0 and the Leech lattice. In addition to the groups $M_{22}$, $M_{23}$, $M_{24}$, also $Mc$, $HS$, $J_2$, and $Suz$ arise in a natural way as subgroups of .0. For example, .0 acts transitively on $\Lambda_5$ and on $\Lambda_7$. If .5 and .7 denote the corresponding one-point stabilizers, then $.5 \cong \mathrm{Aut}(Mc)$ and $.7 \cong HS$. Furthermore, .0 has an element of order 3 which acts fixed-point-free on $\Lambda$. Its centralizer in .0 is the 3-fold cover of $Suz$. Since $J_2 \leqslant Suz$, also $J_2 \leqslant .0$. There is a natural geometric connection between some of these groups and .0, namely, .1 acts on $\Lambda/2\Lambda$, $(\widehat{Suz}) \cdot 2$ acts on $\Lambda/3\Lambda$, and $(\hat{J}_2) \cdot 2$ acts on $\Lambda/5\Lambda$.

Thus, if Conway had studied the Leech lattice some five years earlier, he would have discovered a total of seven new simple groups! Unfortunately he had to settle for three. However, as consolation, his paper on .0 will stand as one of the most elegant achievements of mathematics.

As promised earlier, we conclude this section with a brief description of Thompson's construction of the group $F_3$. Having determined the character table of a simple group $G$ of type $F_3$ (a highly nontrivial task), Thompson concluded that $G$ possesses exactly one irreducible rational-valued character $\chi$ of degree 248. Since the complex Lie group $E_8(\mathbb{C})$ possesses a 248-dimensional representation, it was natural for Thompson to try to relate $G$ to the group $E_8(\mathbb{C})$ [295, 296].

As a further indication of this connection, if $G$ exists, then $G$ must contain a subgroup $D$ which is a nonsplit extension of $E_{32}$ by $L_5(2)$. Ulrich Dempwolff had studied groups $D$ with such a structure and shown that at most one such group $D$ exists (up to isomorphism) and, in addition, every faithful representation of $D$ has degree at least 248 [79]. Furthermore, one can show that the subgroup $D$ of the presumed group $G$ must also contain a subgroup $D_0$, which is a nonsplit extension of $Z_4 \times Z_4 \times Z_4 \times Z_4$ by $\hat{A}_8$. Moreover, if $V$ is a vector space over $\mathbb{C}$ which affords the character $\chi$ (of degree 248) and if one sets $V_0 = C_V(O_2(D_0))$, Thompson determined that $V_0$ is 8-dimensional and that $D_0$, in its action on $V_0$, preserves a certain set of 240 vectors which generate a lattice in $V_0$ isometric to that of a root lattice associated with the Lie algebra $E_8$. As a consequence, $D_0$ can be considered a subgroup of $E_8(\mathbb{C})$.

With such a striking conclusion, Thompson was convinced that $E_8(\mathbb{C})$ must contain a subgroup of the form $D$. With P. Smith's assistance on the computer, he was able to produce an extra element of $E_8(\mathbb{C})$, which together with $D_0$ generated $D$. Thus $E_8(\mathbb{C})$ does indeed contain a subgroup of the form $D$; in particular, this proved the existence of the Dempwolff group. An independent construction of $D$, without the use of computers, has been given by Griess [150]. Moreover, this showed that there existed a module $V$ of dimension 248, affording the character $\chi|_D$ and that $D$ preserved a certain lattice $\Lambda$ of $V$ [whence $D \leqslant \mathrm{Aut}(\Lambda)$]. Thompson went on to prove that $\mathrm{Aut}(\Lambda) \cong Z_2 \times G$, thus establishing the existence of a unique simple group of type $F_3$.

The quadratic form $K$ on $V$ preserved by the action of $D$, not surprisingly, turned out to be the Killing form associated with the complex Lie algebra $E_8$. Thompson showed that $\mathrm{Aut}(\Lambda)$ preserves $K$. Using $K$ to give $V$ the structure of a Lie algebra of type $E_8$, Thompson also proved that $D$ was the largest subgroup of $F_3$ preserving this Lie multiplication. Finally as $\Lambda$ is a lattice, one can "reduce mod $p$"; and Thompson showed in the case $p=3$ that $F_3$ preserves the corresponding Lie multiplication over $GF(3)$. Thus, in fact, $F_3 \leqslant E_8(3)$.

Summarizing, we have

THEOREM 2.88. *There exists a unique simple group of type $F_3$ and it is isomorphic to a subgroup of $E_8(3)$.*

## 2.10. The Fischer–Griess Group $F_1$

As with $J_4$, it was several years after Fischer and Griess's original work before the existence of $F_1$ was established. Soon after the initial "discovery," Griess, Conway, and Norton noticed that every nontrivial irreducible character of a group $G$ of type $F_1$ has degree at least 196,883 and very likely such a group $G$ must have a character of this exact degree. Indeed, on this assumption, Fischer, D. Livingstone, and Thorne eventually computed the full character table of such a group $G$ [204]. (Norton and others at Cambridge had previously done considerable work on the conjugacy classes of $G$.)

In the process of his calculations, Norton computed the character values of this hypothetical character $\chi$ of degree 196,883 and found that it had the following three properties:

(a) $\chi$ *occurs exactly once in its symmetric square.*
(b) $\chi$ *occurs exactly once in its symmetric cube.*
(c) $\chi$ *is self-dual* (*i.e.*, $\chi$ *is real-valued*).

These conditions imply that the underlying vector space $V$ afforded by the representation $\phi$ of $G$ corresponding to $\chi$ can be given a $\phi(G)$-invariant algebra structure having certain additional properties.

Since this "Norton" algebra is the starting point for Griess's construction of $F_1$ [152], we shall describe its general features. First, by definition, an *algebra A* over a field $K$ is a vector space over $K$ on which is defined a distributive multiplication that behaves properly with respect to scalar multiplication by elements of $K$. By definition, a multiplication on $A$ is simply a map $\alpha$ from $A \times A$ to $A$. For $\alpha$ to be distributive means precisely that $\alpha$ "factors" through the tensor product $A \otimes_K A$, i.e., $\alpha = \pi \circ \beta$, where $\pi$ is the natural projection of $A \times A$ on $A \otimes_K A$ and $\beta$ is a map from $A \otimes_K A$ to $A$.

Furthermore, $\alpha$ will be commutative if, in addition, $\beta(v \otimes w) = \beta(w \otimes v)$ for all $v, w \in V$. Hence if $\beta$ is linear, this is equivalent to the condition that each $v \otimes w - w \otimes v$ is in the kernel of $\beta$—in other words, that in turn $\beta$ factors through the symmetric square $A^2$.

Hence, reversing the steps, we see that a linear map $\gamma$ from $A^2$ to $A$ induces in a natural way a commutative algebra structure on the vector space $A$ over $K$.

We apply this to the representation space $V$ of our hypothetical representation $\phi$ of $G$. Because $\phi$ is a constituent of $V^2$, we can write $V^2 = V' \oplus W$, where $V'$ and $W$ are $\phi(G)$-invariant and the representation of $G$ on $V'$ is equivalent to $\phi$ (using Maschke's theorem). For simplicity, identify $V'$ with $V$. Hence, if we define $\gamma$ to be the projection map of $V^2$ to $V$ induced by the direct-sum decomposition $V^2 = V \oplus W$, we see that $\gamma$ commutes with the action of $\phi(G)$. It is immediate from this that the algebra structure induced on $V$ by $\gamma$ and the corresponding product $\alpha$ is invariant under $\phi(G)$—i.e., if we write $v \cdot w$ for $\alpha(v, w)$, then for each $g \in G$

and $v, w \in G$ we have

$$(\phi(g)v)\cdot(\phi(g)w)=\phi(g)(v\cdot w). \tag{2.19}$$

Thus there exists a nontrivial $\phi(G)$-invariant commutative algebra structure on $V$.

But $\phi$ is also a constituent of multiplicity 1 in the symmetric cube of $A$. By the same reasoning as above and since $A^3=(A^2)A=A(A^2)$, it follows that the resulting algebra also admits a nontrivial $\phi(G)$-invariant "associative" form on $V$, i.e., there exists a map $f$ from $V\times V$ to $V$ that commutes with $\phi(G)$ and has the property

$$f(u\cdot v, w)=f(u, v\cdot w) \tag{2.20}$$

for all $u, v, w\in V$.

Furthermore, whenever one has a representation $\psi$ of a group $X$ on a vector space $W$ over $\mathbb{C}$, then the map $h$ from $W\times W^*$ to $\mathbb{C}$,

$$h(w, \delta)=\delta(w), \tag{2.21}$$

for $w\in W$ and $\delta\in W^*$, the dual space of $W$, is invariant under $\psi(X)$. Moreover, if $\psi$ is irreducible and its character self-dual, then by Proposition 1.4(ii) the representation $\psi$ of $G$ on $W$ is equivalent to that of the dual representation $\psi^*$ of $G$ on $W^*$, in which case $W^*$ can be identified with $W$. In that case, $h$ becomes a bilinear map of $W$ into $\mathbb{C}$ invariant under $\psi(X)$. Since our presumed representation $\phi$ of $G$ is, in fact, irreducible and its character $\chi$ is self-dual, our algebra is thus equipped with a nontrivial $\phi(G)$-invariant bilinear form.

Together these conditions serve to define a *Norton* algebra on the representation space $V$: it is a commutative algebra having both a nontrivial bilinear and an associative form. Furthermore, both the algebra and each of the forms are invariant under $\phi(G)$.

In addition to this Norton algebra structure, Griess also makes use of S. Smith's Theorem 4.10, which allows him to define $G$ as a simple group in which the centralizer of some involution is isomorphic to $.1/(D_8)^{12}$. Let $z$ be such an involution and $C$ its centralizer (we follow Griess's notation in our discussion). The group $G$ is to be obtained as a group generated by $C$ and a second 2-local subgroup $N$ (known to exist in a group of type $F_1$ and to be described below). As we shall see, there is a certain involution $\sigma$ in $N$ with $N=\langle N\cap C, \sigma\rangle$, so that $G=\langle C, \sigma\rangle$.

For technical reasons, Griess finds it preferable to work with a vector space $B$ of complex dimension 196,884 ($B=V+V_0$, where $V_0$ is 1-dimensional, is the space afforded by the sum $\phi+1$ of the hypothetical representation $\phi$ and the trivial representation). He first defines an action of $C$ on $B$ (determined from the restriction of $\phi+1$ to $C$). He then examines all $C$-invariant commutative algebra structures $\mathcal{G}(B)$ satisfying conditions (2.20) and (2.21). It turns out that they can all be described in terms of *six* rational parameters. Of course, Griess is seeking one of these algebras that is not only $C$-invariant but also $G$-invariant, equivalently $\sigma$-invariant. From the way $\sigma$ is defined, this places considerable constraints on these parameters. However, his initial analysis does not pin them down uniquely, but only *up to sign*. There then remains the unbelievably difficult task of specifying both $\mathcal{G}(B)$ and the action of $\sigma$ on $B$ in such a way that $\sigma$ becomes an automorphism of $\mathcal{G}(B)$. The calculations required to carry this out are truly prodigious. Ultimately they reduce to a number of horrendous formal identities involving the Conway group .0 and the Leech lattice $\Lambda$. (Since $G=\langle C,\sigma\rangle$ with $\sigma$ normalizing a large subgroup of $C$ and since .0 is directly linked to the structure of $C$, this should not be completely surprising.) The fact that choices of the parameter signs specifying $\mathcal{G}(B)$ have to be made in the very process of constructing the desired involutory automorphism of $\mathcal{G}(B)$ adds a further dimension to the complexities of the problem.

At this point Griess has constructed a Norton algebra $\mathcal{G}(B)$ and a group $G=\langle C,\sigma\rangle$ which acts on $\mathcal{G}(B)$. However, he does not yet know that $G$ is *finite*, let alone of type $F_1$. To prove this, he observes first that the structure constants of $\mathcal{G}(B)$ (with respect to the basis of $B$ used in defining the algebra structure of $B$) lie in $\mathbb{Z}[\frac{1}{6}]$ and that the matrix entries for the elements of $G$ (with respect to this same basis) lie in $\mathbb{Z}[\frac{1}{2}]$. Hence, for all primes $p\geqslant 5$, he can "reduce mod $p$" and obtain a group $G(p)$, generated by the corresponding matrices over $GF(p)$. Thus $G(p)\leqslant GL(196{,}884, p)$, whence $G(p)$ is finite for all primes $p\geqslant 5$. Using techniques of finite group theory, notably Goldschmidt's classification of finite groups with a strongly closed abelian 2-subgroup (see Section 4.8), Griess proves that $C_{G(p)}(z(p))=C(p)$, where $z(p)$ and $C(p)$ denote the images (mod $p$) of the representations of $z$ and $C$ on $B$, respectively. Furthermore, it can be checked that $C(p)\cong C$ and $G(p)\neq O(G(p))C(p)$. Now Smith's theorem (which does not actually require simplicity but only these conditions) yields that each $G(p)$ is a simple group of type $F_1$, so by Griess's theorem, Theorem 2.38, the groups $G(p)$ all have the same order. Since the set of allowable primes $p$ is infinite, it follows finally that $G\cong G(p)$ for all such $p$, and thus $G$ is the desired simple group of type $F_1$.

The fact that Griess was able to make the entire construction by hand not only testifies to his remarkable insight and determination but also "proves" that the group $F_1$, rather than being a "monster," has some natural "geometric" interpretation, reflected in its considerable internal regularity, and so Griess renamed $F_1$ the "friendly giant." A satisfactory explanation of the geometric nature of $F_1$, related in some as yet unknown way to automorphic forms, infinite-dimensional Lie algebra, and/or algebraic geometry, is certainly one of the most intriguing problems concerning simple groups that is untouched by the classification. (Again we refer the reader to [70, 195, 297, 298].)

To complete this brief discussion of $F_1$, we give the definition of $N$ and $\sigma$ and the action of $C$ on $B$. A description of the algebra $\mathcal{G}(B)$ and of the action of $\sigma$ on $B$ are too technical to present here [as are any of the formal identities required for verification that $\sigma$ is an automorphism of $\mathcal{G}(B)$]. Obviously a more conceptual definition of $\mathcal{G}(B)$, if such exists, would be of great interest.

Set $Q=O_2(C)$, so that $Q\cong(D_8)^{12}$. It is known that a group $G$ of type $F_1$ contains a four subgroup $Z$ with $z\in Z\leqslant Q$ such that $N=N_G(Z)$ has the following properties:

(a) $O_2(N)\cap Q\cong(D_8)^{11}$.
(b) $O_2(N)\leqslant C$ and $O_2(N)/O_2(N)\cap Q\cong E_{2^{11}}$.
(c) $(N\cap C)/O_2(N)\cong M_{24}\times Z_2$.
(d) $N/O_2(N)\cong M_{24}\times\Sigma_3$.

Thus $|N: N\cap C|=3$, $N_1=(N')'\leqslant C$, $N_1\cong M_{24}/(D_8)^{11}$, and $N/N_1\cong\Sigma_3$. Furthermore, $Q$ contains an involution $\tau$ not in $O_2(N_1)$ whose image in $\bar{N}=N/O_2(N)$ lies in the $\Sigma_3$ factor $\bar{D}$ of $\bar{N}$. Moreover, one can choose a conjugate $\sigma$ of $\tau$ in $N$ such that $\langle\sigma,\tau\rangle\cong\Sigma_3$ and $\overline{\langle\sigma,\tau\rangle}=\bar{D}$. This gives a description of $N$ and $\sigma$.

Note that once the action of $C$ on $B$ is specified, so is the action of $\tau$. The fact that $\sigma$ is a conjugate of $\tau$ by an element of $G$ of order 3 which normalizes $N_1$ ($\leqslant C$) indicates the severe restriction of the possible actions of $\sigma$ on $B$.

Now to describe the action of $C$ on the vector space $B$, which is to be obtained as the direct sum of three $C$-invariant vector spaces $X, Y, Z$ of respective dimensions 300, 98,280, and 98,304 (summing to 196,884).

First, as .0 acts as a group of automorphisms of the Leech lattice $\Lambda$, the 24-dimensional vector space $U=\mathbb{Q}\otimes_{\mathbb{Z}}\Lambda$ affords a rational-valued representation of .0. The symmetric square $\phi$ of this representation has dimension 300 over $\mathbb{Q}$ and $\phi(Z(.0))$ is trivial. Thus $\phi$ determines a representation of

$.1 = .0/Z(.0)$. We can view $\phi$ as a representation of $C$ in which $\phi(Q)$ is trivial, inasmuch as $C/Q \cong .1$. We denote the corresponding vector space (over $\mathbb{C}$) by $X$. Thus $\dim_{\mathbb{C}} X = 300$ and $C$ acts "rationally" on $X$.

Next let $C_2$ be the preimage in $C$ of a .2 subgroup of $C/Q$. Then $|C:C_2| = 98{,}280$. Because the $GF(2)$-representation of $C_2$ on $Q/\langle z\rangle$ (of dimension 24) is the direct sum of a 1-dimensional and a 23-dimensional representation, $C_2$ has a normal subgroup $C_0$ of index 2. If $\alpha$ denotes the nontrivial homomorphism of $C_2$ into $\mathbb{C}$ whose kernel is $C_0$, then $\alpha$ is a linear (i.e., 1-dimensional) representation of $C_2$. Inducing $\alpha$ to $C$ thus yields a representation $\alpha^*$ of $C$ of dimension 98,280. We let $Y$ denote the corresponding vector space (over $\mathbb{C}$). With respect to an appropriate basis, the matrices of the elements of $C$ can be seen to have entries $+1$, $-1$, or 0.

Finally by general results on representations of direct and central products of groups [130, Section 3.7], $Q \cong (D_8)^{12}$ has a $2^{12}$-dimensional faithful irreducible complex representation inasmuch as $D_8$ has such a 2-dimensional representation. If $T$ denotes the corresponding representation space, then there is a natural way of defining a faithful representation on the space $T \otimes_{\mathbb{Z}} \Lambda$ (of dimension $98{,}304 = 2^{12} \cdot 24$). It is this space that we take as $Z$ with respect to an appropriate basis. The matrix elements of $C$ on $Z$ lie in the field $\mathbb{Q}(\sqrt{-1})$.

The group $C$ acts irreducibly on both $Y$ and $Z$. However, $X$ can be written as $X_1 \oplus X_0$, where $X_1$ is 1-dimensional, $X_1$ and $X_0$ are .0-invariant, and .0 acts irreducibly on $X_0$. Thus $C$ acts on the space $V = X_0 \oplus Y \oplus Z$ of dimension 196,883 as well as on $B$ of dimension 196,884. This is all we shall say about Griess's remarkable construction of $F_1$.

We are left then with the uniqueness problem for a group of type $F_1$. This was carried out in two stages, the first by Thompson [297], the second by Norton [225]. Indeed, Thompson showed that there could exist at most one simple group $G$ of type $F_1$ which possesses an irreducible complex representation of degree 196,883. (As noted in Section 2.7, Norton's proof of the uniqueness of $J_4$ followed the same pattern.)

We have that $G = \langle C, N\rangle$, $|N : C \cap N| = 3$, and $N = \langle C \cap N, a\rangle$ where $|a| = 3$ and $a$ normalizes $N_1 = (N')'$ of index 6 in $N$. Again let $V$ be the 196,883-dimensional vector space on which $G$ acts. Thompson shows first that the actions of $C$ and $N$ on $V$ are uniquely determined up to similarity, then fixes a basis for $V$, with respect to which the action of $C$ can be considered to be unique. Thus the action of $C \cap N$ on $V$ is unique. Since $a$ normalizes $N_1$ and $a$ is determined up to similarity, Thompson then argues that these conditions suffice to force the action of $a$ on $V$.

Thus uniqueness is reduced to the question of whether a simple group $G$ of type $F_1$ must possess an irreducible representation of the specified degree. Norton has shown that this is indeed the case, his result constituting the final chronological step in the classification of the finite simple groups!

Norton's argument is very nice and is based on general connections between the constituents of a primitive permutation representation of a group and its so-called "commuting algebra" of linear transformations, a basis for which is obtained from suitable "incidence matrices" associated with the given permutation representation (a systematic exposition of these developments can be found in D. Higman [168]). In the present situation, Norton works with the permutation representation of an arbitrary simple group $G$ of type $F_1$ determined by the conjugacy class $\mathcal{Y}$ of non-2-central involutions of $G$ (thus for $y \in \mathcal{Y}$, $C_y \cong \hat{F}_2$), and its associated so-called "commuting graph" $\Delta$. By definition, the vertices of $\Delta$ are the elements of $\mathcal{Y}$, with two vertices connected by an edge if and only if they commute. Clearly $G$ acts transitively on $\Delta$ by conjugation, so $G$ can be viewed as a transitive permutation group on the set $\Delta$. In this representation, $C_y$ is the stabilizer of the vertex $y \in \mathcal{Y}$. Since $C_y$ is known to be a maximal subgroup of $G$, $G$ is, in fact, primitive. Note also that $|\Delta| = |G: C_y|$ for $y \in \mathcal{Y}$, which is approximately $10^{20}$, so that the degree of $G$ as a permutation group is very large.

Let me briefly describe the general theory of commuting algebras and incidence matrices. Consider then a primitive permutation group $X$ on a set $\Omega$, of cardinality $n$. The elements of $\Omega$ can be viewed as a basis of a complex vector space $V$, in which case $G$ becomes a group of linear transformations on $V$ and hence a group of $n \times n$ matrices with respect to the given basis. The set of $n \times n$ complex matrices $A$ that commute elementwise with $X$ form an algebra $A(X)$, called the *commuting algebra* of $X$. The dimension of $A(X)$ over $\mathbb{C}$ is precisely the number of irreducible constituents in the (linear) representation of $G$ on $V$, and this in turn is the same as the rank of $X$ as a permutation group on $\Omega$ (i.e., the number of orbits of a one-point stabilizer).

There is a natural way of obtaining a basis for $A(X)$. Consider the action of $X$ on the set $\Omega \times \Omega$ and let $\mathcal{O}_1, \mathcal{O}_2, \ldots, \mathcal{O}_r$ be the distinct orbits of $X$ on $\Omega \times \Omega$. With each $\mathcal{O}_i$, we associate an $n \times n$ *incidence matrix* $A_i$ whose rows and columns are indexed by $\Omega$, with a 1 or 0 in the $j$th row and $k$th column according as the element $(j, k) \in \Omega \times \Omega$ is or is not in $\mathcal{O}_i$. Since $\mathcal{O}_i$ is an orbit under $X$, it is immediate that $A_i$ commutes with $X$ and so is an element of the commuting algebra $A(X)$. Furthermore, it is not difficult to show that the matrices $A_i$, $1 \leqslant i \leqslant r$, form a basis for $A(X)$. In particular, then, $X$ is a rank $r$ permutation group.

There exists a direct relationship between the commuting algebra of $X$ and its group algebra $\mathcal{G}(X)$ over $\mathbb{C}$. Both algebras are "semisimple" and so can be written as the direct sum of "simple" subalgebras (i.e., full matrix algebras over $\mathbb{C}$). Of critical importance for us here is the fact that the *degrees* of the irreducible constituents of the representation of $X$ on $V$ [equivalently, the dimensions of the simple subalgebras in the decomposition of $\mathcal{G}(X)$] are equal to the *multiplicities* (in a suitable order) of the simple constituents in the decomposition of $A(X)$; conversely, the dimensions of the simple constituents in the decomposition of $A(X)$ are equal (in a suitable order) to the multiplicities of the irreducible constituents of the representation of $X$ on $V$. In particular, if the latter representation is "multiplicity free", as is often the case (and is true in the present situation for our group $G$ of type $F_1$), $A(X)$ decomposes into one-dimensional subalgebras and so is, in fact, a commutative algebra.

Another important property of the $A_i$ is the following: for $i>1$, the subalgebra (over $\mathbb{C}$) generated by $A_i$ is $A(X)$ itself. In other words, the full commuting algebra is completely determined by a single $A_i$ and hence by any nontrivial $X$-invariant graph on $\Omega$.

Clearly $X$ leaves invariant the "diagonal" $(j, j)$ of $\Omega\times\Omega$, and this diagonal is a single $X$-orbit (as $X$ is transitive on $\Omega$), which we take to be $\mathcal{O}_1$. Then the corresponding matrix $A_1$ is the identity matrix. (This of course, reflects the fact that the trivial representation occurs as a constituent of the representation of $X$ on $V$.)

There is a direct connection between the incidence matrices $A_i$ and $X$-invariant graphs on $\Omega$. Indeed, for a given $A_i$, define a graph $\Omega_i$ on $\Omega$ as follows: connect the vertices $j, k\in\Omega$ by an edge if and only if the $jk$th entry of $A_i$ is 1. Since $A_i$ commutes with $X$, it is immediate that $X$ acts on the graph $\Omega_i$. Conversely for each such $X$-invariant graph on $\Omega$, one can similarly associate an incidence matrix, equal to one obtained from an orbit of $X$ on $\Omega\times\Omega$. Thus, in fact, there is a one-to-one correspondence between the $A_i$, $1\leq i\leq r$, and $X$-invariant graphs on $\Omega$. [Of course, $A_1$ corresponds to the (trivial) graph with no edges.]

Finally there is yet another set of matrices associated with $A(X)$—the so-called "Higman matrices," which are determined from the basis $A_1, A_2,\ldots,A_r$. Indeed, one has for all $i$, $j$

$$A_iA_j=\sum_{k=1}^{r} H_{ijk}A_k, \tag{2.22}$$

where the $H_{ijk}$ are suitable nonnegative integers. For a given $i$, the matrix $H^{(i)}=(H_{ijk})$ is thus an $r\times r$ integer matrix. The $r$ matrices $H^{(1)}, H^{(2)},\ldots, H^{(r)}$ are known as the *Higman matrices* of the permutation group $X$. Their importance lies in the fact that the multiplicities of the simple subalgebras in the decomposition of $A(X)$, and consequently the degrees of the irreducible constituents of the representation of $X$ on $V$, are completely determined by these matrices $H^{(i)}$, $1 \leqslant i \leqslant r$.

Return now to our simple group $G$ of type $F_1$ and the commuting graph $\Delta$. It can be shown that the rank of $G$ as a (primitive) permutation group on $\Delta$ is exactly 8. Thus in this case the Higman matrices $H^{(i)}$ consist of eight $8 \times 8$ integer matrices.

Of course, this applies, in particular, to Griess's "friendly giant" $F_1$ (as it is a simple group of type $F_1$). Since $F_1$ was constructed as a complex matrix group of degree 196,883, this integer obviously occurs among the degrees of the irreducible representations of $F_1$. Hence, by the previous discussion, it is determined from the Higman matrices $H^{(i)}$ associated with the corresponding commuting graph for $F_1$. For the same reason, if one can show that the corresponding Higman matrices for $G$ are *identical* to those for $F_1$, it will follow that $G$ must possess an irreducible representation of degree 196,883 and uniqueness will have been established.

There you have the theory underlying Norton's uniqueness proof. We conclude with a few further comments about his actual analysis. If we take $y, y' \in \mathcal{Y}$ with $[y, y'] = 1$, the corresponding 2-point stabilizer in $G$ is isomorphic to $({}^2E_6(2)/(Z_2 \times Z_2)) \cdot Z_2$ [${}^2E_6(2)$ is known to possess a covering group by $Z_2 \times Z_2$ and also to have an outer automorphism group isomorphic to $\Sigma_3$]. Furthermore, the three vertices $y, y', yy'$ obviously commute pairwise, and it follows that the stabilizer $Y$ in $G$ of this triangle is isomorphic to $({}^2E_6(2)/(Z_2 \times Z_2)) \cdot \Sigma_3$. To obtain information about $\Delta$, Norton must determine the orbit structure of $Y$ on $\Delta$-$\{y, y', yy'\}$. His analysis depends upon very precise information about the local subgroup structure of an arbitrary simple group of type $F_1$, which must be derived first.

Thus, summarizing the entire discussion, we have

THEOREM 2.89. *There exists a unique simple group of type $F_1$.*

## 2.11. The List of Known Simple Groups and Their Orders

Table 2.4 lists simple $K$-groups together with their orders. The listed groups of Lie type may not be simple, but they contain a simple normal subgroup of index $d$, where $d$ is described in the third column of the table.

**Table 2.4.** Known Simple Groups

| Group $G$ | Order of $G$ | $d$ |
|---|---|---|
| | **Groups of Lie type** | |
| $A_n(q)$ | $q^{n(n+1)/2}\prod_{i=1}^{n}(q^{i+1}-1)$ | $(n+1, q-1)$ |
| $B_n(q),\ n>1$ | $q^{n^2}\prod_{i=1}^{n}(q^{2i}-1)$ | $(2, q-1)$ |
| $C_n(q),\ n>2$ | $q^{n^2}\prod_{i=1}^{n}(q^{2i}-1)$ | $(2, q-1)$ |
| $D_n(q),\ n>3$ | $q^{n(n-1)}(q^n-1)\prod_{i=1}^{n-1}(q^{2i}-1)$ | $(4, q^n-1)$ |
| $G_2(q)$ | $q^6(q^6-1)(q^2-1)$ | 1 |
| $F_4(q)$ | $q^{24}(q^{12}-1)(q^8-1)(q^6-1)(q^2-1)$ | 1 |
| $E_6(q)$ | $q^{36}(q^{12}-1)(q^9-1)(q^8-1)(q^6-1)(q^5-1)(q^2-1)$ | $(3, q-1)$ |
| $E_7(q)$ | $q^{63}(q^{18}-1)(q^{14}-1)(q^{12}-1)(q^{10}-1)(q^8-1)(q^6-1)(q^2-1)$ | $(2, q-1)$ |
| $E_8(q)$ | $q^{120}(q^{30}-1)(q^{24}-1)(q^{20}-1)(q^{18}-1)(q^{14}-1)(q^{12}-1)(q^8-1)(q^2-1)$ | 1 |
| ${}^2A_n(q),\ n>1$ | $q^{n(n+1)/2}\prod_{i=1}^{n}(q^{i+1}-(-1)^{i+1})$ | $(n+1, q+1)$ |
| ${}^2B_2(q),\ q=2^{2m+1}$ | $q^2(q^2+1)(q-1)$ | 1 |
| ${}^2D_n(q),\ n>3$ | $q^{n(n-1)}(q^n+1)\prod_{i=1}^{n-1}(q^{2i}-1)$ | $(4, q^n+1)$ |
| ${}^3D_4(q)$ | $q^{12}(q^8+q^4+1)(q^6-1)(q^2-1)$ | 1 |
| ${}^2G_2(q),\ q=3^{2m+1}$ | $q^3(q^3+1)(q-1)$ | 1 |
| ${}^2F_4(q),\ q=2^{2m+1}$ | $q^{12}(q^6+1)(q^4-1)(q^3+1)(q-1)$ | 1 |
| ${}^2E_6(q)$ | $q^{36}(q^{12}-1)(q^9+1)(q^8-1)(q^6-1)(q^5+1)(q^2-1)$ | $(3, q+1)$ |
| | **Alternating groups** | |
| $A_n,\ n\geq 5,$ | $\frac{1}{2}(n!)$ | |
| | **Sporadic groups** | |
| $M_{11}$ | $7920=2^4\cdot 3^2\cdot 5\cdot 11$ | |
| $M_{12}$ | $95040=2^6\cdot 3^3\cdot 5\cdot 11$ | |
| $M_{22}$ | $443520=2^7\cdot 3^2\cdot 5\cdot 7\cdot 11$ | |
| $M_{23}$ | $10200960=2^7\cdot 3^2\cdot 5\cdot 7\cdot 11\cdot 23$ | |
| $M_{24}$ | $244823040=2^{10}\cdot 3^3\cdot 5\cdot 7\cdot 11\cdot 23$ | |
| $J_1$ | $175560=2^3\cdot 3\cdot 5\cdot 7\cdot 11\cdot 19$ | |
| $J_2$ | $2^7\cdot 3^3\cdot 5^2\cdot 7$ | |
| $J_3$ | $2^7\cdot 3^5\cdot 5\cdot 17\cdot 19$ | |
| $J_4$ | $2^{21}\cdot 3^3\cdot 5\cdot 7\cdot 11^3\cdot 23\cdot 29\cdot 31\cdot 37\cdot 43$ | |

Table 2.4 Sporadic Groups—*Continued*

| | |
|---|---|
| $HS$ | $2^9 \cdot 3^2 \cdot 5^3 \cdot 7 \cdot 11$ |
| $Mc$ | $2^7 \cdot 3^6 \cdot 5^3 \cdot 11$ |
| $Suz$ | $2^{13} \cdot 3^7 \cdot 5^2 \cdot 7 \cdot 11 \cdot 13$ |
| $Ru$ | $2^{14} \cdot 3^3 \cdot 5^3 \cdot 7 \cdot 13 \cdot 29$ |
| $He$ | $2^{10} \cdot 3^3 \cdot 5^2 \cdot 7^3 \cdot 17$ |
| $Ly$ | $2^8 \cdot 3^7 \cdot 5^6 \cdot 7 \cdot 11 \cdot 31 \cdot 37 \cdot 67$ |
| $ON$ | $2^9 \cdot 3^4 \cdot 5 \cdot 7^3 \cdot 11 \cdot 19 \cdot 31$ |
| .1 | $2^{21} \cdot 3^9 \cdot 5^4 \cdot 7^2 \cdot 11 \cdot 13 \cdot 23$ |
| .2 | $2^{18} \cdot 3^6 \cdot 5^3 \cdot 7 \cdot 11 \cdot 23$ |
| .3 | $2^{10} \cdot 3^7 \cdot 5^3 \cdot 7 \cdot 11 \cdot 23$ |
| $M(22)$ | $2^{17} \cdot 3^9 \cdot 5^2 \cdot 7 \cdot 11 \cdot 13$ |
| $M(23)$ | $2^{18} \cdot 3^{13} \cdot 5^2 \cdot 7 \cdot 11 \cdot 13 \cdot 17 \cdot 23$ |
| $M(24)'$ | $2^{21} \cdot 3^{16} \cdot 5^2 \cdot 7^3 \cdot 11 \cdot 13 \cdot 23 \cdot 29$ |
| $F_5$ | $2^{15} \cdot 3^{10} \cdot 5^3 \cdot 7^2 \cdot 13 \cdot 19 \cdot 31$ |
| $F_3$ | $2^{14} \cdot 3^6 \cdot 5^6 \cdot 7 \cdot 11 \cdot 19$ |
| $F_2$ | $2^{41} \cdot 3^{13} \cdot 5^6 \cdot 7^2 \cdot 11 \cdot 13 \cdot 17 \cdot 19 \cdot 23 \cdot 31 \cdot 47$ |
| $F_1$ | $2^{46} \cdot 3^{20} \cdot 5^9 \cdot 7^6 \cdot 11^2 \cdot 13^3 \cdot 17 \cdot 19 \cdot 23 \cdot 29 \cdot 31 \cdot 41 \cdot 47 \cdot 59 \cdot 71$ |

## 2.12. Statement of the Main Classification Theorem

In view of the preceding discussion, we can now give a precise formulation of the classification theorem for finite simple groups.

CLASSIFICATION THEOREM. *If $G$ is a finite simple group, then $G$ is isomorphic to one of the simple $K$-groups listed in Table* 2.4.

The remainder of this book as well as its sequel will be devoted to a detailed discussion of the proof of the classification theorem.

The classification theorem thus shows that the "typical" finite simple group is a group of Lie type. Apart from these, there are only the alternating groups and the 26 sporadic groups. Moreover, in a sense the alternating groups can be thought of as a "degenerate" family of groups of Lie type, inasmuch the symmetric groups arise as *Weyl* groups of linear groups.

It should be added that there appears to be no natural explanation of the number 26. There is nothing magical about it—nothing beyond the classification proof itself, which at its completion demonstrates that this is the exact number of exceptional simple groups.

# 3

# Recognition Theorems

As we have described in Section 1.1, the final stage of every classification theorem involves an "identification" of the group $G$ under investigation with some known simple group $G^*$. Moreover, this identification is made by means of a set of *intrinsic* conditions which serve to "characterize" $G^*$ among simple $K$-groups. For example, the description of Conway's groups in terms of the Leech lattice is *extrinsic*, since it involves an action of the groups on a geometry whose definition is given independently of the groups themselves. To obtain an intrinsic characterization of any of these groups, one must either show that it is possible to *reconstruct* the Leech lattice solely from information about their subgroups or else prove that their multiplication tables are *uniquely determined* by their subgroup structure.

The discussion of the known simple groups in the previous chapter has clearly indicated the three primary methods of identifying the simple groups:

A. By a presentation by generators and relations.
B. By the action of the group on a suitable geometry.
C. By a primitive permutation representation.

These methods are not really distinct, for one can usually pass from one to the other with a slight addition or variation in the argument. However, in general, the form of the recognition theorem to be established depends upon the nature of the internal conditions which one expects to impose on the group $G$. We have already seen the connection between groups of Lie rank 1 and doubly transitive permutation groups. Similarly, the Fischer transposition hypothesis leads most naturally to recognition theorems by geometries, especially those related to rank 3 permutation groups. However, even there, in dealing with the exceptional groups over $GF(2)$, Timmesfeld found it preferable to work with generators and relations. On the other hand, for general classification theorems—especially from the point of view of centralizers of involutions—the most useful form of recognition theorem is by generators and relations, particularly for identifying the groups of Lie type.

Every known simple group possesses such an internal characterization. (For many of the sporadic groups, such characterizations have already been described in Chapter 2 as part of the discussion of uniqueness.) In this chapter, we describe some *recognition* theorem for each of the known simple groups.

## 3.1. The Groups of Lie Type

It was Tits who first realized that the conclusions of Theorem 2.15 could be used as a basis for characterizing the groups of Lie type [304]. He introduced the following terminology.

DEFINITION 3.1. A group $G$ is said to be a $(B, N)$-*pair* provided

1. $B, N \leqslant G$ and $G = BNB$.
2. $B \cap N = H \lhd N$.
3. $W = N/H$ is generated by a set of involutions $w_i$, $1 \leqslant i \leqslant m$.
4. If $v_i$ is a representative of $w_i$ in $N$, then for each $v \in N$ and every $i$, $1 \leqslant i \leqslant m$, we have

$$BvBv_iB \leqslant (BvB) \cup (Bvv_iB).$$

5. $B^{v_i} \neq B$, $1 \leqslant i \leqslant m$.

As noted earlier, these conditions imply that $W$ is a Coxeter group with $w_i$, $1 \leqslant i \leqslant m$, as defining set. $W$ is the *Weyl* group of $G$ and $m$ is the *rank* of $G$.

DEFINITION 3.2. A $(B, N)$-pair $G$ is said to be *split* if $B = (B \cap N)U$, where $U$ is a normal nilpotent subgroup of $B$.

Thus in this terminology Theorem 2.15 asserts that every group of Lie type is a split $(B, N)$-pair. On the other hand, the general analysis of $(B, N)$-pairs proceeds without any splitting assumption. In fact, Tits has shown that all $(B, N)$-pairs of rank at least 3 are necessarily split (see Theorem 3.12 below).

The notion of *parabolic* subgroup defined in Section 2.1 extends in the obvious way to an arbitrary $(B,N)$-pair $G$: they are the subgroups of $G$ containing $B$ together with their $G$-conjugates. Likewise these parabolics can be described in terms of the Weyl group $W$ of $G$.

Set $\mathcal{V}=\{v_i|1\leqslant i\leqslant m\}$ and for each subset $\mathcal{T}$ of $\mathcal{V}$, set

$$G_{\mathcal{T}}=\langle B, v_i|v_i\in\mathcal{T}\rangle.$$

Thus $G_{\mathcal{V}}=G$ and $G_{\varnothing}=B$. One obtains the following extension of Proposition 2.17 directly from the definition of $(B,N)$-pair.

PROPOSITION 3.3. (i) *If $B\leqslant Y\leqslant G$, then $Y=G_{\mathcal{T}}$ for some subset $\mathcal{T}$ of $\mathcal{V}$.*

(ii) *If $\mathcal{T},\mathcal{T}'\subseteq\mathcal{V}$ with $\mathcal{T}\neq\mathcal{T}'$, then $G_{\mathcal{T}}\neq G_{\mathcal{T}'}$.*

(iii) *For any subset $\mathcal{T}$ of $\mathcal{V}$ set $N_{\mathcal{T}}=N\cap G_{\mathcal{T}}$ and let $B_{\mathcal{T}}$ be the largest normal subgroup of $G_{\mathcal{T}}$ contained in $B$. Then $\overline{G}_{\mathcal{T}}=G_{\mathcal{T}}/B_{\mathcal{T}}$ is a $(B,N)$-pair relative to its subgroups $\overline{B}$, $\overline{N}_{\mathcal{T}}$, with the images of $\{v_i|v_i\in\mathcal{T}\}$ a generating set of its Weyl group $\overline{N}_{\mathcal{T}}/\overline{N}_{\mathcal{T}}\cap\overline{B}$.*

In particular, if $\mathcal{T}\subseteq\mathcal{V}$ and $|\mathcal{T}|=r$, the proposition shows that there are precisely $2^{m-r}$ (parabolic) subgroups of $G$ containing $G_{\mathcal{T}}$. As in the Lie situation, the subgroups $G_{\mathcal{T}}$ with $|\mathcal{T}|=1$ are called *minimal* parabolics.

There is a natural incidence geometry $\mathcal{G}$ related to the parabolic subgroups $G_{\mathcal{T}}$ of $G$. Indeed the objects of $\mathcal{G}$ are the left cosets $gG_{\mathcal{T}}$ for $\mathcal{T}\subseteq\mathcal{V}$ and $g\in G$. Two objects $X_1, X_2\in\mathcal{G}$ are then said to be *incident* provided $X_1\cap X_2\neq\varnothing$.

Clearly $G$ acts faithfully on the objects of $\mathcal{G}$ by left multiplication, and this action preserves the incidence relation. We say that $G$ induces a *group of automorphisms* of the geometry $\mathcal{G}$. For example, if $G\cong L_n(q)$ for some $q$ and $n$, then $W\cong\Sigma_n$ and it can be shown that the corresponding geometry is derived from that of $\mathcal{P}_n(q)$, projective $n$-space over $GF(q)$. More precisely, this is the geometry of "flags" in $\mathcal{P}_n(q)$, a flag denoting a set of subspaces $V_i$, $1\leqslant i\leqslant n$, of $\mathcal{P}_n(q)$ with $V_i$ of dimension $i$ and $V_i$ incident with $V_{i+1}$ for all $i$. The group $G$ acts as a group of automorphisms of this flag geometry. [Brauer's characterization of the groups $L_3(q)$ by the structure of the centralizer of an involution also involved the construction of the underlying projective 3-space $\mathcal{P}_3(q)$ (but not its flag geometry), which he built up from properties of the involutions of $G$. Ultimately he was able to show that $G$ acted as a group of projective transformations of $\mathcal{P}_3(q)$, which yielded at once the desired isomorphism $G\cong L_3(q)$.]

As the example of $L_n(q)$ indicates, we can view these $(B, N)$-pair geometries as generalizations of the geometries of projective $n$-space. It is known, of course, that projective space of dimension $n \geqslant 3$ is necessarily Desarguesian; moreover, if the space is finite, then the number of points on each line is of the form $p^r+1$ for some prime $p$ and the geometry is that of $\mathcal{P}_n(p^r)$. It is therefore not completely surprising that these $(B, N)$-pair geometries can be classified when the group $G$ has suitably high rank and, furthermore, that such a classification in turn determines the possibilities for $G$.

A complete classification of these incidence geometries of rank at least 3 was obtained by Tits in a fundamental, elegant paper [307] in which these geometries were placed in a general setting based on his notions of "building" and "apartment." Tits's classification has had far-reaching applicability to the entire theory of algebraic groups, well beyond its direct significance for finite group theory. It would take us too far afield to outline his essentially geometric proof; however, we would like to at least present the basic terminology and indicate its relationship to $(B, N)$-pairs, so that the reader will have some picture of the geometric problem which Tits solved. We require a number of preliminary definitions.

DEFINITION 3.4. A *complex* $\Delta$ is a set together with an order relation $\subset$, read "is a face of" or "is contained in," provided the following two conditions are satisfied:

1. For any $A \in \Delta$, the ordered subset of all faces of $A$ is isomorphic (as a set) to the ordered set of subsets of a set.
2. Any two elements $A, B \in \Delta$ have a unique maximum lower bound, denoted by $A \cap B$.

Every complex possesses a unique minimal element which we denote by 0. Two complexes are *isomorphic* if there exists a one-to-one correspondence between their elements preserving the incidence relations.

Henceforth in the discussion $\Delta$ will always denote a complex in which the number of elements is finite.

DEFINITION 3.5. A subset $\Delta'$ of $\Delta$ is called a *subcomplex* provided:

a. $\Delta'$ is a complex under the ordering relation induced from $\Delta$.
b. If $A \in \Delta'$, then any face of $A$ in $\Delta$ is in $\Delta'$.

DEFINITION 3.6. If $A \in \Delta$, the number of faces of $A$ which are minimal subject to being nonzero is called the *rank* of $A$, denoted by $\operatorname{rk}(A)$.

DEFINITION 3.7. For $A \in \Delta$, the set of all elements of $\Delta$ which contain $A$, under the order relation induced from $\Delta$, is a subcomplex called the *star* of $A$, denoted by $\mathrm{st}(A)$. If $B \in \mathrm{st}(A)$, the rank of $B$ in $\mathrm{st}(A)$ is called the *codimension* of $A$ in $B$, denoted by $\mathrm{codim}_B(A)$.

Clearly $A$ is the unique minimal element of $\mathrm{st}(A)$ and $\mathrm{codim}_B(A)$ measures the number of elements of $\Delta$ contained in $B$, minimal subject to properly containing $A$.

DEFINITION 3.8. $\Delta$ is called a *chamber* complex provided:

1. Every element of $\Delta$ is contained in a maximal element; and
2. If $C, C'$ are two maximal elements, there exists a finite sequence $C = C_0, C_1, \ldots, C_m = C$ of maximal elements such that

$$\mathrm{codim}_{C_{i-1}}(C_{i-1} \cap C_i) = \mathrm{codim}_{C_i}(C_{i-1} \cap C_i) \leqslant 1,$$

   for all $i$, $1 \leqslant i \leqslant m$.

The maximal elements of a chamber complex are called *chambers*.

Condition 2 says, in effect, that one can pass from one chamber to another in a certain prescribed fashion.

Henceforth $\Delta$ will denote a finite chamber complex. It is not difficult to prove that an element $A \in \Delta$ has the same codimension in every chamber containing it. This common codimension is called the *codimension* of $A$ in $\Delta$, denoted by $\mathrm{codim}_\Delta A$.

DEFINITION 3.9. $\Delta$ is called *thick* if every element of codimension 1 is contained in at least three chambers; $\Delta$ is called *thin* if every such element is contained in exactly two chambers.

Now we can introduce the main geometric object.

DEFINITION 3.10. Let $\Delta$ be a finite chamber complex and let $\mathcal{Q}$ be a set of subcomplexes of $\Delta$. The pair $(\Delta, \mathcal{Q})$ is called a *building* and the elements of $\mathcal{Q}$ are called *apartments* provided the following conditions hold:

- B1. $\Delta$ is thick.
- B2. The elements of $\mathcal{Q}$ are thin chamber complexes.
- B3 Any two elements of $\Delta$ belong to an apartment.
- B4. If two elements $A, A'$ are contained in two apartments $\Sigma, \Sigma'$, then there exists an isomorphism of $\Sigma$ onto $\Sigma'$ which leaves invariant $A$ and $A'$ and all their faces.

The significance of the definition is given by the following result, which, of course, motivated Tits's definition.

THEOREM 3.11. *Let $G$ be a $(B, N)$-pair and let $\Delta$ be the set of all left cosets of all subgroups of $G$ containing $B$. Define an ordering $\subset$ on $\Delta$ by setting $gY \subset g'Y'$ if $gY \geqslant g'Y'$ for $g, g' \in G$ and $Y, Y'$ subgroups of $G$ containing $B$. Let $G$ operate on $\Delta$ by left multiplication. Let $\Sigma$ be the subset $\{nY | n \in N, B \leqslant Y\}$ of $\Delta$ and let $\mathfrak{A}$ be the set of all transforms of $\Sigma$ by all elements of $G$. Then $(\Delta, \mathfrak{A})$ is a building and $G$ is transitive on the pairs consisting of an apartment and a chamber of $\Delta$ containing it.*

The proof of the theorem is quite direct. That $\Delta$ is a complex is immediate from the fact that for $\mathfrak{T} \subseteq \mathfrak{V}$, there are exactly $2^{m-r}$ parabolics containing $G_{\mathfrak{T}}$, where $|\mathfrak{T}| = r$ (together with the action of $G$ on $\Delta$). Also $gB \leqslant gY$ for $g \in G$ and $B \leqslant Y$, so $gB \supset gY$. Thus $\{gB | g \in G\}$ is the set of maximal elements of $\Delta$ and every element of $\Delta$ lies in a maximal element. Likewise the elements of codimension 1 in $\Delta$ are precisely the minimal parabolics containing $B$ and their $G$-transforms. Using basic properties of Coxeter groups, applied to the Weyl groups $W$ of $G$, one argues next that $\Delta$ is, in fact, a chamber complex and that $\Sigma$ is a thin chamber complex.

Hence to complete the proof of $B1$ and $B2$, it remains only to show that $\Delta$ is thick, equivalently, that every minimal parabolic $Y \geqslant B$ contains at least three left cosets of $B$, i.e., $|Y : B| \geqslant 3$. However, if false, then $B$ would be normal in $Y$ (subgroups of index $\leqslant 2$ are necessarily normal). But $Y = G_{\mathfrak{T}}$, where $\mathfrak{T} = \{v_i\}$ for some $i$, $1 \leqslant i \leqslant m$, so as $B \lhd Y$, $B^{v_i} = B$, contradicting condition 5 in the definition of a $(B, N)$-pair. Thus $G$ is indeed thick.

Next let $gY$, $g'Y'$ be two elements of $\Delta$. To show that they are contained in a common apartment, write

$$g^{-1}g' = bnb',$$

where $b, b' \in B$ and $n \in N$ (which is possible as $G = BNB$) and set $A = gb\Sigma$. By definition, $A$ is an apartment and consists of the elements $gb(n_1 Y_1)$, $n_1 \in N$, $B \leqslant Y_1$. Observe now that both $gY = gb(1Y) \in A$ and $g'Y' = gbnb'Y' = gb(nY') \in A$. Thus $B3$ also holds.

Finally suppose $gY$, $g'Y'$ are contained in two apartments $A$, $A'$. By translation, we can suppose that $A = \Sigma$, whence $g, g' \in N$. Again by translation, we can also assume that $g = 1$. Thus $gY = Y$, $g'Y' = nY'$, for some $n \in N$, are contained in $\Sigma$ and $x\Sigma$ for some $x \in G$; and to establish $B4$, we must produce an isomorphism of $\Sigma$ on $x\Sigma$ leaving both $Y$ and $nY'$ invariant. In view of the rules for $(B,N)$-pair multiplication, the condition $Y \in x\Sigma$ implies that $Y = xn_0Y$ for suitable $n_0 \in N$. Replacing $x$ by $xn_0$ thus yields $Y = xY$. Hence without loss we can also assume that $x \in Y$. In this case, the desired automorphism will be attained by left multiplication of $\Sigma$ by a suitable element of $Y$.

Indeed, the condition $nY' \in x\Sigma$ implies likewise that

$$nY' = xn'Y' \tag{3.1}$$

for suitable $n' \in N$. Since $x \in Y$, it follows that $n \in Yn'Y'$. Write $Y = G_{\mathfrak{T}}$, $Y' = G_{\mathfrak{T}'}$, where $\mathfrak{T}, \mathfrak{T}' \subseteq \mathfrak{V}$. Let $w$, $w'$ be the images of $n$, $n'$ in $W = N/H$. Again the basic rule for $(B,N)$-pair multiplication yields that $w \in \langle \mathfrak{T} \rangle w' \langle \mathfrak{T}' \rangle$, so $w = w_1 w' w_1'$ for suitable $w_1 \in \langle \mathfrak{T} \rangle, w_1' \in \langle \mathfrak{T}' \rangle$. But then if $n_1$ is a representative of $w_1^{-1}$ in $N$, we have $n_1 \in Y$ and $n_1 n \in n'Y'$. Therefore $xn_1Y = Y$ and $xn_1nY' = nY'$ [using equation (3.1)]. Since clearly $xn_1\Sigma = x\Sigma$, we conclude that left multiplication by $xn_1$ induces an isomorphism of $\Sigma$ on $x\Sigma$ leaving invariant the elements $Y$ and $nY'$. Thus B4 also holds and so $(\Delta, \mathcal{Q})$ is indeed a building.

The apartment $g\Sigma$ is entirely contained in the chamber (i.e., maximal element) $gB$, and $gB$ is clearly the unique chamber containing $g\Sigma$. We see then that $G$ does act transitively on the pairs consisting of an apartment and a chamber containing it, so all parts of Theorem 3.11 are proved.

The theorem shows that the apartments of a building are related to Coxeter groups. Thus a "Weyl geometry" is essentially built into the definition of a building.

Tits's main result is a complete classification of buildings of "dimension" at least 3. Combined with Theorem 3.11, it yields the following fundamental result.

THEOREM 3.12. *If $G$ is a simple $(B,N)$-pair of rank at least* 3, *then $G \in Chev(p)$ for some prime $p$.*

Nothing so definitive can be expected when $G$ has rank 1 or 2. In fact, to say that $G$ is a $(B,N)$-pair of rank 1 is the same as asserting that $G$ is double transitive on the cosets of $B$ [Corollary 2.16]. In the rank 2 case, Feit and G. Higman have obtained the following partial result by a difficult combinatorial analysis of the corresponding geometries [92].

THEOREM 3.13. *If $G$ is a $(B,N)$-pair of rank* 2, *then the Weyl group of $G$ has order* 4, 6, 8, 12, *or* 16.

Each of these orders occurs. Indeed, the groups of Lie type of Lie rank 2 are the groups $L_3(q)$, $Psp(4, q)$, $U_4(q)$, $U_5(q)$, $G_2(q)$, ${}^3D_4(q)$, ${}^2F_4(2^n)$ and the corresponding Weyl groups have orders 6, 8, 8, 8, 12, 12, and 16, respectively. The order 4 case is degenerate and occurs if $G = G_1 \times G_2$, where $G_1$, $G_2$ are each $(B, N)$-pairs of rank 1.

Since the groups of Lie type are actually *split* $(B, N)$-pairs, we would certainly be content with the classification of such groups. Obviously Tits's theorem covers the case of rank at least 3. However, the split case has been handled without recourse to his theorem. Indeed, the combined work of Christoph Hering, Kantor, and Seitz gives a classification of split $(B, N)$-pairs of rank 1 [164, 196]. Using this classification theorem, Fong and Seitz determined all split $(B, N)$-pairs of rank 2. Then on the basis of that result together with the so-called Curtis–Steinberg relations for the groups of Lie type (see Theorem 3.16 below), they were able to classify split $(B, N)$-pairs of arbitrary rank. Thus they obtained the following fundamental classification theorem.

THEOREM 3.14. *If $G$ is a simple split $(B, N)$-pair, then $G \in Chev(p)$ for some prime $p$.*

We postpone discussion of the rank 1 case until the next section, but we shall make a few comments here about the rank 2 case. First, the results of [164, 196] for the rank 1 case determine the possible structures of a parabolic subgroup $G$. The bulk of the analysis is then aimed at showing that two parabolics of $G$ "resemble" a pair of parabolics in some group $G^*$ of Lie type (and Lie rank 2). The recognition portion of their argument deals only with the problem of turning resemblance into isomorphism.

The primary method which they use to prove that $G \cong G^*$ is to show that the given resemblance conditions force $G$ to have a *unique* multiplication table. Since $G^*$ is also a group that resembles $G^*$, $G^*$ also has this multiplication table and so $G$ and $G^*$ must be isomorphic. In view of the double coset multiplication formulas for $G$ as a $(B, N)$-pair, it is only necessary to establish the following facts:

(a) The structures of the groups $U$ and $N$ are uniquely determined. (3.2)
(b) The actions of $H$ on $U$ and of $N$ on $H$ are uniquely determined.

Theorem 3.14 covers every (simple) group in $Chev(p)$ of Lie rank 2 with the single exception of the Tits group $T = {}^2F_4(2)'$, which is of index 2 in the $(B, N)$-pair ${}^2F_4(2)$ [306]. In proving the simplicity of $T$, Tits derived a very pretty presentation of the group ${}^2F_4(2)$. David Parrott, in attempting to give a characterization of the group $T$, first used Tits's presentation to obtain one for $T$ by a standard procedure known as the Reidemeister–Schreier method, which is a general technique for finding a presentation of a subgroup from that of a group [221].

Let $P_1, P_2$ be a pair of parabolic subgroups of ${}^2F_4(2)$ (containing the same Sylow 2-subgroup) and let $Q_1, Q_2$ be their intersections with $T$. Using his presentation of $T$, Parrott then proved the following result [221].

THEOREM 3.15. *Let $G$ be a simple group, let $S \in Syl_2(G)$, and let $R_1$, $R_2$ be 2-local subgroups of $G$ containing $S$ such that $R_i \cong Q_i$, $i=1,2$. Then $G \cong T$.*

The given presentation enabled Parrott to show that the subgroup $G_0 = \langle R_1, R_2 \rangle \cong T$. It follows easily from this that either $G_0$ is strongly embedded in $G$ or $G_0 = G$. Bender's theorem (see Section 4.2) now forces $G = G_0 \cong T$.

Actually Theorem 3.15 paraphrases only the very end of Parrott's analysis. The 2-local $Q_1$ is, in fact, the centralizer of a 2-central involution of $T$, $O_2(Q_1)$ is a metabelian group [i.e., $O_2(Q_1)'$ is abelian] of order $2^9$ and $Q_1/O_2(Q_1)$ is a Frobenius group of order 20. Parrott's main theorem is a characterization of $T$ in terms of the approximate structure of $Q_1$. Thus he showed that a simple group $G$ in which the centralizer $R_1$ of a 2-central involution closely resembles $Q_1$ is necessarily isomorphic to $T$. The bulk of the proof involves the construction of the second maximal 2-local subgroup $R_2$ containing a Sylow 2-subgroup of $R_1$ and determination of the precise structure of $R_1$ and $R_2$.

It should be noted that Thompson uses Parrott's result in his classification of $N$-groups [289].

There is a fundamental second method of identifying the groups of Lie type by means of their so-called *Steinberg* presentations by the generators $X_\alpha(t)$ and suitable relations among them. Indeed, one has [272]

THEOREM 3.16. *Let $G^*$ be a universal Chevalley group over $GF(q)$ of Lie rank at least 2 with root system $\Sigma$ and root subgroups $\chi_\alpha = \langle X_\alpha(t) | t \in GF(q) \rangle$ and subgroups $H_\alpha = \langle h_\alpha(t) | t \in GF(q) - \{0\} \rangle$ for $\alpha \in \Sigma - \{0\}$, as defined in Section 2.1. Then the following relations give a presentation for $G^*$:*

(i) *$\chi_\alpha$ is abelian.*
(ii) *$H_\alpha$ is cyclic.*
(iii) *$[X_\alpha(t), X_\beta(u)]$ satisfies the Chevalley commutator formula for all $\alpha, \beta \in \Sigma - \{0\}$ and $t, u \in GF(q)$.*

In the Lie rank 1 case, a similar presentation exists involving (i), (ii), and a suitable replacement of (iii). Furthermore, as part of Steinberg's development of the twisted groups, he shows that they, too, possess a similar presentation, which we shall not state explicitly. Furthermore, Charles Curtis has obtained a very convenient refinement of the Steinberg presentations, which involves relations only among certain pairs of roots [74].

## 3.2. Doubly Transitive Groups

The theory of doubly transitive permutation groups represents one of the most beautiful and deepest chapters of finite simple group theory. Because of its already noted connection with groups of Lie type of rank 1, its role in the classification proof is fundamental. Although the study of doubly transitive groups yielded no sporadic groups (apart from the already known Mathieu groups), the excitement that followed Suzuki's discovery of an entire family of such new simple groups was much the same as that attending the birth of any of the sporadic groups. Indeed, until Ree showed that the Suzuki groups had a Lie-theoretic interpretation, they were themselves regarded as a family of sporadic groups. Furthermore, doubly transitive groups encompass what may well be the single most difficult problem of the entire classification of simple groups—identification of the family of Ree groups ${}^2G_2(3^n)$.

In order to present a coherent picture of the development of the subject, we shall describe both the resemblance and isomorphism phases of the analysis but shall stress the latter portion as it is the part directly related to recognition. However, even with this restriction, the elaborateness of the theory requires a rather lengthy discussion.

First of all, it is interesting that the Frobenius conjecture on fixed-point-free automorphisms of prime order has strong implications for a very important class of doubly transitive groups—the so-called "Zassenhaus" groups. Indeed, the Frobenius problem arose initially from the study of transitive permutation groups $G$ on a set $\Omega$ in which only the identity fixes more than one point. Using character-theoretic arguments, Frobenius proved that the stabilizer $H$ of a point has a normal complement $U$ whose nonidentity elements are precisely those permutations in $G$ fixing no points of $\Omega$ [130, Theorem 4.5.1]. In particular, it follows that $U$ is transitive on $\Omega$. Hence, according to the definition of the term, $U$ is a *regular* (normal) subgroup of $G$. These conditions also imply that under conjugation, the elements of $H^{\#}$ induce fixed-point-free automorphisms of $U$. Thus $G=HU$ is a Frobenius group with kernel $U$ and complement $H$, as this notion was defined in Section 1.1. We see then that Frobenius's conjecture is related to the nilpotency of this regular normal subgroup.

Hence as an immediate consequence of Thompson's proof of the Frobenius conjecture, one obtains the following result about doubly transitive groups.

THEOREM 3.17. *If $G$ is a doubly transitive permutation group on a set $\Omega$ and only the identity element of $G$ fixes three points, then $G$ is a split $(B, N)$-pair of rank 1 with $B$ the stabilizer of a point $a \in \Omega$ and $B$ is a Frobenius group. Moreover, the Frobenius kernel $U$ of $B$ acts regularly on $\Omega - \{a\}$.*

Indeed, if $B$ is the subgroup fixing a point $a \in \Omega$, the hypothesis implies that $B$ is transitive on $\Omega - \{a\}$ and only the identity of $B$ fixes more than one point of $\Omega - \{a\}$. Hence by the Frobenius–Thompson theorem, $B$ is a Frobenius group with nilpotent kernel $U$ and complement $H$ (where $H$ is a 1-point stabilizer of $B$ on $\Omega - \{a\}$ and hence a 2-point stabilizer of $G$ on $\Omega$). Clearly then $U$ acts regularly on $\Omega - \{a\}$.

Note that $|G: B| = |\Omega| =$ degree of $G$ as a permutation group.

In the mid-1930s, Hans Zassenhaus proved the following characterization of certain doubly transitive groups satisfying the conditions of Theorem 3.17 [324].

THEOREM 3.18. *Let $G$ be a simple doubly transitive permutation group in which only the identity fixes three points and let $B$ be the stabilizer of a point. If the Frobenius kernel of $B$ is abelian, then $G \cong L_2(p^n)$ for some prime $p$ with $p^n > 3$. In particular, this is the case if a two-point stabilizer has even order.*

Zassenhaus's theorem was the first recognition theorem to be established. Almost twenty years was to pass before the true significance of such theorems for the classification of simple groups was to be fully realized. Zassenhaus's original argument had a geometric flavor; however, in [130, Theorem 13.3.5], I have presented a generator-relation type proof, as this is more suggestive of the subsequent developments. I shall comment briefly below on the proof.

In view of Zassenhaus's work, it was natural to introduce the following terminology.

DEFINITION 3.19. A doubly transitive permutation group in which only the identity fixes three points is called a *Zassenhaus* group or, for brevity, a *Z-group*.

By Theorem 3.17, in a $Z$-group $G$, a 1-point stabilizer $B$ is a Frobenius group, so that $B=HU$, where $H$, $U$ denote the Frobenius complement and kernel of $B$, respectively. In particular, $H\cap U=1$. Thus not only is $G$ a split $(B,N)$-pair, but, in addition, the two factors $H=B\cap N$ and $U$ are *disjoint*. Hence there is a natural interest in this subclass of split $(B,N)$-pairs.

DEFINITION 3.20. A split $(B,N)$-pair is said to be *strongly split* if $B$ possesses a normal nilpotent subgroup which is disjoint from $B\cap N$.

In this terminology, a $Z$-group is a strongly split $(B,N)$-pair of rank 1.

The classification of simple split $(B,N)$-pairs of rank 1 in fact proceeds in three stages: first, the $Z$-group case, then the strongly split, non-$Z$-group case, and finally a reduction of the split case to the strongly split case. This describes the logical sequence of the analysis, not its chronological sequence, which followed a quite irregular pattern.

Kantor and Seitz were responsible for the reduction to the strongly split case [196].

THEOREM 3.21. *Let $G$ be a split $(B,N)$-pair of rank* 1, *so that $B=HU$, where $U$ is normal and nilpotent in $B$ and $H=B\cap N$. Then $U$ contains a subgroup $U_0$ with $U_0$ normal in $B$ such that $B=HU_0$ and $H\cap U_0=1$.*

Since $U_0\leqslant U$, also $U_0$ is nilpotent, so $G$ is, in fact, strongly split. Thus the theorem indeed reduces the classification of split $(B,N)$-pairs of rank 1 to the strongly split case.

Let then $G$ be a simple strongly split $(B,N)$-pair of rank 1, so that $B=HU$ with $H=B\cap N$ and $U$ nilpotent and normal in $B$ with $H\cap U=1$. If $v$ is a representative of the generating involution of $W=N/H$, then every element $g\in G-B$ has a *unique* representation of the form

$$g=xu_1vu_2, \qquad \text{where } u_1,u_2\in U \text{ and } x\in H. \tag{3.3}$$

We call (3.3) the *canonical* representation of $g$.

As in the rank 2 case, the multiplication table of $G$ is completely determined by the following data [cf. (3.2) of the previous section; we note, however, that Fong and Seitz show that the analogue of (c) in the rank 2 case is a consequence of (a) and (b)].

(a) The structures of $N$ and $U$.
(b) The actions of $H$ on $U$ and $v$ on $H$. (3.4)
(c) The canonical representation of the element $vuv$ for each $u\in U^{\#}$.

For $u \in U^{\#}$, we can write

$$vuv = h(u)g(u)vf(u),$$

where $h(u) \in H$, and $f(u), g(u) \in U$. A solution of (c) amounts to a determination of the functions $f, g, h$. Using the associative law for $G$, one obtains several, unfortunately *implicit*, relations among $f, g, h$. To obtain a solution of (c), we must, in effect, solve these implicit equations and find explicit expressions for $f$, $g$, and $h$.

Of course, once the structure of $H$ and $U$ are determined, together with the action of $H$ on $U$ and $v$ on $H$, it suffices to determine the values of $f$, $g$, and $h$ on some representative in each orbit of $H$ on $U^{\#}$.

We preserve the above notation and first consider the case of $Z$-groups. An elementary argument yields

PROPOSITION 3.22. *If $G$ is a simple $Z$-group, then*

(i) *$H$ is cyclic.*
(ii) *$v$ can be taken to be an involution.*
(iii) *$v$ inverts $H$.*

Thus the structure of $N$ is uniquely determined from that of $H$, while the structure of $H$ is uniquely determined from its order.

Deeper conclusions about $Z$-groups can be obtained from the Brauer–Suzuki–Wall theorem on groups with dihedral Sylow 2-subgroups (for the case $|H|$ even [49]) and from a theorem of Feit (for the case $|H|$ odd [86]) to which he was led from Suzuki's work on simple groups in which the centralizer of an involution is a 2-group (to be mentioned below).

THEOREM 3.23. *If $G$ is a simple $Z$-group, then*

(i) *$U$ is a $p$-group for some prime $p$.*
(ii) *One of the following holds*:
 (1) $|H| = \varepsilon(|U|-1)$, *where* $\varepsilon = 1$ *or* $\frac{1}{2}$, *and $U$ is elementary abelian.*
 (2) *$U$ is nonabelian and* $|U:U'| \leq 4|H|^2+1$.

It was in his portion of this theorem that Feit developed the basic ideas about "coherent" sets of characters which were to play such a fundamental role later in the proof of the odd-order theorem. (A proof of Feit's theorem as well as the Brauer–Suzuki–Wall theorem appears in my book [130, Theorems 4.6.5, 15.4.1].)

In the abelian case, one easily establishes the following additional result.

**Proposition 3.24.** *If $G$ is a simple $Z$-group in which $U$ is abelian (and hence an elementary $p$-group for some prime $p$), then according to the value of $\varepsilon=1$ or $\frac{1}{2}$, the structure of $B=HU$ is uniquely determined up to isomorphism.*

In view of conditions (a), (b), (c) of (3.4), the preceding three results show that the classification of simple $Z$-groups in which $U$ is abelian is reduced to a determination of the *multiplication functions* $f$, $g$, and $h$. The content of Zassenhaus's theorem is that, in fact, $f$, $g$, and $h$ are uniquely determined (and are, of course, identical to the corresponding multiplication functions of the groups $L_2(p^n)$].

For example, in the case $p=2$, one proves the following:

(1) $\varepsilon=1$ (*whence $H$ has a single orbit on $U^{\#}$*).

(2) *There is a uniquely determined element $u_0 \in U^{\#}$ such that*

$$vu_0v=u_0vu_0. \tag{3.5}$$

(3) *For $u \in U^{\#}$, if $u=u_0^y$ for $y \in H$, then*

$$h(u)=y^2.$$

(4) *For $u \in U^{\#}$, $g(u)=u^{-1}$ and $f(u)=u^{h(u)^{-1}}$.*

Similar conclusions hold when $p$ is odd; however, they are slightly more involved because then $\varepsilon=\frac{1}{2}$ and hence $H$ has two orbits on $U^{\#}$ [(4) continues to hold, while (2) holds for a unique *pair* of elements $u_0, u_0^{-1}$].

In a brilliant piece of work, Suzuki completely analyzed the nonabelian case for $p=2$ [278]. This problem arose from his prior study of simple groups in which the centralizer of every involution is a 2-group. Indeed, he had shown that a simple group $G$ of this type satisfies one of the following conditions:

(1) *$G \cong L_3(4)$ or $L_2(q)$, $q$ a Fermat or Mersenne prime or* 9.

(2) *$G$ is a $Z$-group whose Frobenius kernel $U$ is a 2-group.*

In the first step of his $Z$-group analysis, Suzuki proved, by a very lovely counting argument, that the cyclic group $H$ must permute the involution of $U$ *transitively*. Suzuki asked G. Higman what one could say about the structure of such a 2-group $U$. Using so-called "Lie ring methods" (cf. [130, Section 5.6]), Higman obtained a complete classification of such groups

[170]. Although his answer gave four possible "types," Suzuki was able to show that only one of these could occur in the $Z$-group situation. In particular, $|U|=2^{2n}$ for some odd integer $n$ and $|G:B|=2^{2n}+1$.

For each automorphism $\theta$ of $GF(2^n)$, $n$ odd, let $N(\theta)$ be the group of order $(2^n-1)2^{2n}$ whose elements are the set of all triples $(t, x, y)$, where $t, x, y \in GF(2^n)$ and $t \neq 0$ with multiplication defined by the rule

$$(t, x, y)\cdot(t_1, x_1, y_1)=\left(tt_1, xt_1+x_1, xt_1x_1^{\theta}+yt_1^{1+\theta}+y_1\right).$$

Note that the elements $(1, x, y)$ generate a normal subgroup $U(\theta)$ of order $2^{2n}$. Suppressing the first component, we see that multiplication table in $U(\theta)$ is given by the rule

$$(x, y)\cdot(x_1, y_1)=\left(x+x_1, xx_1^{\theta}+y+y_1\right).$$

Suzuki proves

THEOREM 3.25. *If $G$ is a simple $Z$-group in which the Frobenius kernel $U$ of $B$ is a nonabelian 2-group, then $|G: B|=2^{2n}+1$, $n$ odd, $n>1$, and $B\cong N(\theta)$ for some automorphism $\theta$ of $GF(2^n)$. In particular, $|G|=(2^n-1)2^{2n}(2^{2n}+1)$.*

Clearly also $U\cong U(\theta)$.

The situation here is more complicated than in the abelian case, since there the structure of $B$ was uniquely determined, while now it depends upon the value of the parameter $\theta$. Thus Suzuki must now determine not only the multiplication functions $f$, $g$, and $h$ but also the allowable possibilities for $\theta$.

Crucial to the solution in the abelian case (with $p=2$) was the existence of the (unique) element $u_0$ such that $vu_0v=u_0vu_0$. Since $v$ and $u_0$ are involutions, this is equivalent to the assertion $(vu_0)^3=1$. Suzuki terms either of these identities the *structure equation* for $G$. Likewise the solution in the nonabelian case depends upon a determination of the corresponding structure equation. Suzuki proves the following:

*There is a unique element $u_0$ of order 4 in $U$ such that*

$$vu_0v=u_0vu_0^2;$$

*moreover, this relation is equivalent to the assertion*

$$\left(vu_0^2\right)^5=1.$$

Now Suzuki is able to pin down $f$, $g$, $h$, and $\theta$ *uniquely*, by ingeniously manipulating the functional relations implied by the associative law.

We shall only give the expression for the function $f$. In any $Z$-group, the function $g$ is determined from $f$ by the relation $g(u)=f(u^{-1})^{-1}$ for $u\in U^{\#}$. Also $H$ has $2^n+1$ orbits on $U^{\#}$ and the value of the function $h$ on $U^{\#}$ is determined from its value on representatives of each orbit. Moreover, Suzuki shows that these orbits are represented by the elements $u_0$, $u_0^{-1}$, $u_0^2$, and $u_0t^{-1}u_0^{-1}t$, where $u_0$ is an element of $U$ determined by the structure equation and $t$ ranges over $H^{\#}$.

We identify $U$ with $U(\theta)$, so that each $u\in U$ has the form $u=(x,y)$, $x,y\in GF(2^n)$. Suzuki shows that one can take $(0,1)$ as $u_0$, $(1,0)$ as $u_0^2$, and $(1,1)$ as $u_0^{-1}$. The $H$-orbits of these elements correspond to degenerate cases and Suzuki proves:

$$f(0,y)=\left((y^{-1})^{(1+\theta)^{-1}},0\right)$$

$$f(x,0)=(0,x^{-(1+\theta)})$$

$$f(x,x^{1+\theta})=(x^{-1},x^{-(1+\theta)}).$$

Thus we can assume $x\neq 0$, $y\neq 0$, $y\neq x^{1+\theta}$. To describe $f$ in this case, first set

$$k=k(x,y)=\left(y(x^{1+\theta}+y)^{-1}\right)^{\theta-1},$$

(so that $k\in GF(q)^{\#}$), and set

$$c(k)=(1+k)(k^{-1}+d^{-1}),$$

where $d$ is uniquely determined from the equation

$$k^{1+\theta}+k^{2(1+\theta)}=d^{(1+\theta)}.$$

With these definitions of $k$ and $c$, Suzuki shows that

$$f(x,y)=\left(c(k(x,y))x^{-1},c(k(x,y))y^{-1}\right).$$

This will give some indication of the subtlety of the analysis. There is one further important point which must be mentioned. Suzuki does not reach the conclusion $\theta^2=2$ until *after* he has determined $f$, $g$, and $h$. Thus the exact value of $\theta$ appears to be a "global" question rather than a "local" one.

Hence, for each odd value of $n>1$, Suzuki's analysis has led to a unique solution; the only question left unanswered is the following:

*Does there exist an actual group that satisfies these conditions for a given odd $n>1$?*

In other words, at this point, Suzuki has solved the *uniqueness*, but not the *existence* problem. Part of his great achievement lies in the fact that, on the basis of the internal structure of the hypothetical group $G$ which he had determined, he was eventually able to construct a family of simple groups (parametrized by the odd integer $n$) which satisfy his initial $Z$-group conditions.

Their definition is as follows: consider the $4\times4$ matrices over $GF(2^n)$, $n$ odd,

$$\begin{bmatrix} 1 & 0 & 0 & 0 \\ a & 1 & 0 & 0 \\ a^{1+\theta}+b & a^{\theta} & 1 & 0 \\ a^{2+\theta}+ab+b^{\theta} & b & a & 1 \end{bmatrix}, \begin{bmatrix} c^{1+\theta^{-1}} & 0 & 0 & 0 \\ 0 & c^{\theta-1} & 0 & 0 \\ 0 & 0 & c^{-\theta^{-1}} & 0 \\ 0 & 0 & 0 & c^{-1-\theta^{-1}} \end{bmatrix}, \begin{bmatrix} 0 & 0 & 0 & 1 \\ 0 & 0 & 1 & 0 \\ 0 & 1 & 0 & 0 \\ 1 & 0 & 0 & 0 \end{bmatrix},$$

where $a, b, c\in GF(2^n)$, $n$ odd, $n>1$, $c\neq0$, and $\theta$ is the automorphism of $GF(2^n)$ with $\theta^2=2$. Suzuki shows that the group $G(n)$ generated by the given matrices is a simple $Z$-group in which a 1-point stabilizer $B(n)$ has a nonabelian Frobenius kernel $U(n)$ of order $2^{2n}$. This is the family of Suzuki simple groups, now denoted by $Sz(2^n)$.

Combined with Zassenhaus's theorem, Suzuki's results can be expressed as follows:

THEOREM 3.26. *If $G$ is a simple $Z$-group of odd degree (i.e., in which $|G: B|$ is odd), then $G\cong L_2(2^n)$ or $Sz(2^n)$ for some integer $n$.*

The case of nonabelian $U$ and $p$ odd was treated by Noboru Ito, who showed, using modular character and block theory, that there exist no simple $Z$-groups satisfying these conditions [183]. Ito's original proof has been simplified, first by George Glauberman and subsequently by David Sibley, so that now the analysis depends only on ordinary character theory.

Ito's theorem completes the classification of $Z$-groups:

THEOREM 3.27. *If $G$ is a simple $Z$-group, then $G \cong L_2(p^n)$ for some prime $p$ or $Sz(2^n)$.*

After completing the centralizer-of-involution 2-group classification, Suzuki attacked the more general problem of determining all groups in which the centralizer of every involution has a normal Sylow 2-subgroup [278] (obviously this condition is satisfied in the 2-group case). Again the first part of the analysis is a reduction to double transitivity; and Suzuki establishes a direct generalization of the corresponding result in the 2-group case. Indeed, if $G$ is a simple group having the specified property, he shows that one of the following holds:

1. $G \cong L_3(2^n)$, $n \geq 2$, *or* $L_2(q)$, *where $q$ is a Mersenne or Fermat prime or* 9. (3.6)
2. *$G$ is a strongly split $(B, N)$-pair of rank* 1 *in which both $|G: B|$ and $|H| = |B \cap N|$ are odd.*

For brevity, let us use the term *C-group* for any group $G$ satisfying part 2 of (3.6). The classification of $C$-groups has turned out to be of absolutely fundamental importance for simple group theory. Indeed, Bender's proof of the classification of groups with a strongly embedded subgroup (to be discussed in Section 4.2) consists of a reduction to the $C$-group case.

In view of the importance of $C$-groups for simple group theory, we shall briefly describe Suzuki's analysis of them, even though only the final phase of his argument constitutes a recognition theorem. We limit ourselves to the simple case, noting that Suzuki's inductive analysis requires a solution of the general problem.

Suzuki's goal is to establish the following result:

THEOREM 3.28. *If $G$ is a simple $C$-group, then one of the following holds*:

(i) *$G$ is a $Z$-group* [*whence $G \cong L_2(2^n)$ or $Sz(2^n)$*].
(ii) $G \cong U_3(2^n)$, $n \geq 2$.

[Without the assumption of simplicity, either $G$ is an extension of $L_2(2^n)$, $Sz(2^n)$, or $U_3(2^n)$ by a cyclic group of automorphisms of odd order or else $G$ has a regular normal subgroup (and a very precise structure)].

Clearly Suzuki can assume that $G$ is not a $Z$-group, so that his task is to show that $G$ is a unitary group. To see what is involved in this effort, let me describe the $C$-group structure of the groups $U_3(2^n)$, $n \geq 2$, as rank 1 $(B, N)$-pairs.

1. $|U| = 2^{3n}$, $H$ is cyclic of order $(2^{2n}-1)/d$, where $d=3$ or 1 according as 3 does or does not divide $2^n+1$, and $|U_3(2^n)| = (1/d)(2^{2n}-1)2^{3n}(2^{3n}+1)$.
2. $H = H_0H_1$, where $H_0$ of order $2^n - 1$ is inverted by $v$ and $H_1$ of order $(2^n+1)/d$ is centralized by $v$.
3. $H_0U$ is a Frobenius group and $H_0$ acts transitively on the involutions of $U$.
4. $C_{U_3(2^n)}(H_1) = H_1 \times L$, where $L \cong L_2(2^n)$ and $L \cap U = Z(U)$ is elementary of order $2^n$ (and Sylow in $L$).

At the outset, Suzuki does not know the structure of $H$ [except that it is (solvable) of odd order] nor the structure of $U$ [except that it is nilpotent] nor the action of $v$ on $H$. Moreover, he has no information about the centralizers in $G$ of subgroups of $H$. All these must be pinned down before he can hope to determine the isomorphism type of $G$. Note that as $|H|$ and $|G:B|$ are odd, he does know that $U$ contains a Sylow 2-subgroup $S$ of $G$.

As in the $Z$-group cases, a structure identity again plays a key role; indeed, early in the argument, Suzuki proves the existence of a unique pair of elements $u_0, u_1 \in U$ with $u_0$ an involution satisfying the relation

$$vu_0v = u_1^{-1}vu_1.$$

Moreover, he can also show that

$$|u_0v| = p \qquad \text{for some prime } p \geqslant 3.$$

The value of $p$ turns out to be critical. Recall from the previous discussion that $p=3$ in the groups $L_2(2^n)$ and $p=5$ in the groups $Sz(2^n)$; also one easily checks that $p=3$ in the groups $U_3(2^n)$. This suggests a natural three-case subdivision according as $p>5$, $p=5$, and $p=3$; this is indeed the path that Suzuki follows.

However, he must first establish some preliminary results about the structure of $H$ and its embedding in $G$. He proves:

(3.7)

1. *The subset $H_0$ of elements of $H$ inverted by $v$ is a cyclic normal subgroup of $H$ and its order is precisely the number of involutions of $U$.*
2. *$H_0$ transitively permutes the involutions of $S$ and $H_0U$ is a Frobenius group.*
3. *If $H_1 = C_H(v)$, then $H = H_0H_1$ with $H_0 \cap H_1 = 1$.*
4. *If $X$ is any subgroup of $H_1$, then $C_G(X)$ possesses a normal subgroup $Y$ of odd order such that $C_G(X)/Y$ is a $C$-group (not necessarily simple) whose 1-point stabilizer is the image of $C_B(X)$ and whose structure identity involves the same prime $p$ as $G$.*

In view of part 2 of (3.7), the structure of $S$ is determined from Higman's theorem [thus for a given $n$, $S$ is one of four possible types (or else $S$ is abelian)]. Furthermore, we have $H_1 \neq 1$, otherwise $B = H_0U$ is a Frobenius group, whence $G$ is a $Z$-group, contrary to assumption. In addition, part 4 shows how induction enters into the analysis in a natural way. In particular, if $p > 5$ (and $X \neq 1$), then $C_G(X)/Y$ must have a regular normal subgroup, while if $p = 5$ [and $C_G(X)/Y$ does not have a regular normal subgroup], then $C_G(X)/Y$ has a normal subgroup of odd index isomorphic to $Sz(2^n)$ for some $n$.

In each of the cases $p > 5$, $p = 5$, and $p = 3$ with $S$ elementary abelian, Suzuki proves (as $G$ is not a $Z$-group) that $G$ has a normal subgroup $N$ of suitable odd prime index $q$, contrary to the simplicity of $G$ (more precisely to the fact that $G$ is a minimal counterexample to the more general theorem he is proving). The argument is very delicate. He must first determine the structure of each of the following groups: $H$, $C_G(X_1)$ for $1 \neq X_1 \leqslant H$, $N_B(Q)$ for $Q \in Syl_q(G)$, and $S$ [when $p = 5$, $S$ turns out to be either abelian or of "Suzuki type," i.e., isomorphic to a Sylow 2-subgroup of $Sz(2^n)$ for some $n$]. The existence of $N$ then follows by invoking a suitable transfer theorem.

Thus $p = 3$ and $S$ is not elementary abelian. By yet a further subtle inductive argument (together with Higman's 2-group theorem), Suzuki next forces $S$ to be of "unitary type," i.e., isomorphic to a Sylow 2-subgroup of $U_3(2^n)$, $n \geqslant 2$. This is followed by a character theory argument, similar in spirit to that used in the proof of Feit's theorem [86], which yields the conclusion $S = U$, so that $U$ is a 2-group of a particular shape. Next comes a careful analysis of the structure and embedding of $H$ in $G$, the aim of which is to show that the same conditions prevail as in the groups $U_3(2^n)$, as described above.

Now Suzuki has a good picture of $B = HU = HS$, of the order $(1/d)(2^{2n} - 1)2^{3n}$. Indeed, as in the $Z$-group case, $B$ can be described by triples $(t, x, y)$ [where now $t \in GF(2^{2n})$, $t \neq 0$, and again $x, y \in GF(2^n)$], together with a suitable rule for multiplication of these triples. Again for a given value of $n$, the isomorphism type of $B$ is determined only up to an automorphism $\theta$ of $GF(2^n)$. In the actual groups $U_3(2^n)$, this parameter $\theta$ is the trivial automorphism.

Thus, as before, Suzuki's final task is to prove that $\theta = 1$ and to show that the multiplication functions $f$, $g$, and $h$ of $G$ are uniquely determined. Again this is accomplished by an analysis of the functional relations satisfied by $f$, $g$, and $h$. However, the argument is very much more difficult than in the former case, especially when $d = 3$ as then $H$ has *three* orbits on the elements of $(U/Z(U))^{\#}$ (in contrast to a single orbit in the Suzuki case).

An extremely ingenious argument is required to eliminate an unpleasant subcase. Once this is achieved, Suzuki is able to prove $\theta=1$ and to obtain explicit expressions for $f$, $g$, and $h$ (which we shall not describe here).

Having characterized the unitary groups $U_3(2^n)$ in terms of their strongly split $(B, N)$-pair structure, it was natural for Suzuki to attempt a similar characterization of the groups $U_3(p^n)$, *p odd* [280]. Moreover, as the latter groups have quasi-dihedral or wreathed Sylow 2-subgroups [according as $p^n \equiv 1$ or $-1 \pmod 4$], such a characterization was essential before it would be possible to determine all simple groups with such Sylow 2-subgroups. However, this time because of the greater number of $H$-orbits on $U^\#$ (which depend upon $p$ as well as on $n$), Suzuki was able to unravel the functional relations for $f$, $g$, and $h$ (and then force $\theta=1$) only when $p^n+1$ is prime to 3 (i.e., in the case $d=1$).

For convenience, let us say that a strongly split $(B, N)$-pair of rank 1 is of *odd unitary type* if $|G:B|=p^{3n}+1$, $p$ an odd prime, and $B=HU$, where $|U|=p^{3n}$ and $H$ is cyclic of order $(p^{2n}-1)/d$, where $d=3$ or 1 according as $d$ does or does not divide $p^n+1$.

Thus Suzuki proved

THEOREM 3.29. *If $G$ is a simple group of odd unitary type in which $|G:B|=p^{3n}+1$, $p$ odd, and $p^n+1$ is prime to 3, then $G \cong U_3(p^n)$.*

The case $d=3$ remained untouched for five years after Suzuki's theorem appeared (1964), when O'Nan considered it in his doctoral thesis [226, 227]. O'Nan followed Zassenhaus's geometric point of view rather than Suzuki's generator-relation approach; his analysis (which covers both cases $d=1$ and $d=3$) was positively spectacular. O'Nan's very first result was startling, for he was able to force the conclusion $\theta=1$ solely from an analysis of the structure of $B$. Thus, in the unitary case, the automorphism $\theta$ is "locally" determined. [Since it does not seem possible to prove this in the case of the $Z$-groups $Sz(2^n)$, it had never occurred to Suzuki to look for a local argument in the unitary case.] Thus O'Nan proved:

THEOREM 3.30. *If $G$ is a group of odd unitary type in which $|G:B| = p^{3n}+1$, $p$ an odd prime, then according as $U$ is abelian or nonabelian, the structure of $B$ is uniquely determined; and in the latter case, $B$ is isomorphic to a Borel subgroup of $U_3(p^n)$.*

Next came a brilliant characterization of the unitary groups in terms of their so-called "block design"—at that time a well-known open question in finite geometry. Let $V$ be a three-dimensional vector space over $GF(q^2)$, $q$ a prime power, and let $f$ be a "nondegenerate *Hermitian*" bilinear form on $V$, with respect to the Frobenius automorphism $x \mapsto x^q$, $x \in GF(q^2)$. [*Hermitian* means $f(w,v)=f(v,w)^q$ for all $v, w \in V$ and *nondegenerate* means that if $f(v,w)=0$ for all $v \in V$ and some $w \in V$, then necessarily $w=0$.] Let $I$ denote the family of "isotropic" one-dimensional subspaces of $V$ with respect to $f$ [$v \in V$ is *isotropic* if $f(v,v)=0$]. (See Section 4.14 for a fuller discussion of the geometry of the classical groups.)

One verifies that $I$ has $q^3+1$ points and that the three-dimensional unitary groups $U_3(q)$, $PGU(3,q)$, and $\mathrm{Aut}(U_3(q))$ act naturally on $I$ as doubly transitive permutation groups. By definition, a *block* of $I$ is the set of isotropic one-dimensional subspaces contained in a fixed "nonisotropic" two-dimensional subspace of $V$ (i.e., a subspace not consisting entirely of isotropic vectors). Thus a block consists of $q+1$ elements of $I$. Moreover, there are exactly $q^4-q^3+q^2$ blocks, every point of $I$ is contained in exactly $q^2$ blocks, and any two distinct points of $I$ are contained in a unique block. Together the set of points and blocks of $I$ form the (unitary) block design associated with $U_3(q)$. We denote it by $\mathfrak{U}_q$. An automorphism of $\mathfrak{U}_q$ is any one-to-one transformation of the points of $I$ which transforms blocks into blocks. From its action of $I$, we see at once that $\mathrm{Aut}(U_3(q))$ is a group of automorphisms of $\mathfrak{U}_q$ [whence $\mathrm{Aut}(U_3(q)) \leqslant \mathrm{Aut}(\mathfrak{U}_q)$]. O'Nan proved

THEOREM 3.31. *If $\mathfrak{U}_q$ denotes the natural block design associated with the group $U_3(q)$, $q$ a prime power, then* $\mathrm{Aut}(\mathfrak{U}_q)=\mathrm{Aut}(U_3(q))$.

O'Nan associates a geometry of "circles" with the blocks containing a given point $\infty$ of $I$ and by an analysis of this geometry he ultimately shows that the stabilizer $\mathrm{Aut}(\mathfrak{U}_q)_\infty$ of $\infty$ in $\mathrm{Aut}(U_q)$ is the same as its stabilizer in $\mathrm{Aut}(U_3(q))$, which suffices to establish the theorem. It turns out that the complement in $I$ of a full set of circles having a common tangent is a *line* of $V$, so that $\mathrm{Aut}(\mathfrak{U}_q)_\infty$ consists of line-preserving transformations. This is the key to the proof, for then by a well-known theorem of Dembowski [77], O'Nan can conclude that $\mathrm{Aut}(\mathfrak{U}_q)_\infty$ consists of "semilinear" affine transformations on $V$.

O'Nan's main result is the following, which clearly includes Theorem 3.29 as a special case.

THEOREM 3.32. *If $G$ is a simple group of odd unitary type in which $|G:B| \cong p^{3n}+1$, $p$ odd, then $G \cong U_3(p^n)$.*

He associates a block design $\mathfrak{U}(G)$ with $G$ in such a way that $G$ acts as a group of automorphisms of $\mathfrak{U}(G)$. Indeed, $G$ is doubly transitive on the set $\Omega$ of right cosets of $B$ in $G$, $B$ is a one-point stabilizer, and $H$ is a two-point stabilizer. $H$ is cyclic [of order $(p^{2n}-1)/d$] and so $H$ contains a unique involution $t$. O'Nan shows that $C_U(t)\neq 1$ and that the set $\Omega_t$ of points of $\Omega$ fixed by $t$ has cardinality $1+|C_U(t)|$. A general theorem of Witt [321] now yields that the fixed-point sets $\Omega_{t^g}$, as $g$ ranges over $G$, form a block design on $\Omega$. This is the block design $\mathfrak{U}(G)$. Clearly

$$G \leqslant \operatorname{Aut}(\mathfrak{U}(G)).$$

O'Nan shows by a counting argument that each block of $\mathfrak{U}(G)$ has exactly $p^n+1$ points and that two distinct points of $\Omega$ lie in exactly one block. Thus $\mathfrak{U}(G)$ has some "resemblance" to the block design $\mathfrak{U}_{p^n}$. Clearly O'Nan's goal must be to show that, in fact, $\mathfrak{U}(G)$ and $U_{p^n}$ are *isomorphic* block designs [i.e., there exists a one-to-one transformation of the points of $\mathfrak{U}(G)$ to the points of $\mathfrak{U}_{p^n}$ taking blocks into blocks], for then it will follow by the previous theorem that $G\cong U_3(p^n)$ (as $G$ is simple).

The significance of Theorem 3.30 is now apparent. Indeed, once it is shown that $U$ is nonabelian, it will follow [here, for uniformity, $\infty$ denotes the point of $\Omega$ fixed by $B$ (i.e., the coset $B$ itself)] that the subdesigns $\mathfrak{U}(G)_\infty$ and $(\mathfrak{U}_{p^n})_\infty$ are isomorphic, so that one can assume without loss that they are identical.

In both the abelian and nonabelian cases, $U$ can be described by pairs $(x, y)$, $x, y\in GF(p^n)$, with a suitable rule for multiplication and prescribed action for $H$. By conjugating by elements of $G$, the blocks of $\mathfrak{U}(G)$ can be expressed in terms of suitable pairs of field elements. Such calculations lead fairly quickly to a contradiction when $U$ is abelian. Furthermore, when $d=1$ and $U$ is nonabelian, they similarly yield that $\mathfrak{U}(G)_0$ and $(\mathfrak{U}_{p^n})_0$ have a common block, where $(0,\infty)$ are the two points of $\Omega$ fixed by $H$. However, O'Nan shows that the existence of such a common block is sufficient to guarantee that $\mathfrak{U}(G)\cong\mathfrak{U}_{p^n}$.

Hence at this point O'Nan has obtained an alternate proof of Suzuki's theorem and is left to deal with the more difficult case $d=3$. He first proves the existence of a block in $\mathfrak{U}(G)$ of the form

$$\left\{\left(\omega^{3i},-\tfrac{1}{2}\right),\left(\lambda\omega^{3i+1},-\tfrac{1}{2}+u\right),\left(\eta\omega^{3i+2},-\tfrac{1}{2}+v\right)\right\},\qquad 0\leqslant i\leqslant(p^n+1)/3,$$

where $\omega$ is a primitive $(p^n+1)$st root of unity in $GF(p^{2n})$ and $\lambda$, $\eta$, $u$, $v$ are elements of $GF(p^{2n})$ which must be determined. In the block design $\mathfrak{U}_{p^n}$, these parameters have the values

$$\lambda=1, \qquad \eta=1, \qquad u=0, \qquad v=0.$$

Clearly the goal of the analysis must be to force essentially the same parameter values for $\mathfrak{U}(G)$. The proof is elegant, very ingenious, and extremely intricate. First, O'Nan associates a circle geometry $\mathfrak{D}$ with $\mathfrak{U}(G)$ by analogy with the definition of the circle geometry $\mathcal{C}$ of $\mathfrak{U}_{p^n}$. Moreover, he shows that $\mathfrak{U}_{p^n}$ and $\mathfrak{U}(G)$ are related by a "local isomorphism" (we omit the definition) which enables him to transfer information about $\mathcal{C}$ to $\mathfrak{D}$.

The circle geometry $\mathfrak{D}$ turns out to depend only upon the values of $x=\lambda^{(p^n+1)/3}$ and $y=\eta^{(p^n+1)/3}$, so the conclusion $\mathcal{C}=\mathfrak{D}$ will follow provided $x=y=1$. O'Nan uses the qualitative properties of $\mathfrak{D}$ so far established to ultimately obtain the following two polynomial equations which $x$ and $y$ must satisfy:

$$x^3y+2xy^2+x^2y-y^3-3xy+2x^2+y^2+4y-5x-2=0,$$

$$xy^3+2x^2y+xy^2-x^3-3xy+2y^2+x^2+4x-5y-2=0.$$

When $p\neq 5$, $(x, y)=(1,1)$ is the unique solution; however, when $p=5$, the pairs $(1,2)$, $(2,1)$, and $(3,3)$ are also solutions and further analysis is required to eliminate these possibilities. The most difficult case occurs when $p=5$ and $n=1$. Since Harada had earlier given a characterization of the group $U_3(5)$ [161], O'Nan chooses to invoke Harada's result, thereby enabling him to omit the calculations for this case.

Finally O'Nan forces the values $u=v=0$. Together with the fact that $\mathcal{C}=\mathfrak{D}$, this allows him to conclude that $(\mathfrak{U}_{p^n})_0$ and $\mathfrak{U}(G)_0$ have a common block, so that by his prior sufficiency criterion, the desired conclusion $\mathfrak{U}_{p^n}=\mathfrak{U}(G)$ follows. There is one remarkable feature of the calculations which must be mentioned: they involve the only use of *contour integration* in finite group theory of which I am aware! At one point, O'Nan must evaluate a very complicated rational expression in $\omega$. Treating $\omega$ as an *independent* complex variable, and computing residues in the standard way, he finally ends up with a relationship over $\mathbb{Z}[\omega]$. Factoring out by an appropriate prime ideal in this ring, he obtains a needed relationship for $\omega$ in $GF(p^{2n})$, where now $\omega$ is the original primitive $(p^n+1)$st root in $GF(p^{2n})$.

Return now to the Suzuki groups $Sz(2^n)$, $n$ odd, which we view as acting on a 4-dimensional vector space $V$ over $GF(2^n)$ with respect to a suitable basis $v_1, v_2, v_3, v_4$. If $f$ is the bilinear form on $V$ whose associated matrix

$$(f(v_i, v_j)) = \begin{bmatrix} 0 & 0 & 0 & 1 \\ 0 & 0 & 1 & 0 \\ 0 & 1 & 0 & 0 \\ 1 & 0 & 0 & 0 \end{bmatrix},$$

one checks that the generating matrices of $Sz(2^n)$ leave $f$ invariant. Since $f$ is a nondegenerate alternating form, $Sz(2^n)$ is therefore a subgroup of the symplectic group $C_2(2^n)$. Since $C_2(2^n) \cong B_2(2^n)$, the Suzuki groups are thus also subgroups of the orthogonal groups $B_2(2^n)$.

It took someone such as Ree, who was fully versed in the Lie theory, to realize that these subgroups could also be described in terms of the fixed points of a suitable automorphism of period 2 by means of the Steinberg twisting process. It will be instructive to describe these calculations, for they provide a good illustration of Steinberg's basic ideas.

The associated Lie algebra $B_2$ of $B_2(2^n)$ has four positive roots $\alpha, \beta, \alpha+\beta, \alpha+2\beta$. Thus $B_2(2^n)$ is generated by its Sylow 2-subgroups $U, V$, where

$$U = \langle X_\alpha(t), X_\beta(u), X_{\alpha+\beta}(v), X_{\alpha+2\beta}(w) \rangle,$$

$$V = \langle X_{-\alpha}(t), X_{-\beta}(u), X_{-\alpha-\beta}(v), X_{-\alpha-2\beta}(w) \rangle,$$

and the $X_\gamma(t)$ are the usual root subgroups, $t, u, v, w \in GF(2^n)$.

In characteristic 2, both subgroups $X_{\alpha+\beta}(t)$ and $X_{\alpha+2\beta}(t)$ lie in the center of $U$ and the Steinberg commutator formulas in $U$ reduce to the single identity

$$[X_\alpha(t), X_\beta(u)] = X_{\alpha+\beta}(tu) X_{\alpha+2\beta}(tu^2). \tag{3.8}$$

The "unweighted" Dynkin diagram of $B_2$ has a symmetry of period 2 which interchanges the roots $\alpha, \beta$ and the roots $\alpha+\beta, \alpha+2\beta$. In *characteristic 2* and *only in this characteristic*, this symmetry lifts to an automorphism $\tau$ of $B_2(2^n)$ given by

$$\tau: \begin{array}{ll} X_\alpha(t) \mapsto X_\beta(t), & X_\beta(t) \mapsto X_\alpha(t^2), \\ X_{\alpha+\beta}(t) \mapsto X_{\alpha+2\beta}(t), & X_{\alpha+2\beta}(t) \mapsto X_{\alpha+\beta}(t^2), \end{array}$$

with a similar definition of $\tau$ on $V$.

One checks that the automorphism $\tau^2$ is a generator of the cyclic group of order $n$ of "field" automorphisms of $B_2(2^n)$ induced from the Galois group of $GF(2^n)$. Thus $\langle\tau\rangle$ is cyclic of order $2n$, and as $n$ is odd, it contains a unique involution. Thus there is a unique automorphism $\phi$ of $GF(2^n)$ such that if $\tau_\phi$ is the induced field automorphism of $B_2(2^n)$, then $\sigma=\tau\cdot\tau_\phi$ has period 2. Direct calculation yields that

$$2\phi^2=1 \tag{3.9}$$

and that

$$\sigma:\quad \begin{array}{ll} X_\alpha(t)\mapsto X_\beta(t^\phi), & X_\beta(t)\mapsto X_\alpha(t^{2\phi}) \\ X_{\alpha+\beta}(t)\mapsto X_{\alpha+2\beta}(t^{2\phi}), & X_{\alpha+2\beta}(t)\mapsto X_{\alpha+\beta}(t^\phi). \end{array}$$

Note that as $\phi(2\phi)=1$, $\phi^{-1}=2\phi$, so if we set $\theta=\phi^{-1}$, we have

$$\theta^2=1 \quad \text{and} \quad \theta=2\phi. \tag{3.10}$$

Steinberg's general theory asserts that the subgroup $B_2(2^n)_\sigma$ of fixed points of $\sigma$ on $B_2(2^n)$ is generated by its Sylow subgroups $U_\sigma$, $V_\sigma$, where $U_\sigma=C_U(\sigma)$, $V_\sigma=C_V(\sigma)$ (cf. Section 4.14). Thus $U_\sigma$ is the set of elements of the form

$$X_\alpha(t)X_\beta(u)X_{\alpha+\beta}(v)X_{\alpha+2\beta}(w), \qquad t,u,v,w\in GF(2^n),$$

which are fixed by $\sigma$. Applying $\sigma$ to this expression and using the commutator identity (3.8) together with the fact that $\langle X_{\alpha+\beta}(v), X_{\alpha+2\beta}(w)\rangle\leq Z(U)$, and $2\phi^2=1$, one finds that

$$U_\sigma=\langle X_\alpha(u^{2\phi})X_\beta(u)X_{\alpha+\beta}(w^\phi+u^{2\phi+1})X_{\alpha+2\beta}(w)\,|\,u,w\in GF(2^n)\rangle. \tag{3.11}$$

We must now interpret (3.11) in terms of $4\times4$ matrices over $GF(2^n)$. To do so, we use the natural four-dimensional representation of the Lie algebra $B_2$ to construct our Chevalley group $B_2(2^n)$. We note that the universal Chevalley group $B_2(K)$ over any field $K$ of characteristic 2 has trivial center, so the construction leads to the group $B_2(2^n)$ itself.

In the given representation, one finds that the root elements $x_\alpha, x_\beta, x_{\alpha+\beta}, x_{\alpha+2\beta}$ are represented by the following matrices:

$$x_\alpha: \begin{bmatrix} 0 & 0 & 0 & 0 \\ 0 & 0 & 0 & 0 \\ 0 & 1 & 0 & 0 \\ 0 & 0 & 0 & 0 \end{bmatrix}, \qquad x_\beta: \begin{bmatrix} 0 & 0 & 0 & 0 \\ 1 & 0 & 0 & 0 \\ 0 & 0 & 0 & 0 \\ 0 & 0 & 1 & 0 \end{bmatrix},$$

$$x_{\alpha+\beta}: \begin{bmatrix} 0 & 0 & 0 & 0 \\ 0 & 0 & 0 & 0 \\ 1 & 0 & 0 & 0 \\ 0 & 1 & 0 & 0 \end{bmatrix}, \qquad x_{\alpha+2\beta}: \begin{bmatrix} 0 & 0 & 0 & 0 \\ 0 & 0 & 0 & 0 \\ 0 & 0 & 0 & 0 \\ 1 & 0 & 0 & 0 \end{bmatrix}.$$

(See [186, Chapter 7] for a discussion of the representation theory of simple Lie algebras.) Each of these matrices has square 0. Hence, calculating $X_\gamma(t)=\exp(1+tx_\gamma)$, $\gamma$ a root, $t\in GF(2^n)$, we find that the root subgroups $X_\alpha(t)$, $X_\beta(t)$, $X_{\alpha+\beta}(t)$, $X_{\alpha+2\beta}(t)$ are represented as follows:

$$X_\alpha(t): \begin{bmatrix} 1 & 0 & 0 & 0 \\ 0 & 1 & 0 & 0 \\ 0 & t & 1 & 0 \\ 0 & 0 & 0 & 1 \end{bmatrix}, \qquad X_\beta(t): \begin{bmatrix} 1 & 0 & 0 & 0 \\ t & 1 & 0 & 0 \\ 0 & 0 & 1 & 0 \\ 0 & 0 & t & 1 \end{bmatrix},$$

$$X_{\alpha+\beta}(t): \begin{bmatrix} 1 & 0 & 0 & 0 \\ 0 & 1 & 0 & 0 \\ t & 0 & 1 & 0 \\ 0 & t & 0 & 1 \end{bmatrix}, \qquad X_{\alpha+2\beta}(t): \begin{bmatrix} 1 & 0 & 0 & 0 \\ 0 & 1 & 0 & 0 \\ 0 & 0 & 1 & 0 \\ t & 0 & 0 & 1 \end{bmatrix}. \tag{3.12}$$

Finally put $a=u$ and $b=w^\phi+u^{2\phi+1}$ in (3.11) and note then that

$$w=(b+a^{2\phi+1})^{\phi^{-1}}=b^{2\phi}+a^{2\phi+2}. \tag{3.13}$$

Now using (3.10)–(3.13), we see that $U_\sigma$ is represented by the matrices

$$U_\sigma: \left\langle \begin{bmatrix} 1 & 0 & 0 & 0 \\ 0 & 1 & 0 & 0 \\ 0 & a & 1 & 0 \\ 0 & 0 & 0 & 1 \end{bmatrix} \begin{bmatrix} 1 & 0 & 0 & 0 \\ a & 1 & 0 & 0 \\ 0 & 0 & 1 & 0 \\ 0 & 0 & a & 1 \end{bmatrix} \begin{bmatrix} 1 & 0 & 0 & 0 \\ 0 & 1 & 0 & 0 \\ b & 0 & 1 & 0 \\ 0 & b & 0 & 1 \end{bmatrix} \begin{bmatrix} 1 & 0 & 0 & 0 \\ 0 & 1 & 0 & 0 \\ 0 & 0 & 1 & 0 \\ a^{\theta+2}+b^\theta & 0 & 0 & 1 \end{bmatrix} \middle| a,b\in GF(2^n) \right\rangle,$$

whence

$$U_\sigma: \left\langle \begin{bmatrix} 1 & 0 & 0 & 0 \\ a & 1 & 0 & 0 \\ a^{1+\theta} & a^\theta & 1 & 0 \\ a^{2+\theta}+ab+b^\theta & b & a & 1 \end{bmatrix} \middle| a, b \in GF(2^n) \right\rangle.$$

Thus the representation of $U_\sigma$ by $4\times4$ matrices is identical to a Sylow 2-subgroup of the group $Sz(2^n)$ defined by Suzuki. Similarly the representation of $V_\sigma$ is identical to a Sylow 2-subgroup of $Sz(2^n)$ given by upper triangular matrices. We therefore conclude that

$$Sz(2^n) = \langle U_\sigma, V_\sigma \rangle.$$

As pointed out earlier, the groups $G_2(3^n)$ and $F_4(2^n)$ also have an *extra* automorphism, analogous to that of $B_2(2^n)$ described above. Once Ree understood how the Suzuki groups could be constructed from the groups $B_2(2^n)$, he knew that he could repeat the Steinberg process for $G_2(3^n)$ and $F_4(2^n)$ when $n$ is odd, thus obtaining two further (new) families of simple groups [238, 239]. In the first case, as $G_2$ has Lie rank 2, the fixed points of the corresponding automorphism $\sigma$ form a group of Lie rank 1 and hence a strongly split $(B, N)$-pair of rank 1. In conformity with the Lie notation, the resulting groups are denoted by the symbol ${}^2G_2(3^n)$, $n$ odd. [Similarly Ree's second family is denoted by ${}^2F_4(2^n)$; however, its members have Lie rank 2.]

Here we are interested only in the groups ${}^2G_2(3^n)$ [as the groups ${}^2F_4(2^n)$ are not of Lie rank 1]. Summarizing Ree's results, we have

THEOREM 3.33. *The groups* ${}^2G_2(3^n)$, *$n$ odd, are simple for $n>1$ and have the following $(B, N)$-pair structure*:

(i) $B = HU$, *where* $|U| = 3^{3n}$ *and $H$ is cyclic of order* $3^n - 1$.

(ii) $|{}^2G_2(3^n)| = (3^n-1)3^{3n}(3^{3n}+1)$.

(iii) $|O(H)| = \frac{1}{2}(3^n-1)$ *and $O(H)U$ is a Frobenius group.*

(iv) *If $a$ is the involution of $H$, then*

$$C_{{}^2G_2(3^n)}(a) \cong Z_2 \times L_2(3^n).$$

(v) ${}^2G_2(3^n)$ *has elementary abelian Sylow 2-subgroups of order* 8.

[The group ${}^2G_2(3)$ is isomorphic to $\mathrm{Aut}(L_2(8))$, which has a (simple) normal subgroup of index 3.]

The theorem shows that there is a definite parallel with the unitary case, although now $U$ has class 3 rather than class 2. Note also that in ${}^2G_2(3^n)$, regarded as a doubly transitive group of degree $3^{3n}+1$, a subgroup fixing three letters has order 2, so that these groups are "almost" $Z$-groups.

Our concern is with the converse of Theorem 3.33: Are the Ree groups "characterized" by the five conditions of the theorem? In other words, if we call an arbitrary strongly split $(B, N)$-pair satisfying these conditions a group of *Ree type*, we would like to assert that the Ree groups themselves are the only groups of Ree type.

As already remarked, the proof of this assertion has involved enormous difficulties; and its complete resolution has taken approximately fifteen years. Thompson became interested in the problem soon after the discovery of the Ree groups, for he saw that the classification of simple groups with abelian Sylow 2-subgroups could not be completed without first classifying the groups of Ree type. Over a ten-year period, Thompson made a heroic effort to show that the only simple groups of Ree type were the Ree groups ${}^2G_2(3^n)$ themselves. His analysis followed the same general functional equation pattern as in the linear, Suzuki, and unitary cases. However, even in the actual Ree groups the explicit form of the multiplication functions $f$ and $g$ are unbelievably complicated, involving literally hundreds of terms (in complete contrast to the other cases). In fact, Ree managed to construct his groups without ever giving a full description of $f$ and $g$, but only their values on certain very special $H$-orbits of $U$.

Altogether Thompson wrote three papers on groups of Ree type, ultimately reducing their classification to a problem concerning the solutions of a certain system of equations with coefficients in $GF(3^n)$ [290]. In the first paper he determines the possible structure of the subgroup $B$. Although more difficult than in the Suzuki case, the final answer is entirely analogous. Indeed, for each automorphism $\theta$ of $GF(3^n)$, $n$ odd, let $N(\theta)$ be the group of order $(3^n-1)3^{3n}$, whose elements are the quadruples $(t, x, y, z)$, where $t, x, y, z \in GF(3^n)$ and $t \neq 0$, with multiplication defined by the rule

$$(t,x,y,z)\cdot(t_1,x_1,y_1,z_1)=\big(tt_1,\, xt_1+x_1,\, yt_1^{\theta+1}+y_1+xt_1\cdot x_1^{\theta}-x^{\theta}t_1^{\theta}x_1,$$
$$zt_1^{\theta+2}+z_1+yt_1^{\theta+1}x_1+x^{\theta}t_1^{\theta}\cdot x_1^2+xt_1x_1^{\theta+1}-x^2t_1^2x_1^{\theta}\big).$$

Thompson proves

THEOREM 3.34. *If $G$ is a simple group of Ree type in which $|G:B| = 3^{3n}+1$, $n$ odd, then $B \cong N(\theta)$ for some automorphism $\theta$ of $GF(3^n)$.*

Just as $\theta^2 = 2$ in the Suzuki groups, so in the actual Ree groups one has

$$\theta^2 = 3,$$

where now 3 denotes the Frobenius automorphism $x \mapsto x^3$ for $x \in GF(3^n)$. Thus again the classification problem is reduced to a determination of $f$, $g$, $h$ and $\theta$.

In his second paper, Thompson attempts to unravel the implicit functional equations for $f$, $g$, $h$, and $\theta$. With much effort he determines a system of polynomial equations over $GF(3^n)$ which must be satisfied by $\theta$ (but with $\theta$ occurring in the exponents). Specifically he proves

THEOREM 3.35. *Let $G$ be a simple group of Ree type with $B \cong N(\theta)$ for some automorphism $\theta$ of $GF(3^n)$, $n$ odd. Then we have*

(i) *$\theta$ is a generator of the cyclic group* $\mathrm{Aut}(GF(3^n))$.
(ii) *There are two integers $a$ and $s$ with $a$ even and $s$ odd such that $x^{(\theta+1)a} = x^2$ and $x^{(\theta+2)s} = x$ for all $x \in GF(3^n)$.*
(iii) *Let $E$ be the set of triples $(z, y, u)$ with $z, y, u \in GF(3^n)$ and $zy \neq 0$ such that*

$$z^{\theta+2} - y^{\theta+2} = -1,$$

$$z^{\theta+1} - y^{\theta+1} = u.$$

*Then the following identity holds on $E$:*

$$z(u-1)^a - (z-y+1)u^a - y(u+1)^a + (u^2-1)^a = 0.$$

Despite the complexity of this system of relations, Thompson felt that they captured the structure of a group of Ree type. However, he was never able to demonstrate this fact except in the important special case in which *by assumption* $\theta^2 = 3$. Again with great effort, using Theorem 3.35 (as well as other results from his second paper), Thompson was able to show that the multiplication table of $G$ was uniquely determined. Thus he proved

THEOREM 3.36. *If $G$ is a simple group of Ree type with $B \cong N(\theta)$ and $\theta^2 = 3$, then $G \cong {}^2G_2(3^n)$ for some odd $n$.*

[Implicit in his proof of Theorem 3.36 is the fact that for each choice of generator $\theta$ of $\mathrm{Aut}(GF(3^n))$, there is at most one group of Ree type $G$ with $B = N(\sigma)$.]

Thompson's papers are dated 1967, 1971, and 1977, indicating the long period of time this problem occupied his attention. Shortly after the third paper Suzuki's student Mark Hopkins showed in his doctoral thesis, using a computer at the University of Illinois, that for $n \leqslant 29$ one indeed must have $\theta^2 = 3$ [180]. However, the problem of proving $\theta^2 = 3$ in general remained open until Enrico Bombieri took it up in the spring of 1979, precisely where Thompson left off. I was extremely gratified to learn that Bombieri's interest had been stimulated by a remark in the earlier survey article [132], which has been expanded into this book, to the effect that the Ree group problem was "accessible to the nonspecialist [in finite group theory]."

Bombieri treated Thompson's equations from the perspective of classical elimination theory [36]. By successive elimination of the variables $a$, $u$, and $y$, he eventually obtained a relation of the form

$$H(z, z^{\theta}, z^{\theta^2}, \dots, z^{\theta^k}) = 0,$$

where $H(z_0, z_1, \dots, z_k)$ is a polynomial in $k+1$ variables, not identically 0, with the following bounds:

$$k = 4 \quad \text{and} \quad \deg_{z_i}(H) < 3^{10}, \qquad 0 \leqslant i \leqslant k.$$

Despite the limited information about the exact form of the polynomial $H$, Bombieri was able to use it to obtain precise information about the automorphism $\theta$, namely, he proved that there are integers $m$ and $\lambda$ satisfying

$$1 \leqslant m \leqslant k \quad \text{and} \quad 3^{|\lambda|} \leqslant \max\{\deg_{z_i}(H)\}$$

such that

$$\theta^m = 3^{\lambda}.$$

In view of the previous bounds on $k$ and $\deg_{z_i}(H)$, this yields

$$m \leqslant 4 \quad \text{and} \quad |\lambda| \leqslant 9.$$

Note that this gives a bound on the number of possibilities for $\theta$, independent of the number of elements in the underlying field $GF(3^n)$.

Now with these explicit values of $\theta^m$, Bombieri repeats the entire elimination process, obtaining a new polynomial relationship

$$K(z, z^{\theta}, \dots, z^{\theta^{m-1}}) = 0.$$

This algorithmic procedure yields either a bound for $n$, a polynomial $K$ which is identically 0, or a solution for $\theta$ of the form $\theta=3^{\lambda}$, $|\lambda|\leqslant 9$. If $3^n$ is sufficiently large, Bombieri excludes the last possibility with the aid of Bézout's classical theorem concerning the number of intersections of two algebraic curves in projective 2-space. Furthermore, he can show in most cases that $K$ is not identically 0. Altogether there remain (apart from the case $\theta^2=3$) exactly 178 cases unaccounted for, in each of which $n\leqslant 83$.

Andrew Odlyzko and Hunt independently checked these 178 cases (on computers at Bell Laboratories and the University of New South Wales, Australia, respectively) and obtained contradictions in each case [36].

Thus the determination of $\theta$ was complete:

THEOREM 3.37. *If $G$ is a simple group of Ree type with $B\cong N(\sigma)$, then* $\theta^2=3$.

Combined with Theorem 3.35, the desired characterization of the Ree groups was finally achieved.

THEOREM 3.38. *If $G$ is a simple split $(B,N)$-pair of Ree type, then $G\cong {}^2G_2(3^n)$ for some odd integer $n$.*

An interesting question is whether O'Nan's geometric approach to the unitary groups could be adapted to groups of Ree type to provide an alternate (possibly simpler) proof of Theorem 3.38. The problem is, of course, much more complicated since in the Ree group case the number of orbits of $H$ on $U$ is much greater than in the unitary case, $U$ has class 3 rather than class 2, and the parameter $\theta$ on which the structure of $B$ depends cannot be determined in advance.

On the basis of these specific classification theorems, one could now consider the general strongly split $(B,N)$-pair rank 1 problem. The determination of all such groups has itself been quite difficult and, in particular, has involved local reduction arguments of the same general flavor as in Suzuki's unitary case analysis. Bender, Hering, Kantor, O'Nan, Seitz, Ernest Shult, and Suzuki considered some aspect of the problem. Their combined effort (together with Theorem 3.21) yields a complete classification of split $(B,N)$-pairs of rank 1.

We state their result only in the simple case.

THEOREM 3.39. *If $G$ is a simple split $(B,N)$-pair of rank 1, then $G\cong L_2(q)$, $U_3(q)$, $Sz(2^n)$, or ${}^2G_2(3^n)$, $n$ odd, $n>1$.*

We conclude this discussion with a statement of O'Nan's fundamental structure theorem for arbitrary doubly transitive groups [229]. Here if $X$ is a simple group and $X \leqslant Y \leqslant \mathrm{Aut}(X)$, we call $Y$ a *holomorph* of $X$. Moreover, if $X$ is any permutation group on a set $\Omega$, we say that $X$ acts *semiregularly* on $\Omega$ if each orbit of $X$ on $\Omega$ has cardinality $|X|$. In particular, if $X$ has only one orbit (whence $X$ is transitive on $\Omega$), then $X$ acts regularly on $\Omega$.

THEOREM 3.40. *If $G$ is a doubly transitive permutation group acting on a set $\Omega$, then one of the following holds:*

(i) *A one-point stabilizer of $G$ on $\Omega$ is a local subgroup of $G$.*
(ii) *A one-point stabilizer of $G$ on $\Omega$ is a holomorph of a simple group.*

Furthermore, theorems of Derek Holt and O'Nan determine the possibilities for $G$ in many of the local cases [178, 228]:

THEOREM 3.41. *Let $G$ be a doubly transitive permutation group on a set $\Omega$ in which no abelian normal subgroup acts regularly on $\Omega$. Suppose a one-point stabilizer $G_1$ of $G$ on $\Omega$ is a local subgroup of $G$ and assume one of the following conditions holds:*

(a) $|\Omega|$ $(=|G:G_1|)$ *is odd.*
(b) $G_1$ *is solvable.*
(c) $G_1$ *has an abelian normal subgroup which does not act semiregularly on $\Omega-\{1\}$.*

*Then $G \cong L_n(q)$ for some $n$ and $q$, $U_3(q)$ for some $q$, $Sz(2^m)$, or ${}^2G_2(3^m)$ for some odd $m$.*

Assuming the local and regular normal subgroup cases can be completed, Theorem 3.40 gives strong indication that the determination of all simple groups yields as a consequence a complete classification of all doubly transitive permutation groups.

## 3.3. The Alternating Groups

There is a classical presentation of the alternating groups, which can be used to identify them in any given classification problem. It is built up in a natural way from the symmetric group of two lower degree, using the involutions $(12)(34), (12)(45), \ldots, (12)((n-1)n)$ and adjoining the 3-cycle

(123) to this set. Thus we have

THEOREM 3.42. *If the group $G$ is generated by elements $x_1, x_2, \ldots, x_{n-2}$ subject only to the relations $x_1^3 = 1$, $x_i^2 = 1$, $2 \leqslant i \leqslant n-2$, $(x_i x_{i+1})^3 = 1$, $1 \leqslant i \leqslant n-3$, and $(x_i x_j)^2 = 1$, $1 \leqslant i \leqslant n-4$, $i+1 < j$, then $G \cong A_n$.*

A proof of this theorem as well as of the corresponding presentation of $\Sigma_n$ can be found in Bertram Huppert's book [181, pp. 137–139].

## 3.4. The Sporadic Groups

The discussion of the last chapter shows that those sporadic groups constructed from the centralizer of an involution are satisfactorily characterized once both existence and uniqueness are established. Likewise the rank 3 sporadic groups are satisfactorily characterized by their one-point stabilizers with specified action on the three orbits.

Thus we are left with only the five Mathieu groups and three Conway groups to consider.

### A. The Mathieu Groups

Remarkably the first characterization of $M_{11}$ is due to Camille Jordan in 1872, who proved [194]:

THEOREM 3.43. *If $G$ is a quadruply transitive permutation group in which only the identity element fixes four letters, then $G \cong \Sigma_4$, $\Sigma_5$, $A_6$, or $M_{11}$.*

This result has been extended by M. Hall to the case in which the subgroup fixing four letters had odd order, the group $A_7$ representing an added possibility (see [152, Theorem 5.8.1]). The proof divides into the cases $n \leqslant 7$ and $n \geqslant 8$, where $n$ is the number of letters on which $G$ acts. In the latter case, $n = 11$ is forced and $G$ is shown to be generated by a specific set of permutations. This means that $G$ is uniquely determined from the given conditions (when $n \geqslant 8$); and as $M_{11}$ satisfies these conditions, it follows that $G \cong M_{11}$.

The Mathieu groups have been characterized by both the structure of the centralizer of a 2-central involution (Brauer for $M_{11}$ [40, 42]; Brauer–Fong for $M_{12}$ [45]; Janko for $M_{22}$ and $M_{23}$ [190]; and Held for $M_{24}$ [165]) and their orders (Ralph Stanton, a student of Brauer's in the late 1940s, for

$M_{12}$ and $M_{24}$ [267]; Parrott and W. Wong for $M_{11}$ [232, 323]; Parrott and Anita Bryce, in their doctoral theses under Janko, $M_{22}$ and $M_{23}$, respectively, [232, 50]).

We state only the latter results.

THEOREM 3.44. *If $G$ is a simple group of order $M_n$, $n=11$, 12, 22, 23, or* 24, *then $G \cong M_n$.*

One can follow essentially the same proof in each case. From the order of $G$ and its simplicity, one first obtains its complete local structure and conjugacy classes of elements. From this one derives its character table. Now, using the Brauer trick, one produces a subgroup $G_1$ of index $n$. The permutation character on the cosets of $G_1$ has the form $1_G + \chi$, where $\chi$ is irreducible, which implies that $G$ is doubly transitive on the cosets of $G_1$. Finally, using this permutation representation and the character table, one forces certain elements of $G$ to be represented by specific permutations, which means that the multiplication table of $G$ is again uniquely determined. Thus $G$ is uniquely determined by its order; and as $M_n$ is simple of the same order as $G$, we conclude that $G \cong M_n$.

## B. The Conway Groups

Characterizations of the Conway groups by the centralizers of their involutions have involved two distinct types of recognition theorems:

(a) By construction of an associated graph.
(b) By construction of a lattice from internal properties of the group.

Frederick Smith has proved [259]:

THEOREM 3.45. *If $G$ is a group with $O(G)=Z(G)=1$ in which the centralizer $C$ of an involution is 2-constrained with $O_2(C)$ extra-special of order $2^9$ and $C/O_2(C) \cong Sp(6,2)$, then $G \cong .2$.*

Smith's approach parallels the Parrott–S. K. Wong centralizer-of-involution characterization of $Mc$ described in Section 2.4. By general methods of 2-local analysis, Smith determines the involution fusion pattern and the order of $G$ (using the Thompson order formula). Then by a delicate generator-relation argument, he shows that $G$ must contain a $(B, N)$-pair subgroup $H_0$, which from known theorems he is able to identify as $U_6(2)$. The group $H = N_G(H_0)$ is shown to be a unique extension of $U_6(2)$ by an outer automorphism of order 2. Using only local considerations, he then proves

PROPOSITION 3.46. *$G$ is a primitive rank 3 permutation group on the cosets of $H$ with subdegrees 891 and 1,408.*

Finally Smith proves the uniqueness of the resulting graph.

Nicholas Patterson proceeds in a similar way to characterize .1 [235].

THEOREM 3.47. *If $G$ is a group with $O(G)=Z(G)=1$ in which the centralizer $C$ of an involution is 2-constrained with $O_2(C)$ extra-special of order $2^9$ and $C/O_2(C)\cong\Omega_8^+(2)$ [of index 2 in $O_8^+(2)$], then $G\cong.1$.*

In this case the corresponding subgroup $H$ is isomorphic to $\widehat{Suz}$, the cover of $Suz$ by $Z_3$, which Patterson constructs inside $G$ by generators and relations and identifies from its associated rank 3 graph. Likewise he determines the subdegrees of the permutation representation of $G$ on the cosets of $H$ (it is no longer of rank 3) and proves that the resulting graph is uniquely determined.

Perhaps a similar approach is possible for .3, but Daniel Fendel, in his dissertation under Feit [95], followed a more character-theoretic path, ultimately reducing the recognition problem to a theorem of Feit concerning groups having a rational-valued representation of degree 23 [89]. Fendel proved

THEOREM 3.48. *If $G$ is a group with $O(G)=Z(G)=1$ in which the centralizer $C$ of an involution is isomorphic to that of a 2-central involution of .3 [thus $C\cong\widehat{Sp}(6,2)$], then $G\cong.3$.*

[Here $\widehat{Sp}(6,2)$ denotes the covering group of $Sp(6,2)$ by $Z_2$.]

After completing the local analysis, Fendel obtains the character table of $G$ and then argues that $G$ possesses an irreducible rational representation of degree 23. Now he can invoke the following theorem of Feit to complete the proof.

THEOREM 3.49. *If $G$ is a group having a faithful irreducible rational-valued representation of degree 23 and $G$ has no subgroups of index 23 or 24, then $G$ is isomorphic to a subgroup of $Z_2\times.2$ or $Z_2\times.3$.*

Feit's paper on integral representations of groups, of which this theorem is only one of the important results, is one of the deepest papers ever written in the representation theory of finite groups, intertwining algebraic number theory, modular character theory, and group actions on integral lattices. In it, Feit establishes criteria for the action of a group $G$ on a lattice $\mathfrak{L}$ to force the uniqueness of $\mathfrak{L}$ up to isometry. Under the assumptions of Theorem 3.49, he argues that there are exactly three possibilities for $\mathfrak{L}$, each corresponding to a specific sublattice of the Leech lattice.

### C. Sporadic Groups by Centralizers of Involutions

In Section 2.4, we have listed the 11 centralizers of involutions that have given rise to sporadic simple groups; and we have just described how the Conway groups are determined from the centralizer of one of their involutions. Since $M_{24}$ (which has the same centralizer of a 2-central involution as $He$) occurs as one of the possible conclusions of Held's analysis, we have thus described 15 of the sporadic groups by centralizers of involutions and have discussed characterizations of 13 of them in terms of these centralizers.

Each of the remaining 11 sporadic groups have similar characterizations. Parrott and S. K. Wong's centralizer-of-involution characterization of $Mc$ has already been described in Section 2.4. Although such results involve more than solely recognition theorems, we shall list them in Table 3.1 to

**Table 3.1.** Centralizers of Involutions in Some Sporadic Groups

| Group | Centralizer of an involution[a] |
|---|---|
| $M_{11}$ | $GL(2,3) \cong \Sigma_3/Q_3$, trivial core |
| $M_{12}$ | $\Sigma_3/Q_8 * Q_8$, trivial core |
| $M_{22}$ | $\Sigma_4/E_{16}$, $\Sigma_4$ acting faithfully |
| $M_{23}$ | $L_3(2)/E_{16}$, 2-constrained |
| $M_{24}$ | $L_3(2)/(D_8)^3$ |
| $HS$ | $\Sigma_5/Q_8 * Q_8 * Z_4$, 2-constrained |
| $Mc$ | $\hat{A}_8$ |
| $Suz$ | $PSp(4,3)/D_8 * D_8 * D_8$, 2-constrained |
| $Ru$ | $\Sigma_5/X$, where $X \cong E_{16}/Q_8 \times E_{16}$, 2-constrained |
| $M(22)$ | $\hat{U}_6(2)$ |
| $M(23)$ | $\widehat{M(22)}$ |
| $M(24)'$ | $\hat{U}_4(3)\cdot 2/(D_8)^6$, 2-constrained |
| .3 | $\widehat{Sp}(6,2)$ |
| .2 | $Sp(6,2)/(D_8)^4$ |
| .1 | $\Omega_8^+(2)/(D_8)^4$ |

[a] Here $\hat{U}_4(3)$ denotes the covering group of $U_4(3)$ by $Z_3$.

round out the discussion. In each case, the involution in question is 2-central, with the exception of $M(22)$.

The principal result in this connection is the following:

THEOREM 3.50. *If $G$ is a simple group in which the centralizer of an involution is isomorphic to one of the 29 groups listed in either Table 3.1 or in Tables 2.1, 2.2 (note that the groups $F_1$, $F_2$, and $F_5$ appear twice in the tables of Section 2.4), then one of the following holds:*

(i) *$G$ is isomorphic to one of the known sporadic groups.*
(ii) *$G \cong L_3(3)$ or $L_5(2)$.*

The last two groups arise from $\Sigma_3/Q_8$ and $L_3(2)/(D_8)^3$, respectively.

Additional comments concerning the proof of Theorem 3.50 in the 11 cases not yet discussed will be given in the sequel, which will contain a full analysis of centralizers of involutions in simple groups.

Many of the sporadic groups possess more than one conjugacy class of involutions; and for the classification proof, it has been necessary to characterize them in terms of other centralizers of involutions than solely those of Theorem 3.50, specifically the centralizers of involutions listed in Table 3.2.

Here $\widehat{L_3}(4)$ and $\hat{U}_6(2)$ denote the perfect central extension of $L_3(4)$ by $Z_2 \times Z_2$ and of $U_6(2)$ by $Z_2$, respectively.

These "secondary" centralizers occur in the context of obtaining a complete solution of so-called "standard-form" problems, which constitutes a fundamental chapter of finite simple group theory. In practice, the goal of the analysis in each of these five centralizer-of-involution problems is to force the structure of the centralizer of the corresponding "primary" involution and then to invoke Theorem 3.50 to identify the group under investiga-

**Table 3.2.** Centralizers of Non-2-Central Involutions in Some Sporadic Groups

| Group | Centralizer of involution |
|---|---|
| $J_2$ | $Z_2 \times Z_2 \times A_5$ |
| $HS$ | $Z_2 \times \mathrm{Aut}(A_6)$ |
| $He$ | $\widehat{L_3}(4) \cdot 2$ |
| $Ru$ | $Z_2 \times Z_2 \times Sz(8)$ |
| $Suz$ | $(Z_2 \times Z_2 \times L_3(4)) \cdot 2$ |

tion. A more detailed description of the procedure will be given in the sequel.

The classification proof requires similar "standard form" characterizations of the groups of Lie type and of the alternating groups; but likewise we postpone discussion of the needed results until the next book.

# 4

# General Techniques of Local Analysis

In this long chapter, we shall describe and illustrate the principal methods and results that underlie local group-theoretic analysis. (We shall not attempt a systematic discussion of ordinary and modular character theory, but shall limit our comments on those techniques to the few specific places in the text in which they are needed.)

It is often difficult to distinguish between a "technique" and a "classification theorem," for once a result of the latter type is proved, it becomes a tool for all subsequent classification theorems. We have not troubled ourselves here with this distinction; rather our aim is to lay out for the reader the most important general ideas of local and internal geometric analysis which have provided the basis for the major results that have been established within the four phases of the classification of simple groups.

To keep at least some bounds on the length of the discussion, we shall be selective in the degree of detail with which we treat a particular topic. The proofs and outlines to be presented have all been chosen with a view to illuminating the basic ideas of local analysis.

## 4.1. Solvable Groups

It is impossible to develop an understanding of local analysis without first seeing how it works in the solvable case. Indeed, in a very real sense, solvable groups represent a test case for the general theory (which should be abundantly clear from the stress we have already placed on the odd-order and $N$-group analyses). The general local subgroup of a simple group is "built up" from "solvable pieces" and "nonsolvable pieces" [e.g., the generalized Fitting subgroup $F^*(X)$ of the group $X$ is the product of the

Fitting subgroup $F(X)$, which is a nilpotent and hence solvable group, and the layer $L(X)$, which is a semisimple and hence nonsolvable group (provided that $L(X) \neq 1$)]. Moreover, arguments that work for solvable local subgroups often carry over to the "solvable piece" of an arbitrary local subgroup. Beyond this, the arguments and ideas from the solvable case have provided the basis for many broad extensions to general local analysis within arbitrary simple groups.

For these reasons, we shall indicate some of the main features of solvable local analysis together with key properties of solvable groups which underlie it. Since the notion of constraint is a direct generalization of a property of solvable groups (Proposition 1.25), a number of our statements will hold for this wider class of groups.

Local analysis is a complex process; however, if one were forced to state a single quality that comes closest to capturing its essence, I think it would have to be "proof by contraction." Questions concerning the structure of a local subgroup $H$ of the simple group $G$ under investigation are continually being *reduced* to more manageable questions about sections $K$ of $H$ of some specified shape (often $K$ even has a natural identification with a group of linear transformations of some vector space associated with $H$). In the solvable situation, the procedure works especially well because one has strong control over the subgroup structure of a solvable group. The latter fact is best exemplified by P. Hall's well-known *extended Sylow* theorems [130, Theorem 6.4.1].

THEOREM 4.1. *Let $X$ be a solvable group and $\pi$ a set of primes. Then we have*

(i) *$X$ possesses a Hall $\pi$-subgroup (i.e., a subgroup $Y$ such that $|Y|$ is divisible only by primes in $\pi$ and $|X:Y|$ only by primes in $\pi'$).*
(ii) *Any two Hall $\pi$-subgroups of $X$ are conjugate in $X$.*
(iii) *Every $\pi$-subgroup of $X$ (i.e., of order divisible only by primes in $\pi$) is contained in a Hall $\pi$-subgroup of $X$.*

Because of Theorem 4.1, many questions about the structure of $X$ needed for local analysis can be reduced to questions concerning the Hall $\{p, q\}$-subgroups of $X$ as $p, q$ range over suitable pairs of prime divisors of $X$.

When Theorem 4.1 is combined with the Schur–Zassenhaus theorem (footnote on page 51), one obtains the following consequence:

THEOREM 4.2. *Suppose the group $A$ acts on the group $X$ with $A$ and $X$ of relatively prime orders. If either $A$ or $X$ is solvable, then for any set of* *primes $\pi$, we have*

(i) *A leaves invariant some Hall $\pi$-subgroup of X.*
(ii) *Any two A-invariant Hall $\pi$-subgroups of X are conjugate by an element of $C_X(A)$.*

The theorem is often used when $A$ is a $p$-group for some prime $p$. For example, if $H$ is a local subgroup of $G$ and $A$ a $p$-subgroup of $H$, it can be applied with $O_{p'}(H)$ as $X$. In particular, it follows that $A$ leaves invariant a Sylow $q$-subgroup of $O_{p'}(H)$ for each prime $q$.

Perhaps the single most quoted result of local analysis is the so-called "Frattini argument," an elementary consequence of Sylow's theorem, valid in all groups [130, Theorem 1.3.7].

PROPOSITION 4.3. *Let Y be a normal subgroup of the group X and P a Sylow p-subgroup of Y for any prime p. Then*

$$X = YN_X(P).$$

The Frattini argument is used to reduce questions about $X$ to questions about $X_1 = N_X(P)$, a group which has a normal $p$-subgroup (namely $P$). Combining this with Proposition 1.12(iii) one can often further reduce to a question of the action of $X_1$ on the Frattini factor group $\bar{P} = P/\phi(P)$. Since $\bar{P}$ is elementary abelian, it can be viewed as a vector space over $GF(p)$, in which case the group $\bar{X}_1 = X_1/C_{X_1}(\bar{P})$ becomes a group of linear transformations on $\bar{P}$, thus reducing us ultimately to questions about groups of linear transformations.

This last reduction process can be expressed very succinctly in $p$-constrained groups.

PROPOSITION 4.4. *Let X be a p-constrained group with $O_{p'}(X) = 1$, p a prime [equivalently with $F^*(X) = O_p(X)$]. Then $X/O_p(X)$ acts faithfully as a group of linear transformations of $O_p(X)/\phi(O_p(X))$ regarded as a vector space over $GF(p)$.*

A basic result of P. Hall and G. Higman [130, Theorem 5.3.7] allows one to reduce actions of $p'$-groups on $p$-groups to actions on subgroups. We state it only for abelian $p'$-groups.

THEOREM 4.5. *Let A be an abelian p'-group acting nontrivially on the p-group P, p a prime. If Q is a minimal A-invariant subgroup of P on which A acts nontrivially, then*

(i) *$Q$ is a special group (see Section 1.4 for definition).*
(ii) *$A$ acts trivially on $\phi(Q)$ and irreducibly on $Q/\phi(Q)$ [regarded as a vector space over $GF(p)$].*

Thus properties of the action of $A$ on $P$ are reduced to actions on the subgroup $Q$ of $P$ of *class at most* 2. However, some of the nontrivial action of $A$ on $P$ may be "lost" in this reduction since some elements of $A$ that act nontrivially on $P$ may well act trivially on $Q$. Furthermore, the theorem gives no information about the embedding of $Q$ in $P$. The following result of Thompson's from the odd-order paper shows how far one can go to rectify this situation [130, Theorem 5.3.11].

THEOREM 4.6. *A $p$-group $P$, $p$ a prime, contains a characteristic subgroup $Q$ with the following properties:*

(i) *$Q/Z(Q)$ is elementary abelian (in particular, $Q$ has class $\leqslant 2$).*
(ii) *$Q/Z(Q)$ is in the center of $P/Z(Q)$ [whence $[P, Q] \leqslant Z(Q)$].*
(iii) *If $\alpha$ is a nontrivial automorphism of $P$ of order prime to $p$, then the restriction of $\alpha$ to $Q$ is nontrivial.*

By condition (iii), if $A$ is a $p'$-group acting faithfully on $P$, then the restriction of $A$ to such a characteristic subgroup $Q$ acts faithfully on $Q$.

One can often drop down to an even smaller characteristic subgroup in view of the following result.

PROPOSITION 4.7. *If $P$ is a $p$-group such that either $P$ is abelian or $p$ is odd, then any nontrivial automorphism of $P$ of order prime to $p$ restricts nontrivially to $\Omega_1(P)$.*

There is often one further step in the dropping-down process, which does not require solvability [130, Theorem 5.3.6; 143, Lemma 1]. Note that if $A$, $X$ are subgroups of any group, then $[A, X]$ is normal in the group $\langle A, X \rangle$ [130, Theorem 2.2.1(iii)].

PROPOSITION 4.8. *Let the group $A$ act on the group $X$ and assume that either $A$ is perfect or $A$ and $X$ have relatively prime orders. Then $[X, A, A] = [X, A]$. In particular, if $A$ centralizes $[X, A]$, then $A$ centralizes $X$.*

The effect of the proposition is that under such conditions one can replace the group $X$ by its subgroup $Y = [X, A]$, which thus has the added property $Y = [Y, A]$.

Theorem 4.6 may appear to be stronger than Theorem 4.5; however, the critical fact that $A$ acts irreducibly on $Q/\phi(Q)$, which is part of Theorem 4.5, need not hold in Theorem 4.6. Hence which one of these two results is preferable in a given situation depends upon the nature of the subsequent analysis. The importance of the condition of irreducibility in Theorem 4.5 can be seen from the following observations. Set $\bar{Q}=Q/\phi(Q)$ and $A_0=C_A(\bar{Q})$. Since $A$ is abelian by assumption in Theorem 4.5, $A/A_0$ is cyclic by Theorem 1.2. But as $A_0$ acts trivially on $Q/\phi(Q)$, $A_0$ acts trivially on $Q$ by Proposition 1.12(iii). Thus, in fact, $A/C_A(Q)=A/A_0$ is cyclic.

Often one reaches such a configuration $AQ$ with $A/A_0$ cyclic and $A$ a $q$-group for some prime $q$, in conjunction with Proposition 4.4 (applied with $q$ in place of $p$), so that in addition the group $AQ$ can be considered to be a group of linear transformations of some vector space $V$ over $GF(q)$. In general, it is difficult to analyze groups of linear transformations over fields of characteristic $q$ which possess normal $q$-subgroups. The subgroup $A_0=C_A(Q)$ is just such a normal $q$-subgroup of $AQ$ [since $A$ is abelian, we have, in fact, $A_0 \leqslant Z(AQ)$], so if possible we would prefer to "get rid" of $A_0$. The natural way to do so is, of course, to consider the subspace $W=C_V(A_0)$, for then $AQ$ and hence $\overline{AQ}=AQ/A_0$ act on $W$. However, for this reduction to be useful, we must know that the action of $\overline{AQ}$ on $W$ is *faithful*—i.e., that $\overline{AQ}$ is itself a group of linear transformations on $W$. Fortunately this is guaranteed by Thompson's well-known ($A\times$B)-*lemma*, another result from the odd-order paper. For uniformity of notation, we interchange the roles of $p$ and $q$ in its statement.

PROPOSITION 4.9. *Let $A\times B$ act on the $p$-group $P$, $p$ a prime, where $A$ is a $p'$-group and $B$ is a $p$-group. Then any element of $A$ which acts nontrivially on $P$ acts nontrivially on $C_P(B)$.*

Thus, replacing $AQ$ by $AQ/A_0$, and again interchanging the primes $p$ and $q$, we reach the following situation: a group $X=AQ$, where $Q$ is a special $q$-group and $A$ is a cyclic $p$-group which acts trivially on $\phi(Q)$ and faithfully and irreducibly on $Q/\phi(Q)$, with $X$ acting *faithfully* on a vector space $W$ over $GF(p)$. However, we are still missing one desirable property: *irreducibility* of the action of $X$. To achieve this, we must make yet a further reduction, namely, we must pass to a composition factor of $W$ under the action of $X$. The effectiveness of this reduction depends upon the following general property of so-called "chain stabilizers."

PROPOSITION 4.10. *Let $P$ be a $p$-group, $p$ a prime, and $A$ a $p'$-group which acts on $P$, and let $1=P_1\leqslant P_2\leqslant \cdots \leqslant P_n=P$ be a chain of $A$-invariant* *subgroups of $P$ with each $P_i \lhd P_{i+1}$, $1 \leq i \leq n-1$. If $A$ acts trivially on each factor group $P_{i+1}/P_i$, $1 \leq i \leq n-1$, then $A$ acts trivially on $P$.*

In view of the proposition, we can find a composition factor $U$ of $W$ under the action of $X$ on which $Q$ acts *nontrivially*. Thus $X$ is irreducible and $Q$ is nontrivial on $U$. However, this time we may well have lost the property of faithfulness which held for $X$ in its action on $W$. But this can be dealt with, since we can easily show that the kernel $K$ of the representation of $X$ on $U$ is a proper subgroup of $Q$ and that $\bar{Q}=Q/K$ is also a special $q$-group with $A$ acting trivially on $\phi(\bar{Q})$ and irreducibly on $\bar{Q}/\phi(\bar{Q})$.

To continue the discussion, replace $X$ by $X/K$ and $U$ by $V$. Thus $X=AQ$ is an irreducible group of linear transformations of the vector space $V$ over $GF(p)$ with $Q$ a special $q$-group, $A$ a cyclic $p$-group acting trivially on $\phi(Q)$ and faithfully and irreducibly on $Q/\phi(Q)$. The condition of irreducibility imposes yet a further restriction on $Q$. Indeed, $\phi(Q) \leq Z(Q)$ as $Q$ is special, so as $A$ acts trivially on $\phi(Q)$, it follows that $\phi(Q) \leq Z(X)$. But now as $X$ acts irreducibly on $V$, Schur's lemma (cf. Theorem 1.2) implies that $\phi(Q)$ must be *cyclic*. We conclude therefore from the definitions that $Q$ is either *elementary abelian* or *extra-special*.

It turns out that several very important questions about the structure of solvable groups can be reduced by just this series of steps to questions concerning the *degree* of the *minimal polynomial* of a generator $a$ of the subgroup $A$ of our irreducible group $X$ of linear transformations of $V$. Because $a$ has order $p^m$ for some $m$ and $V$ is defined over $GF(p)$, 1 is the only characteristic root of $a$ as a linear transformation of $V$, so the minimal polynomial of $a$ has the form

$$(x-1)^r=0.$$

Obviously $r \leq |a|$. We are interested in both the minimal possible value of $r$ as well as conditions under which equality holds. When $Q$ is abelian, equality is easily proved by using Clifford's theorem (Theorem 1.5). However, in the extra-special case, the result is deep and depends on a beautiful theorem of P. Hall and G. Higman [130, Theorem 11.1.1], a result that had great influence on Thompson and helped to shape his approach to the analysis of groups of odd order. Their theorem has since been extended in a variety of directions; in particular, it is the basis for many developments in the general theory of solvable groups. We shall limit ourselves here to conditions that imply equality.

THEOREM 4.11. *Let $X$ be an irreducible group of linear transformations of the vector space $V$ over $GF(p)$ with $X=AQ$, where $Q$ is either an elementary abelian or extra-special $q$-group for some prime $q \neq p$ and $A$ is a cyclic $p$-group that acts trivially on $\phi(Q)$ and faithfully and irreducibly on $Q/\phi(Q)$. If $a$ is a generator of $A$, then the minimal polynomial of $a$ on $V$ has the form*

$$(x-1)^{|a|}=0$$

*under any of the following conditions:*

(i) *$p$ is not a Fermat prime or 2.*
(ii) *$p$ and $q$ are both odd.*
(iii) *$Q$ is abelian.*

That equality does not hold in general can be seen from the example $X=SL(2,3)=AQ$, where $Q \cong Q_8$ and $\langle a \rangle \cong Z_3$ in its natural 2-dimensional representation on a vector space $V$ over $GF(3)$. Since $V$ has dimension 2 over $GF(3)$ and minimal polynomials always divide the corresponding characteristic polynomials (which have the same degree as the vector space dimension), clearly $(x-1)^2=0$ is the minimal polynomial of $a$.

The total line of argument just outlined allows one to predict the embedding of certain abelian subgroups of Sylow $p$-subgroups of solvable groups—namely, that they fall "near the bottom" of the group. Since Theorem 4.11 leads to exceptions when $p$ or $q$ is 2, we restrict ourselves to the odd-order case for simplicity.

THEOREM 4.12. *Let $X$ be a solvable group of odd order with $O_{p'}(X)=1$, $p$ a prime, and let $A$ be an abelian $p$-subgroup of $X$. If $[O_p(X), A, A]=[[O_p(X), A], A]=1$, then $A \leq O_p(X)$.*

Any abelian normal subgroup $A$ of a Sylow $p$-subgroup $P$ of $X$ satisfies the hypothesis of the theorem. Indeed, normality implies that $[P, A] \leq A$; and as $A$ is abelian, it follows that $[P, A, A]=[[P, A], A]=1$. But $O_p(X) \leq P$ by Sylow's theorem as $O_p(X) \lhd X$, so $[O_p(X), A, A]=1$. In some situations, one can show that certain abelian subgroups of $P$ of maximal order or rank (but not necessarily normal in $P$) also have the required property.

Theorem 4.12 is very important for local analysis; in fact, several of the basic topics to be discussed in this chapter—"$p$-stability," "factorizations," "failure of factorization," and the theory of "pushing up"—all have their origins in considerations of this nature.

There is another important kind of reduction in local analysis of a different character. It stems from the following *generational* result.

THEOREM 4.13. *Let $A$ be a noncyclic abelian $p$-group, $p$ a prime, which acts on the $p'$-group $X$. Then*

$$X=\langle C_X(B)|B\leqslant A, |A:B|\leqslant p\rangle.$$

*In particular,*

$$X=\langle C_X(a)|a\in A^{\#}\rangle.$$

Using Theorem 4.2, one reduces the proof to the case that $X$ is a $q$-group for some prime $q\neq p$. Then, again using Proposition 1.12(iii), one can show that it suffices to prove the theorem with $\bar{X}=X/\phi(X)$ in place of $X$. However, as $\bar{X}$ is elementary abelian and hence a vector space over $GF(q)$, the desired conclusion follows in this case from Maschke's theorem (Theorem 1.1).

The point of Theorem 4.13 is that if $A$ is a noncyclic abelian $p$-subgroup of a simple group $G$, it reduces the study of the $A$-invariant $p'$-subgroups $X$ of $G$ to *$p$-local* questions inasmuch as each of the groups $C_X(B)$ lies in the $p$-local subgroup $N_G(B)$ of $G$. It was from considerations of this type that Thompson was able to establish his fundamental transitivity theorem [130, Theorem 8.5.4].

THEOREM 4.14. *Let $G$ be a group in which every $p$-local subgroup is $p$-constrained (in particular, solvable), $p$ a prime. Let $P\in Syl_p(G)$ and let $A$ be an abelian normal subgroup of $P$ maximal under inclusion. If $m_p(A)\geqslant 3$, then for any prime $q\neq p$, any two maximal $A$-invariant $q$-subgroups of $G$ are conjugate by an element of $C_G(A)$.*

The condition $m_p(A)\geqslant 3$ is critical, for it allows one to "compare $p$-locally" any two maximal $A$-invariant $q$-subgroups $Q_1$ and $Q_2$ of $G$. Indeed, by Theorem 4.13, there is $B_i\leqslant A$ with $|A:B_i|\leqslant p$ such that $C_{Q_i}(B)\neq 1$, $i=1, 2$. Since $m_p(A)\geqslant 3$, $B_1\cap B_2\neq 1$, so $N_G(B_1\cap B_2)$ is a $p$-local subgroup of $G$ containing $C_{Q_i}(B)$ for *both* $i=1$ and 2.

A consequence of the condition that $A$ is a maximal abelian normal subgroup of $P$ is the fact that $C_G(A)=A\times O_{p'}(C_G(A))$, which clearly implies that $A$ *contains every element of order $p$ in its centralizer* (cf. [130, Theorem 7.6.5]). In the case of *odd $p$*, Thompson was able to extend the transitivity theorem to any abelian $p$-subgroup $A$ of $G$ (again of rank $\geqslant 3$) with the latter property, in particular, to abelian $p$-subgroups of $G$ of maximal rank [assuming $m_p(G)\geqslant 3$]. Thompson's proof depends upon the following result, an analogue of which he had previously established in the maximal abelian normal case [292] (an alternate proof of the theorem was given later by Bender [31]).

THEOREM 4.15. *Let $X$ be a $p$-constrained group, $p$ an odd prime, and let $A$ be an abelian $p$-subgroup of $X$ which contains every element of order $p$ in $C_X(A)$. Then every $A$-invariant $p'$-subgroup of $X$ lies in $O_{p'}(X)$.*

The result is false for $p=2$, the group $X=SL(2,3)$ with $A=Z(X)$ (of order 2) providing a counterexample [any subgroup of $X$ of order 3 is $A$-invariant, but does not lie in $O_{2'}(X)=O(X)$].

In the odd-order and $N$-group contexts, the transitivity theorem represented the first step in Thompson's proof of so-called "uniqueness" theorems, which were of central importance for determining the structure of the critical maximal local subgroups of $G$. Because the primes 2 and 3 are exceptional in the latter case [the possible involvement of $SL(2,3)$ in $G$ makes 3 exceptional], we shall state Thompson's results only in the odd-order case.

THEOREM 4.16. *Let $G$ be a simple group of odd order with all proper subgroups solvable, let $P\in Syl_p(G)$ for some prime $p$, and let $A$ be an abelian normal subgroup of $P$ maximal under inclusion.* If $m_p(A)\geqslant 3$, *then the set of all $A$-invariant $p'$-subgroups of $G$ generates a $p'$-group.*

COROLLARY 4.17. *Under the assumptions and notation of Theorem* 4.16, $\langle O_{p'}(C_G(a))|a\in A^{\#}\rangle$ *is a $p'$-group.*

Since $A\lhd P$ by hypothesis, $P$ permutes the set of all $A$-invariant $p'$-subgroups of $G$ under conjugation, so normalizes their join. Since every $P$-invariant $p'$-subgroup of $G$ is certainly $A$-invariant, we obtain the following further corollary, which was the goal of this phase of Thompson's analysis.

COROLLARY 4.18. *Under the assumption and notation of Theorem* 4.16, *$G$ possesses a unique maximal $P$-invariant $p'$-subgroup.*

Thompson referred to any $P$-invariant $p'$-subgroup of $G$ as a *$P$-signalizer*, and later the term was extended to arbitrary $p$-subgroups of $G$. For any

$p$-subgroup $R$ of $G$, Thompson introduced the symbol $\mathcal{N}_G(R)$ [for brevity $\mathcal{N}(R)$] for the set of $R$-signalizers in $G$—i.e., for the set of $R$-invariant $p'$-subgroups of $G$. A substantial portion of local analysis deals with the set $\mathcal{N}(R)$ (or some specified subset of it), especially when $R$ is abelian.

Finally I should like to mention a very elementary result, known as the *three subgroups lemma*; it is used frequently throughout both solvable and general local analysis [130, Theorem 2.2.3].

PROPOSITION 4.19. *If $H, K, L$ are subgroups of the group $X$ such that $[K, L, H]=1$ and $[L, H, K]=1$, then $[H, K, L]=1$.*

As an illustration of its use, suppose $\alpha$ is an automorphism of a perfect group $X$ such that $\alpha$ acts trivially on $X/Z(X)$. Then the three subgroups lemma implies that $\alpha$ acts trivially on $X$ (i.e., $\alpha=1$). Indeed, as $\alpha$ centralizes $X/Z(X)$, $[X, \alpha] \leqslant Z(X)$, whence $[X, \alpha, X] \leqslant [Z(X), X]=1$. Hence, by the proposition, $[X, X, \alpha]=1$. But $[X, X]=X$ as $X$ is perfect, so, in fact, $[X, \alpha]=1$, i.e., $\alpha$ acts trivially on $X$.

I hope this entire discussion gives some feeling for the nature of solvable local analysis. The ideas we have touched on here will be further reinforced and developed throughout the chapter, as well as in the sequel.

## 4.2. Strong Embedding

As we have already pointed out, the simplicity of $G$ must somehow be used to force $G$ to have an internal structure resembling that of a simple $K$-group $G^*$. The single most general tool for accomplishing this is by means of the construction of a *strongly embedded* subgroup of $G$, followed by invocation of Bender's complete classification of groups possessing a strongly embedded subgroup (almost always in order to obtain a contradiction).

There is a certain analogy between the nonexistence of strongly embedded subgroups in the general finite simple group and the nondegeneracy of the Killing form in the study of semisimple Lie algebras $L$. The latter enables one to conclude that $L$ has a trivial radical, which has a profound effect on its internal structure. The appropriate analogue for finite groups $G$ of the radical of $L$ is $Sol(G)$, the largest normal solvable subgroup of $G$. By the Feit–Thompson theorem, the core $O(G)$ of $G$ is solvable, so $O(G) \leqslant Sol(G)$. If $G$ is simple, then, of course, $Sol(G)$ and, in particular, $O(G)$ is trivial. The nonexistence of a strongly embedded subgroup, when combined with the condition $O(G)=1$, yields internal structural properties of $G$.

Typically, the argument goes as follows. To show that the core of the centralizer $C$ of an involution of $G$ "resembles" that of some known simple group, one assumes the contrary and on the basis of this assumption constructs inside $G$ a subgroup $M$ with nontrivial core which is strongly embedded in $G$. But a consequence of Bender's theorem (Corollary 4.27 below) implies that $O(M) \leqslant O(G)$. However, $O(G)=1$ as $G$ is simple, contrary to the fact that $O(M) \neq 1$ by construction. Thus we can deduce significant properties of $O(C)$ from the nonexistence of a strongly embedded subgroup in $G$.

On the other hand, there are fundamental uses of Bender's theorem unrelated to cores, which occur at a later stage of the analysis, after one has shown that $G$ resembles internally some simple $K$-group $G^*$. This time one constructs *inside* $G$ by purely group-theoretic means a subgroup $G_0$ isomorphic to $G^*$. To obtain the desired conclusion that $G \cong G^*$, one must obviously prove that $G_0 = G$. Assuming this to be false, one argues that $N_G(G_0)$ is strongly embedded in $G$. Bender's theorem (Theorem 4.24 below) then yields that $G \cong L_2(2^n)$, $U_3(2^n)$, $Sz(2^n)$ for some $n$. One now simply checks that none of these groups possesses a strongly embedded subgroup of the structure of $G_0$.

There are many equivalent definitions of a strongly embedded subgroup; we take the following:

DEFINITION 4.20. Let $X$ be a finite group and $H$ a proper subgroup of $X$ of even order. $H$ is said to be *strongly embedded* in $X$ provided $N_X(T) \leqslant H$ for every nontrivial 2-subgroup $T$ of $H$.

Clearly by Sylow's theorem, it suffices to impose the condition on subgroups $T$ of a fixed Sylow 2-subgroup $S$ of $H$.

In the terminology of Definition 1.32, it follows that $\Gamma_{S,1}(X) \leqslant H$. Thus the assertion that a group $X$ has a strongly embedded subgroup is equivalent to $X$ having a proper 1-generated core.

As a consequence of the definition, one can easily prove

PROPOSITION 4.21. *If $H$ is a strongly embedded subgroup of $X$, then the following conditions hold*:

(i) *If $S \in Syl_2(H)$, then $S \in Syl_2(X)$.*
(ii) *For any $x \in X - H$, $|H \cap H^x|$ is odd. In particular, $H = N_X(H)$.*
(iii) *In the permutation representation of $X$ on the conjugates of $H$ in $X$, a 1-point stabilizer has even order and every 2-point stabilizer has odd order.*

Further properties of groups with a strongly embedded subgroup $H$ are established in [130, Section 9.2]. In particular, $X$ and $H$ each have only one conjugacy class of involutions, and each coset of $H$ in $X-H$ contains precisely one involution. The latter condition implies that any conjugate $H_1$ of $H$ distinct from $H$ has the form $H_1=H^t$ for some involution $t$ of $X$. In addition, if $y \in \mathcal{I}(H)$, then $H=C_X(y)K$, where $K$ has odd order. Thus the embedding and structure of a strongly embedded subgroup are very restricted.

Suzuki's results on split $(B, N)$-pairs of rank 1 (in particular, Theorem 3.28) imply the following basic result about groups with a strongly embedded subgroup.

THEOREM 4.22. *Let $G$ be a simple group having a strongly embedded subgroup $H$. If the permutation representation of $G$ on the conjugates of $H$ is doubly transitive and if $H$ possesses a normal subgroup that acts regularly on the conjugates other than $H$, then $G \cong L_2(2^n)$, $n \geq 2$, $Sz(2^{2n+1})$, $n \geq 1$, or $U_3(2^n)$, $n \geq 2$.*

Bender took up the general strongly embedded problem at this point. His classification [26] proceeds in two steps. He first proves

THEOREM 4.23. *If $G$ is a group with a strongly embedded subgroup $H$, then one of the following holds*:

(i) *$G$ is doubly transitive on the conjugates of $H$ in $G$.*
(ii) *$G$ has cyclic or quaternion Sylow 2-subgroups and $O(G)$ is transitive on the conjugates of $H$ in $G$.*

Note that as $H=N_G(H)$ by Proposition 4.21, the permutation-theoretic assertion that an element $x \in G$ fixes a given conjugate of $H$ is equivalent to the statement that $x$ lies in that conjugate.

The proof of the theorem is by induction on $|G|$ and involves a careful analysis of the normalizers in $G$ of $p$-subgroups of $H$ that lie in at least three conjugates of $H$, $p$ an odd prime. Ultimately the proof is reduced to the special case in which $G$ has cyclic or quaternion Sylow 2-subgroups [in which case $G=O(G)H$ and $H=C_G(z)$ for $z \in \mathcal{I}(H)$ by the Brauer–Suzuki theorem, to be discussed in Section 4.6].

Bender's main result (stated here only for simple groups) is the following:

THEOREM 4.24. *If $G$ is a simple group with a strongly embedded subgroup $H$, then $G \cong L_2(2^n)$, $n \geq 2$, $Sz(2^{2n+1})$, $n \geq 1$, or $U_3(2^n)$, $n \geq 2$.*

In view of Theorems 4.22 and 4.23, Bender can assume that $G$ is doubly transitive on the conjugates of $H$ in $G$ and that a Sylow 2-subgroup $S$ of $H$ is not normal in $H$. Moreover, he can also take $G$ to be a minimal counterexample (to a more general theorem classifying arbitrary finite groups with a strongly embedded subgroup).

The following lemma enables Bender to apply induction.

LEMMA 4.25. *Let $Y \leqslant H$ and suppose that both $|Y \cap H|$ is even and $|Y \cap H^g|$ is even for some $g \in G - H$. Then $Y \cap H$ is strongly embedded in $Y$.*

With the aid of the lemma, Bender proves the following key result.

PROPOSITION 4.26. *Let $Y$ be a subgroup of $H$ that lies in at least three conjugates of $H$. Then we have*

(i) *$|C_H(Y)|$ is odd.*
(ii) *$C_G(Y)$ transitively permutes the set of conjugates of $H$ containing $Y$.*

To establish the theorem, Bender must derive a contradiction from his assumptions. This is obtained by a careful analysis of a certain nontrivial subgroup $E$ of $H$ of odd order. Let $g \in G - H$ and set $D = H \cap H^g$, so that $|D|$ is odd. By Proposition 4.21, $H = DS$. Since $S$ is not normal in $H$ by assumption, a *normal* subgroup $W$ of $H$ chosen minimal subject to $H = DW$ is necessarily distinct from $S$. This implies that $D \cap W \neq 1$, otherwise $|W| = |S|$ and then $W = S$. The pertinent subgroup $E$ is defined to be $D \cap W$.

Bender's theorem has the following fundamental corollary.

COROLLARY 4.27. *If the group $G$ contains a strongly embedded subgroup $H$, then $O(H) \leqslant O(G)$.*

The corollary can be viewed as a nonsimplicity criterion, since it asserts, under the assumption of strong embedding, that the core of a proper subgroup lies in the core of the entire group.

We come now to Aschbacher's generalization of Bender's strong embedding theorem to groups with a proper 2-generated core, which as we have pointed out earlier yields as a corollary the classification of nonconnected simple groups with a connected Sylow 2-group. We state Aschbacher's result only in the simple case [8].

THEOREM 4.28. *If $G$ is a simple group with a proper 2-generated core, then $G \cong L_2(q)$, $q \geqslant 3$, $Sz(2^{2n+1})$, $n \geqslant 1$, $U_3(2^n)$, $n \geqslant 1$, $M_{11}$, or $J_1$.*

Aschbacher considers a minimal counterexample $G$ (to a more general theorem classifying all groups with a proper 2-generated core). The following lemma is easily proved.

LEMMA 4.29. *If $z$ is an involution of $G$ such that $C_z<G, m_2(C_z)\geq 3$, and $C_z$ has a proper 2-generated core, then $\bar{C}_z=C_z/O(C_z)$ has a normal subgroup $\bar{L}\cong L_2(2^n)$, $U_3(2^n)$, $Sz(2^n)$, $\hat{S}z(8)$, $SL_2(5)$, or $SL(2,5)*SL(2,5)$; and $m_2(C_{\bar{C}_z}(\bar{L}))=1$.*

Here $\hat{S}z(8)$ denotes a covering group of $Sz(8)$ by $Z_2$. Centralizers of involutions of the above general form were first considered by Walter and me in [142], where such involutions were called *exceptional* [actually our definition had not covered the cases in which $\bar{L}\cong SL(2,5)$ or $SL(2,5)*SL(2,5)$]. In particular, we had there proved all but one case of the following result, which Aschbacher handles.

PROPOSITION 4.30. *If $z$ is an exceptional involution of the simple group $X$ and $z$ is 2-central in $X$, then $X\cong J_1$.*

This result depends in part upon Janko's characterization of $J_1$ by the centralizer of an involution of the form $Z_2\times L_2(4)$.

To establish his main theorem, Aschbacher reduces the problem to a second major theorem which gives a basic criterion for a group to possess a strongly embedded subgroup. The reduction is carried out as follows. First, he argues easily that a minimal counterexample $G$ (to the general theorem) is quasisimple with $O(G)=1$ and $|Z(G)|\leq 2$. If $G$ is not simple, a careful analysis of a Sylow 2-subgroup $\bar{S}$ of $\bar{G}=G/Z(G)$ shows that $S$ is isomorphic to a Sylow 2-subgroup of $A_9$ and the fusion pattern of involutions in $\bar{G}$ is the same as that in $A_9$. A theorem by Harada and me [134] is now applicable and yields that $\bar{G}\cong A_9$. It follows at once that $\bar{G}\cong\hat{A}_9$, which is one of Aschbacher's possible conclusions.

Thus the minimal couterexample $G$ is simple. If $m_2(G)\leq 2$, $G$ is determined from known classification theorems [4], and as $G$ has a proper 2-generated core, $G\cong L_2(q)$ for suitable odd $q$ or $M_{11}$, so $G$ is not a counterexample. Hence also $m_2(G)\geq 3$.

Let $S\in Syl_2(G)$ and set $M=\Gamma_{S,2}(G)$, so that $M<G$ by hypothesis. Let $z$ be an involution of $Z(S)$. Then $S\leq C_z$, so $m_2(C_z)=m_2(S)\geq 3$. If $C_z\nleq M$, then $\Gamma_{S,2}(C_z)<C_z$, so $C_z$ has a proper 2-generated core. Hence $z$ is exceptional by Lemma 4.29 and consequently $G\cong J_1$ by Proposition 4.30. Again $G$ is not a counterexample. We thus conclude that $C_z\leq M$. Furthermore, by

definition of $M$, $m_2(M \cap M^g) \leqslant 1$ and so $M \cap M^g$ has cyclic or generalized quaternion Sylow 2-subgroups whenever $M \neq M^g$. Since $C_z \leqslant M$, this immediately yields that $z \in M^g$ if and only if $g \in M$.

Now let $u$ be any involution of $C_z$ such that $C_u \nleqslant M$ and let $H$ be the subgroup of $C_u$ generated by all $G$-conjugates of $z$ that lie in $C_u$. It is easily seen that either $u$ is exceptional or $m_2(C_u) = 2$. Now with the aid of Glauberman's $Z^*$-theorem (see Section 4.6), Aschbacher argues that $H \cap M$ is strongly embedded in $H$. These are precisely the hypotheses of his strong embedding criterion, which enables him to conclude that $G$ has a strongly embedded subgroup [whence $G \cong L_2(2^n)$, $Sz(2^n)$, or $U_3(2^n)$ by Bender's theorem] and so is not a counterexample.

We shall not discuss Aschbacher's strong embedding criterion, except to say that its proof is similar in spirit to Bender's argument. In particular, Aschbacher again shows that a minimal counterexample $G$ is simple and acts doubly transitively on the set $\Omega$ of $G$-conjugates of $M$. A contradiction is obtained by a detailed analysis of the structure of this doubly transitive group $G$ in terms of its action on $\Omega$. The argument utilizes some key ideas of Shult from his classification of groups $G$ which contain an involution $t$ whose "weak closure in $C_G(t)$ with respect to $G$" (see Section 4.10 for definition) is assumed to be abelian [252]. Although Shult's result was never published, it has had considerable influence on simple group theory—in particular, on certain aspects of the work of Goldschmidt and Timmesfeld in addition to Aschbacher.

Subsequently Aschbacher slightly extended his strong embedding criterion, establishing the following result, which has become an important tool of local analysis [10].

THEOREM 4.31. *Let $G$ be a simple group, $M$ a proper subgroup of $G$, and $z$ an involution of $M$. Then $G$ possesses a strongly embedded subgroup under either of the following conditions*:

(a) *If $g \in G$ with $z^g$ centralizing $z$, $z^g \neq z$, and $z^g \in M$, then $g \in M$*; *or*

(b) (1) *$z$ is in the center of a Sylow 2-subgroup of $G$*;
  (2) *If $g \in G$ with $z^g \in M$, then $g \in M$*; *and*
  (3) *If $g \in G$ with $z^g$ centralizing $z$ and $z^g \neq z$, then $C_G(zz^g) \leqslant M$.*

Subsequently Derek Holt [179] and F. Smith [260] independently studied groups satisfying conditions (b) (1) and (b) (2) alone. We state their result in permutation-theoretic language.

THEOREM 4.32. *Let $G$ be a primitive permutation group on a set $\Omega$. If $O(G)=Z(G)=1$ and some 2-central involution of $G$ fixes exactly one point of $\Omega$, then either $G$ has a strongly embedded subgroup or $G \cong A_n$ or $\Sigma_n$, $n$ odd.*

Finally combining Theorem 4.28 with Proposition 1.33, we obtain the following result, which solves one of the major parts of the nonconnected simple group problem. For simplicity, we limit ourselves to the 2-rank at least 3 case.

THEOREM 4.33. *If $G$ is a nonconnected simple group of 2-rank at least 3 with a connected Sylow 2-subgroup, then $G \cong L_2(2^n)$, $n \geqslant 3$, $Sz(2^{2n+1})$, $n \geqslant 1$, $U_3(2^n)$, $n \geqslant 3$, or $J_1$.*

## 4.3. Signalizer Functors

As a first step in proving that the centralizer $C$ of an involution in an arbitrary simple group $G$ resembles the centralizer $C^*$ in some simple $K$-group $G^*$, one would naturally try to prove that the core $O(C)$ bears a close relationship to $O(C^*)$. However, as the following result shows, the structure of $O(C^*)$ is very restricted.

PROPOSITION 4.34. *If $G^*$ is a simple $K$-group, then the core of the centralizer of an involution of $G^*$ is cyclic.*

In view of this result, we clearly need some general method for limiting the structure of the cores of centralizers of involutions in arbitrary simple groups. *Signalizer functors* represent the principal tool that has been developed to accomplish this purpose. Because of the fundamental importance of the "signalizer functor method" in the study of simple groups, I shall spend some time motivating the signalizer functor concept. The notion grew out of an attempt to abstract Thompson's uniqueness results in the odd-order and $N$-group analyses, so that they could be applied more generally to simple groups with nonsolvable local subgroups.

The objective of the $N$-group analysis was not only to show that the cores of the centralizers of involutions were cyclic but also to establish the stronger conclusion that they were *trivial*! Thompson's uniqueness theorems, for $p=2$ (cf. Theorem 4.14–4.16), represented the first step of the proof; the second was achieved by the following result, whose hypotheses are directly linked to the conclusions of his uniqueness theorems.

THEOREM 4.35. *Let $G$ be a simple group, $S$ a Sylow 2-subgroup of $G$, and assume the following*:

(a) $W=\langle O(C_x)|x\in\mathcal{I}(S)\rangle$ *is of odd order.*
(b) $C_W(x)=O(C_x)$ *for every* $x\in\mathcal{I}(S)$.

*Then* $O(C_t)=1$ *for every* $t\in\mathcal{I}(G)$.

It will be instructive to prove this theorem since it involves a typical application of the Bender–Aschbacher results. First, as $G$ is simple, the Brauer–Suzuki theorem (see Section 4.6) implies that $m_2(G)\geqslant 2$. Thus $G$ and hence $S$ contains a four subgroup $U$. For any such $U$, set

$$W_U=\langle O(C_u)|u\in U^{\#}\rangle. \tag{4.1}$$

Then by (4.1) and assumption (b), we have

$$W_U=\langle C_W(u)|u\in U^{\#}\rangle. \tag{4.2}$$

Since $|W|$ is odd and $U$ is noncyclic abelian, it follows from (4.2) and Theorem 4.13 that

$$W=W_{U^g}. \tag{4.3}$$

Furthermore, if $g\in G$ and $U^g\leqslant S$, it is immediate from (4.1) that

$$(W_U)^g=W_{U^g}. \tag{4.4}$$

Now set $N=N_G(W)$. We claim that

$$\Gamma_{S,2}(G)\leqslant N. \tag{4.5}$$

Indeed, let $T\leqslant S$ with $m_2(T)\geqslant 2$. We must show that $N_G(T)\leqslant N$ for any such $T$. It will suffice to prove that $y\in N$ for any $y\in N_G(T)$. Since $m_2(T)\geqslant 2$, $T$ contains a four subgroup $V$. Then $V^y\leqslant T$ and $V^y$ is also a four group. Hence (4.3) holds for both $V$ and $V^y$, whence

$$W=W_V=W_{V^y}. \tag{4.6}$$

But by (4.4), $W_{V^y}=(W_V)^y$, so by (4.6) we have

$$W=W^y. \tag{4.7}$$

Thus $y\in N=N_G(W)$, proving (4.5).

Suppose $W \neq 1$. Since $|W|$ is odd, $W < G$; and as $G$ is simple, it follows that $N = N_G(W) < G$. Hence $\Gamma_{S,2}(G) < G$ by (4.5) and so $G$ *has a proper 2-generated core*. Now Aschbacher's classification theorem yields that $G \cong L_2(q)$, $q > 3$, $Sz(2^{2n+1})$, $n \geqslant 1$, $U_3(2^n)$, $n \geqslant 2$, $M_{11}$, or $J_1$. However, the precise form of the centralizers of involutions in each of these groups is known and in none of them does the subgroup $W$ turn out to be nontrivial of odd order. Hence we reach a contradiction and so we conclude that $W = 1$.

It follows now from (b) that $O(C_x) = 1$ for every $x \in \mathcal{I}(S)$. Hence, by Sylow's theorem, we obtain the desired conclusion $O(C_t) = 1$ for every $t \in \mathcal{I}(G)$.

In essence, Theorem 4.35 asserts (under the given hypothesis) that $O(C_G(t)) = C_{O(G)}(t)$ for every involution $t$ of $S$ (and hence of $G$); since $O(G) = 1$ by assumption, the triviality of each $O(C_G(t))$ then follows.

How can we hope to extend Theorem 4.35? To get some perspective on the question, let us at least suppose that our simple group $G$ under investigation internally resembles a $K$-group $G^*$ (but with $G^*$ as far from simple as you wish). If $t^* \in \mathcal{I}(G^*)$, then clearly $C_{O(G^*)}(t^*) \leqslant O(C_{G^*}(t^*))$. However, there is no reason for equality to hold in general; indeed, that conclusion would require the validity of Proposition 4.34 with "trivial" in place of "cyclic," which is known to be false for the general $K$-group. On the other hand, if $t$ is an involution of $G$ corresponding in the given "resemblance" to $t^*$, there "ought to be" some subgroup of $O(C_G(t)) = O(C_t)$ corresponding to $C_{O(G^*)}(t^*)$; and it is this hypothetical subgroup that we want to get hold of. To conceptualize the situation, let us denote this as yet undetermined subgroup by $\theta(C_t)$ and let us set

$$\theta(G) = \langle \theta(C_t) | t \in \mathcal{I}(G) \rangle.$$

If $G$ truly resembles $G^*$, then as $G$ is simple, so must be $G^*$, whence $O(G^*) = 1$ and hence $O(C_{G^*}(t^*)) = 1$ for each $t^* \in \mathcal{I}(G^*)$. But then we should have $O(C_t) = 1$ and hence $\theta(C_t) = 1$ for each $t \in \mathcal{I}(G)$, equivalently that $\theta(G) = 1$. Thus our objective must be to prove the triviality of $\theta(G)$. On the other hand, if the given resemblance is not quite so precise, we might hope as a reasonable first approximation to show that the hypothetical subgroup $\theta(G)$ at least has odd order, since it is intended to correspond to a subgroup of $O(G^*)$. This latter assertion is, in fact, the essential content of Thompson's uniqueness results (for $p = 2$) in the $N$-group situation.

Again by Sylow's theorem, one can reduce the discussion to a fixed Sylow 2-subgroup $S$ of $G$. Thus it suffices to consider the subgroup

$$\theta(G; S) = \langle \theta(C_x) | x \in \mathcal{I}(S) \rangle;$$

and as a first step in proving the triviality of each $\theta(C_t)$, the aim should be to establish the following conclusions:

$$\begin{aligned}&\text{(a)}\quad \theta(G;S)\ \textit{has odd order}.\\&\text{(b)}\quad C_{\theta(G;S)}(x)=\theta(C_x)\ \textit{for each}\ x\in\mathscr{I}(S).\end{aligned}\tag{4.8}$$

There is a serious technical difficulty in trying to prove such a result directly—namely, if $x\notin Z(S)$, then $S\nleq C_x$, so we do not even know at the outset that our hypothetical subgroup $\theta(C_x)$ is $S$-invariant. This difficulty can be overcome if instead of working with $S$, we limit ourselves to *abelian* subgroups $A$ of $S$, for then if $x\in\mathscr{I}(A)$, we shall at least have $A\leqslant C_x$. Clearly there will be no loss if we restrict ourselves to elementary abelian subgroups of $S$.

Thus we should like to be able to prove, for suitable elementary abelian $A\leqslant S$, that

$$\begin{aligned}&\text{(a)}\quad \theta(G;A)=\langle\theta(C_x)|x\in A^{\#}\rangle\ \textit{has odd order}.\\&\text{(b)}\quad C_{\theta(G;A)}(x)=\theta(C_x)\ \textit{for every}\ x\in A^{\#}.\end{aligned}\tag{4.9}$$

These considerations also explain why Thompson first worked with abelian $p$-subgroups (cf. Theorems 4.14 and 4.15) and only subsequently brought the full Sylow $p$-subgroup $P$ into the picture (Corollary 4.17).

In the presence of connectedness (see Definition 1.31), such a result will be entirely sufficient; and this will help to explain the fundamental importance of this condition. Indeed, we have

PROPOSITION 4.36. *Let $G$ be a connected group (or a group with a connected Sylow 2-subgroup) of 2-rank at least 3 and let $S\in Syl_2(G)$. If (4.9) holds for every abelian subgroup $A$ of $S$ with $m_2(A)\geqslant 3$, then (4.8) holds for $S$.*

Indeed, if $A$ is an elementary subgroup of $S$ with $m_2(A)\geqslant 3$ and $B$ a noncyclic subgroup of $A$, it is immediate from (4.9) that $\theta(G;B)=\theta(G;A)$. Using this equality together with the connectivity of $G$ (or $S$), it is an easy exercise to prove the following statement:

$$\theta(G;U)=\theta(G;V)\tag{4.10}$$

for any pair of four subgroups $U$ and $V$ of $S$. Since any involution of $S$ lies in some four subgroup of $S$, this implies that

$$\theta(G;S)=\theta(G;U)\tag{4.11}$$

for any four subgroup $U$ of $S$. Also connectivity implies that $U\leqslant A$ for some elementary $A \leqslant S$ with $m_2(A) \geqslant 3$, so $|\theta(G;U)|$ is odd by (4.9) and hence $|\theta(G;S)|$ is odd. Since any $x \in \mathscr{I}(S)$ lies in some such $A$, it also follows [using (4.9)] that $C_{\theta(G;S)}(x) = \theta(C_G(x))$, so (4.8) holds.

This entire discussion has been based on the assumption that $G$ resembles a $K$-group $G^*$, so that we have been implicitly assuming that a likely candidate for $\theta(C_t)$, $t \in \mathscr{I}(G)$, is actually available. How then do we identify a "good" candidate in an arbitrary simple group? Again we can use the case that $G$ resembles $G^*$ as a guide. Indeed, if $x^*$, $y^*$ are commuting involutions of $G^*$, the following equality holds:

$$C_{O(G^*)}(x^*) \cap C_{G^*}(y^*) = C_{O(G^*)}(y^*) \cap C_{G^*}(x^*), \tag{4.12}$$

inasmuch as both of these groups are equal to $C_{O(G^*)}(\langle x^*, y^* \rangle)$.

This identity suggests that good candidates for $\theta$ ought to satisfy the following condition for every pair of commuting involutions $x$, $y$ of $G$:

$$\theta(C_x) \cap C_y = \theta(C_y) \cap C_x. \tag{4.13}$$

Considerations of this nature led me to feel that equation (4.13) represented the key to the hypothetical subgroups $\theta(C_t)$; and I made it the basis for the definition of what I called a "signalizer functor," viewing it as an abstract generalization of Thompson's notion of signalizer. Now came the crucial question: Are there circumstances under which conditions (4.9) hold for a signalizer functor? The signalizer functor theorem asserts that this is indeed the case.

My original proof [128, 129] was very complicated and was patterned very closely on Thompson's uniqueness proofs in the odd-order case. However, some time later Goldschmidt made a significant extension and improvement of the signalizer functor theorem, utilizing ideas of Bender in the latter's simplification of Thompson's odd-order uniqueness results [121]. In particular, Goldschmidt was able to reduce the restrictions on the rank of $A$ which I had been forced to make and at the same time simplify and clarify the notion of a signalizer functor. We shall follow Goldschmidt's now standard terminology. Since signalizer functors are also important in studying centralizers of elements of odd prime order, all definitions will be made for arbitrary primes.

Definition 4.37. Let $X$ be a group and $A$ an elementary abelian $p$-subgroup of $X$ for some prime $p$. Suppose that for each $a \in A^{\#}$, there is associated an $A$-invariant $p'$-subgroup (i.e., of order prime to $p$) $\theta(C_X(a))$ of

$C_X(a)$ such that for each $a, b \in A^{\#}$,

$$\theta(C_X(a)) \cap C_X(b) = \theta(C_X(b)) \cap C_X(a).$$

Then $\theta$ is said to be an *A-signalizer functor* on $X$.

We set

$$\theta(X; A) = \langle \theta(C_X(a)) | a \in A^{\#} \rangle$$

and call $\theta(X; A)$ the *closure* of $\theta$. $\theta$ is said to be *complete* provided

(a) $\theta(X; A)$ is a $p'$-group; and (4.14)
(b) $C_{\theta(X;A)}(a) = \theta(C_X(a))$ for $a \in A^{\#}$.

$\theta$ is said to be *solvable* if $\theta(C_X(a))$ is solvable for each $a \in A^{\#}$.

Note that if $p=2$, then each $\theta(C_X(a))$ has odd order and so is solvable by the Feit–Thompson theorem. Hence, in this case, *every* $A$-signalizer functor on $G$ is necessarily solvable.

The solvable signalizer functor theorem asserts:

THEOREM 4.38. *Let $X$ be a group, $p$ a prime, $A$ an elementary abelian $p$-subgroup of $X$ of rank at least* 3, *and $\theta$ a solvable $A$-signalizer functor on $G$. Then $\theta$ is complete.*

In particular, if $A$ is an elementary abelian 2-subgroup of $X$ of rank$\geqslant 3$ and $\theta$ is an $A$-signalizer functor on $X$, then $\theta(X; A)$ is a subgroup of $X$ of odd order.

Goldschmidt covers all cases of the theorem except $p$ odd and $m_p(A)=3$. Subsequently Glauberman proved the theorem in the stated form [116], and Bender gave an alternate proof of Goldschmidt's results [29]. We shall outline Goldschmidt's beautiful rank$\geqslant$4 proof—certainly one of the most elegant papers in all of local analysis. The rank 3 case is more technical and involves additional special arguments; however, our initial results apply to all cases.

In the odd-order and $N$-group situations, Thompson had been able to work with the set $\mathscr{N}(A)$ of *all* $A$-invariant $p'$-subgroups of the given group $G$ (for suitable abelian $p$-subgroups $A$); Theorem 4.15 asserts precisely that the set $\mathscr{N}(A)$ possesses a unique maximal element; in other words, the subgroup of $G$ generated by the elements of $\mathscr{N}(A)$ is itself an element of $\mathscr{N}(A)$.

In the general situation of signalizer functors, one must work instead only with certain $A$-invariant $p'$-subgroups of the group $X$, directly related to the given $A$-signalizer functor $\theta$, namely, we denote by $\mathcal{N}_\theta(A)$ the set of $A$-invariant $p'$-subgroups $Y$ of $X$ with the property

$$C_Y(a) = Y \cap \theta(C_X(a)) \qquad \text{for all } a \in A^{\#}.$$

In view of Theorem 4.12, if $Y \in \mathcal{N}_\theta(A)$, then

$$Y = \langle Y \cap \theta(C_X(a)) | a \in A^{\#} \rangle,$$

so $Y \leqslant \theta(X; A)$. Moreover, if $\theta$ is complete, it is immediate from the definition that then $\theta(X; A) \in \mathcal{N}_\theta(A)$. Thus we see that in this terminology the signalizer functor theorem reduces to the assertion that the set $\mathcal{N}_\theta(A)$ possesses a unique maximal element. This is the abstract reformulation of Thompson's results.

Note that the definition of a signalizer functor implies at once that $\theta(C_X(a)) \in \mathcal{N}_\theta(A)$ and is the unique maximal element of $\mathcal{N}_\theta(A)$ contained in $C_X(a)$ for every $a \in A^{\#}$. Similarly if for $B \leqslant A$ we set

$$\theta(C_X(B)) = \bigcap_{b \in B^{\#}} \theta(C_X(b)),$$

it implies that the corresponding assertions hold for $\theta(C_X(B))$ in place of $\theta(C_X(a))$.

In the proof of the solvable signalizer functor theorem, elements of $\mathcal{N}_\theta(A)$, which are $q$-groups or $\{q, r\}$-groups for various primes $q$ and $r$, play an important role; the set of such $A$-invariant $p'$-groups is denoted by $\mathcal{N}_\theta(A; q)$, $\mathcal{N}_\theta(A; q, r)$, respectively. Partially ordering the elements of $\mathcal{N}_\theta(A; q)$, $\mathcal{N}_\theta(A; q, r)$ by inclusion, the corresponding set of maximal elements is denoted by $\mathcal{N}_\theta^*(A; q)$, $\mathcal{N}_\theta^*(A; q, r)$, respectively.

One easily obtains the following direct extension of Thompson transitivity (Theorem 4.14)

PROPOSITION 4.39. *For every prime $q$, any two elements of $\mathcal{N}_\theta^*(X; q)$ are conjugate by an element of $\theta(C_X(A))$. In particular, Theorem 4.38 holds if $\theta(C_X(a))$ is a $q$-group for every $a \in A^{\#}$.*

One of the key results in the proof is verification of the theorem in the *solvable* case. Goldschmidt proves

PROPOSITION 4.40. *If $X$ is a solvable $p'$-group invariant under $A$, then $\theta$ is complete. In particular, if $X = \langle X \cap \theta(C_X(a)) | a \in A^{\#} \rangle$, then $X \in \mathcal{N}_\theta(A)$.*

The next result, which is not difficult, allows one to argue on $|X|$.

PROPOSITION 4.41. *If $X$ is a minimal counterexample to Theorem 4.38, then $\theta(X; A)$ normalizes no nontrivial element of $\mathscr{M}_\theta(A)$.*

Now we impose the restriction $m_p(A) \geqslant 4$. Goldschmidt uses Proposition 4.40 *in the rank $\geqslant 3$ case* to construct other signalizer functors from the given $\theta$. Indeed, let $\pi(\theta)$ be the set of prime divisors of $|\theta(C_X(a))|$ as $a$ ranges over $A^\#$. For each prime $q \in \pi(\theta)$, and each $D \leqslant A$ with $D \cong E_{p^2}$, set

$$\Delta_q^\theta(D) = \bigcap_{d \in D} O_{q'}(\theta(C_X(d)));$$

and for $a \in A^\#$, set

$$\theta_q(C_X(a)) = \langle \theta(C_X(a)) \cap \Delta_q^\theta(D) | E_{p^2} \cong D \leqslant A \rangle.$$

Goldschmidt proves

PROPOSITION 4.42. *For each $q \in \pi(\theta)$, $\theta_q$ is a solvable $A$-signalizer functor on $X$ and $q \notin \pi(\theta_q)$. In particular, $|\pi(\theta_q)| < |\pi(\theta)|$.*

Clearly the proposition allows Goldschmidt to argue by induction on the size of $|\pi(\theta)|$ in proving the solvable signalizer functor theorem and to conclude that the theorem holds for each $\theta_q$. This immediately implies (i) of the following results; (ii) is easily established.

PROPOSITION 4.43. *For each $q \in \pi(\theta)$, we have*

(i) $\theta_q(X; A) \in \mathscr{M}_\theta(A)$.
(ii) $\theta(X; A)$ *normalizes* $\theta_q(X; A)$.

Combined with Proposition 4.41, this yields

PROPOSITION 4.44. *If $X$ is a minimal counterexample to Theorem 4.38, then $\theta_q$ is trivial for each $q \in \pi(\theta)$, i.e., $\Delta_q^\theta(D) = 1$ for all $E_{p^2} \cong D \leqslant A$ and all $q \in \pi(\theta)$.*

On the other hand, Proposition 4.39 implies

PROPOSITION 4.45. *If $X$ is a minimal counterexample to Theorem 4.38, then $|\pi(\theta)| \geqslant 2$.*

Clearly to establish the theorem, one must show that Propositions 4.44 and 4.45 are mutually incompatible—in other words, the nontriviality of $\theta$ must imply the nontriviality of some $\Delta_q^{\theta}(D)$. Goldschmidt's proof provides an excellent illustration of the ideas behind Bender's simplification of the uniqueness results of the odd-order theorem and of the general group-theoretic technique that it fostered, now known as the "Bender method" (to be further discussed in Section 4.8), so it will be worthwhile to explain Goldschmidt's argument.

Let $q, r \in \pi(\theta)$ with $q \neq r$ (Proposition 4.45) and let $Q \in \text{И}_q^*(A; q)$, $R \in \text{И}_r^*(A; r)$. Also let $Z_q$, $Z_r$ be minimal $A$-invariant subgroups of $\Omega_1(Z(Q))$, $\Omega_1(Z(R))$, respectively, so that $A$ acts irreducibly on $Z_q$ and $Z_r$, regarded as vector spaces over $GF(q)$, $GF(r)$, respectively. Set $A_q = C_A(Z_q)$ and $A_r = C_A(Z_r)$; then by Theorem 1.2, $A/A_q$ and $A/A_r$ are cyclic. Since $m_p(A) \geqslant 4$, there is thus $a \in A^{\#}$ centralizing $Z_q$ and $Z_r$, so as $\theta$ is an $A$-signalizer functor, $\langle Z_q, Z_r \rangle \leqslant \theta(C_X(a))$. Using Theorem 4.2, one proves easily that $\langle Z_q, Z_r^y \rangle$ is a $\{q, r\}$-group for some $y \in C_X(A) \cap \theta(C_X(a))$ $(\leqslant \theta(C_X(A)))$. In view of Proposition 4.39, we can suppose without loss that $y = 1$, whence $\langle Z_q, Z_r \rangle \in \text{И}_\theta(A; q, r)$. Thus we have

LEMMA 4.46. *$\langle Z_q, Z_r \rangle \leqslant H$ for some $H \in \text{И}_\theta^*(A; q, r)$.*

Typical in Bender's approach is a division into two cases according as the Fitting subgroup $F(H)$ of $H$ is or is not of prime power order. Consider the first possibility and, for definiteness, assume that $F(H)$ is an $r$-group, whence $F(H) = O_r(H)$ and $O_q(H) = 1$. Since $H$ is solvable, it is $r$-constrained, so $Z_q$ does not centralize $O_r(H)$. Now $Z_q \times A_q$ acts on $O_r(H)$ and as $m_p(A_q) \geqslant 3$, Theorem 4.13 implies that for some $E_{p^2} \cong D \leqslant A_q$, $Z_q$ does not centralize $S = C_{O_r(H)}(D)$. However, as $Z_q \leqslant Z(Q)$, one can show, using Proposition 4.39, that $Z_q \leqslant O_{q'q}(\theta(C_X(d)))^*$ for each $d \in D^{\#}$, whence $S_0 = [S, Z_q] \leqslant O_{q'}(\theta(C_X(d)))$ for each $d \in D^{\#}$. Hence $S_0 \leqslant \Delta_q^{\theta}(D)$ by definition of this term. Since $S_0 \neq 1$, Proposition 4.44 is therefore contradicted. Thus we have

LEMMA 4.47. *$F(H)$ is not of prime power order.*

The same argument, in fact, shows that $Z_q \leqslant O_q(H)$ and $Z_r \leqslant O_r(H)$, so that $\langle Z_q, Z_r \rangle \leqslant F(H)$.

It is this, the more difficult case, that involves the essence of the Bender method. The key ingredient is always a "uniqueness" criterion for suitable $\{q, r\}$-subgroups $K$ of the group $X$ under investigation. Some distinguished

*For any group $X$ and any prime $q$, $O_{q'q}(X)$ denotes the preimage in $X$ of $O_q(X/O_{q'}(X))$.

set of subgroups of $X$ is specified (often the set of maximal subgroups of $X$), and $\mathscr{S}$ is partitioned into suitable subsets $\mathscr{S}_i$, $1 \leq i \leq n$. One then proves that if $K \leq Y \in \mathscr{S}$, then $Y \in \mathscr{S}_i$ for a *unique* value of $i$, $1 \leq i \leq n$, and it is this uniqueness assertion that allows one to complete the analysis.

In the present instance the set $\mathscr{S} = \mathscr{M}^*_\theta(A; q, r)$ and the subsets $\mathscr{S}_i$ are the distinct orbits of the elements of $\mathscr{S}$ under conjugation by $\theta(C_X(A))$. Moreover, the "uniqueness subgroups" $K$ of $H$ are the subgroups of the form $K = K_q \times K_r$, where $K_q$, $K_r$ are minimal $A$-invariant subgroups of $\Omega_1(Z(O_q(H)))$, $\Omega_1(Z(O_r(H)))$, respectively. For simplicity of notation, let us assume that $Z_q$, $Z_r$ can be taken for $K_q$, $K_r$, so that $\langle Z_q, Z_r \rangle$ can be taken for $K$. Goldschmidt proves

PROPOSITION 4.48. *Suppose $\langle Z_q, Z_r \rangle \leq H^* \in \mathscr{M}^*_\theta(A; q, r)$. Then $H^*$ and $H$ are elements of the same subset $\mathscr{S}_i$ of $\mathscr{S}$.*

The proof of the proposition is similar in spirit to that of Lemma 4.47 but more intricate.

When combined with Proposition 4.39, the proposition has the following corollary, which is the key fact needed to obtain a final contradiction.

COROLLARY 4.49. *If $a \in A^\#$ centralizes $\langle Z_q, Z_r \rangle$, then $H$ contains a Sylow $q$-subgroup and a Sylow $r$-subgroup of $\theta(C_X(a))$.*

To reach the desired contradiction, note once again that as $A/A_q$ and $A/A_r$ are cyclic with $m_p(A) \geq 4$, there is $E_{p^2} \cong D \leq A_q \cap A_r$, so that $\langle Z_q, Z_r \rangle \leq \theta(C_X(d))$ for each $d \in D^\#$. Hence by Corollary 4.49, $H$ contains a Sylow $q$-subgroup $Q_d$ of $\theta(C_X(d))$ for each such $d$. Since $Z_r \leq O_r(H)$, this implies that $[Z_r, Q_d]$ is an $r$-group. Using the $q$-constraint of the solvable group $\theta(C_X(d))$, this immediately yields that $Z_r \leq O_{q'}(\theta(C_X(d)))$ for each $d \in D^\#$. Thus $Z_r \leq \Delta^\theta_q(D)$, once again contradicting the fact that $\Delta^\theta_q(D) = 1$ by Proposition 4.44.

In dealing with signalizer functors when $A$ is an elementary abelian $p$-group for some *odd* prime $p$, the solvable signalizer functor theorem is insufficient. A few years ago, Lyons and I established the first nonsolvable signalizer functor theorem, imposing suitable conditions on the nonsolvable composition factors of the groups $\theta(C_X(a))$ [139]. That result has been considerably extended by Patrick McBride, who has proved [215]:

THEOREM 4.50. *Let $X$ be a group, $p$ a prime, $A$ an elementary abelian $p$-subgroup of $X$ of rank at least* 3, *and $\theta$ a nonsolvable signalizer functor on $X$. If $\theta(C_X(a))$ is a $K$-group for each $a \in A^\#$, then $\theta$ is complete.*

## 4.4. *k*-Balanced Groups

To make effective use of the signalizer functor theorem to study cores of centralizers of involutions in a group $G$, we clearly need methods for constructing good signalizer functors on $G$.

This is easiest to do when $G$ is a group of noncomponent type—equivalently, a group in which the centralizer of every involution is 2-constrained. Using Thompson's $(A\times B)$-lemma (Proposition 4.9) one easily obtains the following result.

PROPOSITION 4.51. *Let $G$ be a group of noncomponent type and $A$ an elementary abelian 2-subgroup of $G$. If for each $a\in A^{\#}$, we set*

$$\theta(C_a)=O(C_a),$$

*then $\theta$ is an $A$-signalizer functor on $G$.*

Indeed, each $O(C_a)$ is an $A$-invariant subgroup of $C_a$ of odd order, so we need only establish the basic compatibility condition (3.13) with $O$ as $\theta$. By symmetry, it suffices to prove that

$$D=O(C_x)\cap C_y\leqslant O(C_y) \tag{4.9}$$

for $x, y\in A^{\#}$. Set $C=C_x\cap C_y$ and also let $Q$ be an $A$-invariant Sylow 2-subgroup of $O_{2'2}(C_y)$. Then clearly $C_Q(x)\leqslant C_x\cap C_y=C$. But by its definition, $D\lhd C$ and $|D|$ is odd, so $D\leqslant O(C)$. Hence, $[D, C_Q(x)]\leqslant O(C)$ and consequently

$$[D, C_Q(x)] \text{ has odd order.} \tag{4.10}$$

On the other hand, if we set $\bar{C}_y=C_y/O(C_y)$, then $\bar{Q}=O_2(\bar{C}_y)$ by definition of $Q$ and so $[\bar{D},\bar{Q}]\leqslant\bar{Q}$. In particular,

$$[\bar{D}, C_{\bar{Q}}(\bar{x})] \text{ is a 2-group.} \tag{4.11}$$

Together (4.10) and (4.11) yield

$$[\bar{D}, C_{\bar{Q}}(\bar{x})]=1. \tag{4.12}$$

Thus $\bar{D}\times\langle\bar{x}\rangle$ acts on $\bar{Q}$ with $\langle\bar{x}\rangle$ a 2-group, $\bar{D}$ a 2′-group and $\bar{D}$ acting trivially on the fixed points of $\bar{x}$ on $\bar{Q}$. Hence by the $(A\times B)$-lemma, $\bar{D}$ acts trivially on $\bar{Q}$. However, as $G$ is of noncomponent type, $C_y$ is 2-constrained, so $C_{\bar{C}_y}(\bar{Q}) \leqslant \bar{Q}$. Thus $\bar{D} \leqslant \bar{Q}$, forcing $\bar{D}=1$ (as $|\bar{D}|$ is odd). We conclude that $D \leqslant O(C_y)$, giving the desired conclusion (4.9).

If, in addition, $G$ is simple and connected of 2-rank at least 3, the signalizer functor theorem shows immediately that the assumptions of Proposition 4.36 are satisfied. Hence so are the assumptions of Theorem 4.35. We can therefore conclude that $O(C_t)=1$ for every $t \in \mathscr{I}(G)$. This shows the power of the signalizer functor method. Furthermore, once one knows that cores of centralizers are trivial, one can continue the analysis, again with the aid of the $(A \times B)$-lemma, to obtain the stronger conclusion that $F^*(H)=O_2(H)$ for every 2-local subgroup $H$ of $G$—equivalently, $H$ is 2-constrained and $O(H)=1$. Hence a connected simple group of noncomponent type of 2-rank $\geqslant 3$ is necessarily of characteristic 2 type, precisely the conclusion of Theorem 1.40.

The proof of the proposition yields as a corollary:

Corollary 4.52. *If $G$ is a group of noncomponent type, then for any pair of commuting involutions $a, b \in G$, we have*

$$O(C_a) \cap C_b = O(C_b) \cap C_a. \tag{4.13}$$

Definition 4.53. A group $G$ that satisfies (4.13) for every $a, b \in \mathscr{I}(G)$ with $[a, b]=1$ is said to be *balanced*.

Thus the statement that $G$ is balanced is equivalent to the assertion that $\theta = O$ defines a signalizer functor on any elementary abelian 2-subgroup of $G$. Hence the results of the preceding section, when rephrased in terms of balanced groups, yield the following extension of Theorem 1.40.

Theorem 4.54. *Let $G$ be a balanced group with $O(G)=1$ and $m_2(G) \geqslant 3$. If either $G$ or a Sylow 2-subgroup of $G$ is connected, then $O(C_t)=1$ for every involution $t$ of $G$.*

If $G$ is of noncomponent type, then $L(C_x/O(C_x))=1$ for every $x \in \mathscr{I}(G)$. When such layers are nontrivial, we cannot expect $G$ to be balanced in general. Indeed, this need not be the case even when $G$ is of known type. Thus the question of balance is related to properties of these components. We shall make this connection more precise.

Definition 4.55. *A simple group $K$ is said to be locally balanced provided whenever $K \leqslant H \leqslant \mathrm{Aut}(K)$ and $x \in \mathcal{I}(H)$, we have $O(C_H(x))=1$.*

The following result shows the breadth of this concept for $K$-groups.

Proposition 4.56. *A simple group $K$ is locally balanced provided* (a), (b), *or* (c) *holds*:

(a) $K \in Chev(2)$ *with* $K \not\cong L_2(4)$ *or* $L_3(4)$.
(b) $K \cong A_n$ *with $n$ even*, $n \geqslant 8$.
(c) *$K$ is sporadic with* $K \not\cong He$.

Note that $L_2(4) \cong A_5$ and a transposition of $\Sigma_5$ has $A_2 \times \Sigma_3$ as its centralizer. Also the unitary automorphism of $L_3(4)$ has $U_3(2)$ as its fixed points, $U_3(2)$ being solvable with $O(U_3(2)) \cong Z_3 \times Z_3$. Similarly, using the isomorphism $A_6 \cong L_2(9)$, we check that the case $m=6$ must be excluded in (b). Likewise $He$ possesses an outer automorphism of order 2 whose fixed points are a nonsplit extension of $\Sigma_7$ by $Z_3$.

Definition 4.57. $\mathfrak{L}_p(G) = \{K \mid K$ is a component of $L(C_x/O_{p'}(C_x))$, $x \in \mathcal{I}_p(G)\}$.

Thus the elements of $\mathfrak{L}_p(G)$ are suitable quasisimple sections of $G$. In particular, if $K \in \mathfrak{L}_p(G)$, then $K/Z(K)$ is a simple group. In practice, these simple factors are of known type.

Proposition 4.51 extends quite easily to the case in which the simple factors of the elements of $\mathfrak{L}_2(G)$ are locally balanced. Thus one has:

Proposition 4.58. *If $G$ is a group in which $K/Z(K)$ is locally balanced for every $K \in \mathfrak{L}_2(G)$, then $G$ is balanced.*

This result provides an excellent illustration of the way in which $K$-properties of proper sections of a group $G$ [here the elements of $\mathfrak{L}_2(G)$] are used to obtain general results about local subgroups of $G$ (here centralizers of involutions), so we shall outline the proof.

We use the same notation as in Proposition 4.51, and again it suffices to establish (4.9), which in turn will follow provided $\bar{D} \leqslant \bar{Q} = O_2(\bar{C}_y)$, where again $\bar{C}_y = C_y/O(C_y)$. First, the same argument as in Proposition 4.51 yields that $\bar{D}$ centralizes $\bar{Q}$. Hence we need only show that $\bar{D}$ also centralizes the layer $\bar{L} = L(\bar{C}_y)$. Indeed, in that case, $\bar{D}$ will centralize $\bar{F}^* = F^*(\bar{C}_y) = \bar{L}\bar{Q} = L(\bar{C}_y)O_2(\bar{C}_y)$, whence $\bar{D} \leqslant \bar{F}^*$ by Proposition 1.27. Thus $\bar{D} \leqslant C_{\bar{F}^*}(\bar{F}^*) \leqslant \bar{Q}$.

(This is a typical use of Proposition 1.27 in general local analysis; it shows very effectively how the verification of local properties of subgroups are reduced to questions about layers.)

Recall that $D \leqslant O(C)$, where $C = C_x \cap C_y$, so $[D, C]$ has odd order. But it is easily seen that $C$ covers $C_{\bar{L}}(\bar{x})$, which immediately implies that

$$[\bar{D}, C_{\bar{L}}(\bar{x})] \text{ has odd order.} \tag{4.14}$$

This is a rather strong conclusion; indeed, using it, one obtains by a direct argument the following two assertions:

(a) $\bar{D}$ leaves invariant each component of $\bar{L}$.
(b) $\bar{D}$ centralizes those components of $\bar{L}$ that are either centralized by $\bar{x}$ or not left invariant by $\bar{x}$. (4.15)

Thus, to complete the proof, it remains to show that $\bar{D}$ centralizes every component $\bar{K}$ of $\bar{L}$ that is normalized but not centralized by $\bar{x}$. By (4.15a), also $\bar{D}$ normalizes $\bar{K}$. Since $\bar{D}\langle\bar{x}\rangle = \bar{D} \times \langle\bar{x}\rangle$ is a group, it follows that $\bar{H} = \bar{K}\bar{D}\langle\bar{x}\rangle$ is a subgroup of $\bar{C}_y$. Set $H^* = \bar{H}/C_{\bar{H}}(\bar{K})$ and denote by $Y^*$ the image in $H^*$ of any subgroup or subset $\bar{Y}$ of $\bar{H}$. We must prove that $D^* = 1$, for then $\bar{D}$ will centralize $\bar{K}$.

Since $\bar{K}$ is quasisimple, $K^* \cong \bar{K}/Z(\bar{K})$ is simple. Furthermore, it is not difficult to show that $C_{H^*}(K^*) = 1$. [The argument is given following the statement of the three subgroups lemma (Proposition 4.19).] Hence $H^*$ can be identified with a subgroup of $\mathrm{Aut}(K^*)$. Moreover, $x^*$ is an involution as $\bar{x}$ does not centralize $\bar{K}$. In addition, as $D \leqslant O(C_x)$, it follows from easily established general results about images of cores, first, that $\bar{D} \leqslant O(\bar{C}_y(\bar{x}))$, whence $\bar{D} \leqslant O(C_{\bar{H}}(\bar{x}))$ (as $\bar{H} \leqslant \bar{C}_y$) and, second, that $D^* \leqslant O(C_{H^*}(x^*))$. However, by hypothesis, $K^*$ is *locally balanced*, so by definition $O(C_{H^*}(x^*)) = 1$. Thus $D^* = 1$, as required,

It turns out that the nonlocally balanced simple $K$-groups satisfy conditions related to local balance. Moreover, these conditions hold for arbitrary primes and not only for the prime 2. To state them, we require a preliminary definition.

Definition 4.59. If $X$ is a group and $A$ an elementary abelian $p$-subgroup of $X$, $p$ a prime, we set

$$\Delta_X(A) = \bigcap_{a \in A^{\#}} O_{p'}(C_X(a)).$$

[Thus if $|A| = p$, $\Delta_X(A) = O_{p'}(C_X(a))$, where $\langle a\rangle = A$.]

DEFINITION 4.60. A simple group $K$ is said to be *locally k-balanced* for the prime $p$, $k$ a positive integer, provided whenever $K \leqslant H \leqslant \mathrm{Aut}(K)$ and $A$ is an elementary abelian $p$-subgroup of $H$ of rank $k$, we have $\Delta_H(A)=1$.

Thus local balance is exactly the same as local 1-balance for the prime 2. Note that if $K$ is locally $k$-balanced, then it is also locally $m$-balanced for all $m \geqslant k$. Hence the groups in Proposition 4.56 are locally $m$-balanced for all positive $m$ (for the prime 2).

Seitz has recently completed the analysis of local $k$-balance for odd primes for the groups of Lie type [250]. The corresponding results for the alternating groups are easy to verify. On the other hand, in the case of the sporadic groups, O'Nan has prepared a systematic list of their various local properties [231], from which Lyons has been able to check local $k$-balance for odd primes. Likewise local $k$-balance for the prime 2 can be established from the already determined structure of the centralizers of involutions acting on simple $K$-groups.

THEOREM 4.61. *If $K$ is a simple $K$-group, then either $K$ is locally 2-balanced for the prime $p$ or one of the following holds:*

(i) $K \cong L_p(q)$, $p \mid q-1$.
(ii) $K \cong U_p(q)$, $p \mid q+1$.
(iii) $K \cong A_n$, *where either $p$ is odd, $n=sp^k+r$ with $2 \leqslant r \leqslant p-1$, or $p=3$ and $n=sp^k+4$, or $p=2$ and $n=sp^k+3$.*
(iv) $p=5$ *and* $K \cong M(22)$.

One also has:

PROPOSITION 4.62. *If $K \cong L_p(q)$, $U_p(q)$, or $M(22)$, then $K$ is locally 3-balanced for the prime $p$.*

The groups $A_n$ are exceptional. Indeed, if $n=sp^k+r$, with either $p$ odd and $2 \leqslant r \leqslant p-1$ or $p=2$ and $r=3$, and we take $H \cong \Sigma_n$; then $H$ contains an elementary abelian $p$-subgroup $A$ of rank $k$ which acts semiregularly on $sp^k$ letters and fixes the remaining $r$ letters (i.e., each element of $A^{\#}$ fixes none of the $sp^k$ letters). Then, for each $a \in A^{\#}$, one checks that $O_{p'}(C_H(a))=Y$, where $Y \cong \Sigma_r$, the symmetric group on the $r$ letters fixed by $A$. Hence $Y=\Delta_H(A)$ if $p$ is odd, while $O_3(Y)=\Delta_H(A)$ if $p=2$; and so $\Delta_H(A) \neq 1$. Thus $A_n$ is not locally $k$-balanced for such a value of $n$.

Just as local balance on components implies global balance, the same is true of local $k$-balance.

DEFINITION 4.63. A group $G$ is said to be *k-balanced* for the prime $p$ if whenever $A$ is an elementary abelian $p$-subgroup of $G$ of rank $k$ and $b$ an element of $G$ of order $p$ which centralizes $A$, we have

$$\Delta_G(A) \cap C_b \leqslant O_{p'}(C_b).$$

It follows that balance is the same as 1-balance for the prime 2. We have the following extension of Proposition 4.58.

PROPOSITION 4.64. *If $G$ is a group in which $K/Z(K)$ is locally $k$-balanced for the prime $p$ for every element of $\mathfrak{L}_p(G)$, then $G$ is $k$-balanced for the prime $p$.*

Just as balance leads immediately to the existence of signalizer functors, so also one can construct signalizer functors when $G$ is $k$-balanced for the prime $p$.

PROPOSITION 4.65. *Let $G$ be a group that is $k$-balanced for some prime $p$ and let $A$ be an elementary abelian $p$-subgroup of $G$ with $m_p(A) \geqslant k+2$. Suppose that $O_{p'}(C_a)$ is a $K$-group (for example, solvable) for each $a \in A^{\#}$. If for $a \in A^{\#}$, we set*

$$\theta(C_a) = \langle \Delta_G(D) \cap C_a | D \leqslant A, m_p(D) = k \rangle,$$

*then $\theta$ is an $A$-signalizer functor on $G$.*

Because of the assumption of $k$-balance, each $\theta(C_a) \leqslant O_{p'}(C_a)$, so $\theta(C_a)$ is an $A$-invariant $p'$-subgroup of $C_a$. Also if $k=1$, it is immediate that each $\theta(C_a) = O_{p'}(C_a)$, whence $G$ is balanced for $p$ and so $O_{p'}$ is an $A$-signalizer functor. If $k \geqslant 2$ and each $O_{p'}(C_a)$ is solvable, one uses Proposition 4.40 to prove that $\theta$ is an $A$-signalizer functor on $G$. McBride has shown that these arguments extend to the general case of the proposition.

The theory of $k$-balance (and $L$-balance in the next section) which Walter and I developed [144], has had a great many applications to the study of $p'$-cores of centralizers of elements of order $p$. Furthermore, the reader will undoubtedly have noted the close relationship between Proposition 4.65 in the case $k=2$ and the definitions of the subsidiary signalizer functors $\theta_q$ which Goldschmidt introduced in the course of his solvable rank $\geqslant 4$ signalizer functor theorem (Theorem 4.38).

We have discussed the case of groups of noncomponent type and, more generally, balanced groups (Theorems 4.54 and 1.40, which in turn depend upon Theorem 4.35) in considerable detail because one can see very clearly in those cases how the signalizer functor method works to "kill" cores of centralizers of involutions. The general study of $p'$-cores of centralizers is much more complicated, even, for example, under the assumption that $G$ is 2-balanced for the prime $p=2$. Indeed, in that case, one would clearly try to extend the proof of Theorem 4.35, using Proposition 4.65 in place of Proposition 4.51. However, to carry this out smoothly, $G$ must have 2-rank at least 4 and must also satisfy a stronger form of connectedness (called "3-connectedness"). If these conditions are satisfied and if the 2-balanced signalizer functor is nontrivial, one can conclude exactly as in Theorem 4.35 that $G$ possesses a proper 3-*generated core*. But as no classification of such groups exists at present, further analysis is needed to reach the stronger conclusion that $G$, in fact, possesses a proper 2-generated core, so that Theorem 4.28 can be applied. Furthermore, even if one ultimately does manage to reach a contradiction from the nontriviality of the 2-balanced functor, one ends up with the following conclusion:

$$\Delta_G(D)=1 \qquad \text{for every four subgroup } D \text{ of } G. \tag{4.16}$$

Hence one is still left with the task of determining the implications of (4.16) for the structure of the cores of the centralizers of involutions.

However, in view of Theorem 4.61 and Proposition 4.64, even the assumption of 2-balance is too restrictive for the general analysis. To deal with some of the difficulties that may arise, Goldschmidt has introduced some variations of the 2-balanced signalizer functor which have turned out to be very effective for many general classification problems. These functors are defined in terms of commutators of suitable subgroups of $A$ on $O_{p'}(C_a)$ for $a\in A^{\#}$. However, in addition to or in place of local balance assumptions on the components of $L(C_a/O_{p'}(C_a))$, they require assumptions on the embedding of $A$ in $C_a$ for $a\in A^{\#}$.

We shall give two examples, limited to the case $p=2$. The first was used by Goldschmidt in the proof of his "product fusion" theorem (see Section 4.9) and later by Richard Foote [104], Ronald Solomon [265], and others. It depends on the following definition.

DEFINITION 4.66. Let $A$ be an elementary subgroup of $G$ of order 16 and write $A=A_1\times A_2$, where $A_1$, $A_2$ are four groups. We say that $G$ is *core-separated* with respect to the given decomposition provided that for any $a\in A^{\#}$ and any component $L$ of $L(C_a/O(C_a))$, either $A_1$ or $A_2$ centralizes $L$.

PROPOSITION 4.67. *Suppose $G$ is core-separated with respect to the decomposition $A=A_1\times A_2$ of the elementary subgroup $A$ of $G$ of order 16. If for $a\in A^{\#}$, we set*

$$\theta(C_a)=\bigcap_{i=1,2}[O(C_a),A_i](O(C_G(A_i))\cap O(C_a)),$$

*then $\theta$ is an $A$-signalizer functor on $G$.*

Note that no local balance conditions are required in this case, but only the core-separation property. The proof again depends on elementary generational statements (concerning the action of $A$ on subgroups of $G$ of odd order). Some time earlier Harada and I, in our study of groups with Sylow 2-groups of the form dihedral$\times$dihedral, had used an analogous functor, but our proof depended upon specific properties of the components of $L(C_a/O(C_a))$. Goldschmidt's realization that such conditions were superfluous was the basis for his completely general product fusion theorem.

The second example was used by Aschbacher in the proof of his fundamental classical involution theorem [13]. It depends on the following definition.

DEFINITION 4.68. A simple group $K$ is said to be *strongly locally 2-balanced* with respect to the four subgroup $A\leqslant \mathrm{Aut}(K)$ provided that for any subgroup $H$ of $\mathrm{Aut}(K)$ containing $KA$, we have $\Delta_H(A)=1$ and $[O(C_H(a)),A]=1$ for each $a\in A^{\#}$.

This commutator condition obviously holds if $K$ is locally balanced, but for groups of Lie type of odd characteristic or alternating groups of odd degree, it depends very much upon the embedding of $A$ in $\mathrm{Aut}(K)$. However, if $K\cong L_2(q)$, $q$ odd, $q$ not a Fermat or Mersenne prime or 9, it fails for *every* choice of $A$.

PROPOSITION 4.69. *Let $A$ be an elementary subgroup of $G$ of order 8. For each pair of elements $a,a'\in A^{\#}$, assume*

(a) *$a'$ centralizes all but at most one component $L$ of $L(C_a/O(C_a))$.*
(b) *If $L$ exists and $\langle a\rangle=C_A(L)$, then $L/Z(L)$ is strongly locally 2-balanced with respect to $A/\langle a\rangle$.*

*Under these conditions, if $B$ is a fixed four subgroup of $A$ and for $a\in A^{\#}$, we set*

$$\theta(C_a)=[O(C_a),B](O(C_G(B))\cap C_a),$$

*then $\theta$ is an $A$-signalizer functor on $G$.*

Note that $A$ leaves $L$ invariant and $A/\langle a \rangle$ is, in fact, a four subgroup of $\mathrm{Aut}(L/Z(L))$ in (b). The proof of the proposition is very similar to that of Proposition 4.67. Ordinarily, the assumption of local 2-balance on components yields a signalizer functor only when $A$ has rank at least 4 (cf. Proposition 4.65). The significance of Aschbacher's result is that one obtains a functor even in the rank 3 case provided one has strong local 2-balance and the embedding assumption (a) of the proposition.

## 4.5. $L$-Balance

So far the discussion has focused on $p'$-cores of centralizers of elements of order $p$ in a group $G$. The elements $\mathfrak{L}_p(G)$ have entered only in relation to a reduction of the question whether $G$ is $k$-balanced for $p$ to properties of the elements of $\mathfrak{L}_p(G)$. However, for the further analysis of the structure of the centralizers of elements of order $p$, it is important to determine the relationship between those elements of $\mathfrak{L}_p(G)$ that occur in $L(C_a/O_{p'}(C_a))$ and those that occur in $L(C_b/O_{p'}(C_b))$ for commuting elements $a$, $b$ of order $p$ in $G$. We shall describe this relationship in the present section. We need some preliminary results and definitions [143, 144].

PROPOSITION 4.70. *Let $X$ be a group, $\pi$ a set of primes, and set $\bar{X} = X/O_{\pi'}(X)$. If $\bar{K}$ is a component of $L(\bar{X})$, then $X$ contains a unique subnormal subgroup $K$ which is minimal subject to mapping on $\bar{K}$. $K$ is perfect and $K$ has no proper normal subgroups of $\pi'$-index.*

DEFINITION 4.71. If $X$, $\bar{X}$, $\bar{K}$, and $K$ are as in the proposition, we call $K$ a *$\pi$-component* of $X$. The product of all $\pi$-components of $X$ is called the *$\pi$-layer* of $X$ and is denoted by $L_{\pi'}(X)$. For completeness, we set $L_{\pi'}(X) = 1$ if $X$ has no $\pi$-components (equivalently, if $L(\bar{X}) = 1$).

PROPOSITION 4.72. *For any group $X$ and any set of primes $\pi$, we have*

(i) *The $\pi$-components of $X$ are characterized as the set of minimal perfect subnormal subgroups of $X$ which are not $\pi'$-groups.*
(ii) *Each $\pi$-component of $X$ is normal in $L_{\pi'}(X)$.*
(iii) *Every element of $X$ induces by conjugation a permutation of the set of $\pi$-components of $X$.*

Propositions 4.70 and 4.72 are quite straightforward and their proof requires only elementary properties of finite groups.

In local analysis, there are two particular cases which are of primary importance: $\pi=\{p\}$ for some prime $p$ and $\pi=\{2'\}$. Correspondingly we speak of *p-components* and *2′-components*, *p-layer* and *2′-layer*, and we write $L_{p'}(X)$, $L_2(X)$ for $L_{\{p\}}(X)$, $L_{\{2'\}}(X)$, respectively. 2-components are basic for the study of centralizers of involutions in simple groups of component type and both $p$-components, $p$-odd, and $2'$-components in the analysis of groups of characteristic 2 type.

In the case $p=2$, a 2-component $K$ of $X$ has the property that $K/O_{2'}(K)=K/O(K)$ is quasisimple; while a $2'$-component $K$ has the property that $K/O_2(K)$ is quasisimple. Moreover, in both cases, $K$ is perfect. In particular, 2-components are never 2-constrained, while $2'$-components are either 2-constrained or quasisimple. Of course, if $X$ is 2-constrained, it has no quasisimple subnormal subgroups, so in that case $2'$-components are necessarily 2-constrained.

We first consider the case of $p$-components. We are interested in the following general question:

*If $P$ is a $p$-subgroup of the group $X$, $p$ a prime, what can we say about the embedding of $L_{p'}(C_X(P))$ in $X$?*

In the case $p=2$, we have the following fundamental result, which is one form of what has come to be called *L-balance* [144].

THEOREM 4.73. *For any group $X$ and any 2-subgroup $T$ of $X$, we have*

$$L_{2'}(C_X(T))\leqslant L_{2'}(X).$$

We outline the proof, which relies ultimately on a deep theorem of Glauberman concerning the automorphism group of a finite group which will be described in the next section (Theorem 4.103). Apart from this use of Glauberman's result, the proof is quite elementary.

Using induction, one easily reduces to the case $O(X)=1$. Now assume false and choose $T$ of maximal order violating the desired conclusion. Set $K=L_{2'}(C_X(T))$, so that $K$ is perfect and $K/O(K)$ is a product of quasisimple groups. Since the theorem fails for $K$, $K\nleqslant L(X)$. Let $J$ denote the product of the 2-components of $K$ which do not lie in $L(X)$, so that $J\neq 1$. Set $Y=L(X)O_2(X)$. Since $Y/L(X)$ is a 2-group, no 2-component of $J$ is contained in $Y$. We conclude at once now from the structure of $J$ that

$$J\cap Y\leqslant O_{2'2}(J).$$

Furthermore, Proposition 1.27 yields

LEMMA 4.74. *No 2-component of $J$ centralizes $Y$.*

Now we apply Glauberman's theorem; in view of Lemma 4.74, it yields:

PROPOSITION 4.75. *$J$ does not centralize a Sylow 2-subgroup of $Y$.*

We use the maximality of $T$ to contradict this conclusion. Let $Q \in Syl_2(N_Y(T))$ and set $R=QT$, so that $R$ is a 2-group. Now $K$ and hence $J$ are invariant under $N_Y(T)$, so as $[J,T]=1$ and $Y \lhd X$,

$$[J,R]=[J,Q] \leqslant J \cap Y \leqslant O_{2'2}(J).$$

Thus, if $\bar{J}=J/O(J)$, it follows that $[\bar{J},R] \leqslant O_2(\bar{J})$. Thus $R$ centralizes $\bar{J}/O_2(\bar{J})$. However, as $J$ is a product of 2-components, $\bar{J}$ is semisimple and so $O_2(\bar{J}) \leqslant Z(\bar{J})$. Since $J$ is perfect, the argument following Proposition 4.19 (the three subgroups lemma) implies that $R$ centralizes $\bar{J}=J/O(J)$.

Now set $I=L_{2'}(C_J(R))$. Since $R$ centralizes $J/O(J)$, $I$ covers $J/O(J)$, i.e., $J=O(J)I$, and we see that none of the 2-components of $I$ lie in $Y$. But as $J$ is a product of 2-components of $C_X(T)$, $I$ is, in fact, a product of 2-components of $C_X(R)$. Thus $J \leqslant L_{2'}(C_X(R))$ and so the theorem fails for $R$. Maximality of $T$ now forces $T=R$.

Finally $R=QT \in Syl_2(N_Y(T)T)$, so $T \in Syl_2(N_Y(T)T)$. Hence by Corollary 1.11, $T \in Syl_2(YT)$. We thus conclude that $T \cap Y \in Syl_2(Y)$. However, $J$ centralizes $T \cap Y$, so $J$ centralizes a Sylow 2-subgroup of $Y$, giving the desired contradiction.

As a corollary of Theorem 4.73, one obtains the more standard form of $L$-balance.

COROLLARY 4.76. *If $a, b$ are commuting involutions of the group $G$, then*

$$L_{2'}(L_{2'}(C_a) \cap C_b) = L_{2'}(L_{2'}(C_b) \cap C_a).$$

No analogue of Glauberman's theorem has been proved for odd primes. However, such an analogue does hold for simple $K$-groups. Hence the proof of Theorem 4.73 can be extended to yield the following two results, which are known as *$L_{p'}$-balance*. (Thus $L$-balance is the same as $L_{2'}$-balance.)

THEOREM 4.77. *If $X$ is a $K$-group and $P$ a $p$-subgroup of $X$ for any prime $p$, then we have*

$$L_{p'}(C_X(P)) \leq L_{p'}(X).$$

COROLLARY 4.78. *Let $G$ be a group in which all $p$-local subgroups are $K$-groups for some prime $p$. If $a, b$ are commuting elements of order $p$ in $G$, then*

$$L_{p'}(L_{p'}(C_a) \cap C_b) = L_{p'}(L_{p'}(C_b) \cap C_a).$$

One can give a slightly sharper form of this last result (and likewise of Corollary 4.76).

THEOREM 4.79. *Let $G$ be a group in which all $p$-local subgroups are $K$-groups for some prime $p$ and let $a, b$ be commuting elements of order $p$ in $G$. Let $J$ be a $p$-component of $L_{p'}(L_{p'}(C_a) \cap C_b)$ and let $K$ be the normal closure of $J$ in $L_{p'}(C_b)$. Then we have*

(i) *$K$ is either a single $\langle a \rangle$-invariant $p$-component of $C_b$ or $K$ is a product of $p$ $p$-components cycled by $a$.*
(ii) *$J$ is a $p$-component of $L_{p'}(C_K(a))$.*

Theorem 4.79 is the key result. Here is how it is used: one starts with a $p$-component $I$ of $L_{p'}(C_a)$, where $a$ is an element of order $p$ of the group $G$ under investigation, and a second element $b$ of order $p$ in $C_a$ which acts on $I$. It is then immediate that $L_{p'}(C_I(b))$ is a product of $p$-components of $L_{p'}(L_{p'}(C_a) \cap C_b)$, so Theorem 4.79 applies to any $p$-component $J$ of $L_{p'}(C_I(b))$. One refers to the normal closure $K$ of $J$ in $L_{p'}(C_b)$ as the "pump-up" of $J$ in $C_b$. The theorem gives us a very good hold on this pump-up. Indeed, for simplicity consider the case in which $K$ is a single $\langle a \rangle$-invariant $p$-component of $C_b$ and set $\bar{C}_b = C_b / O_{p'}(C_b)$. Then $\bar{K}$ is a quasisimple group and $\bar{J}$ is a $p$-component of $L_{p'}(C_{\bar{K}}(\bar{a}))$. When $\bar{K}$ is a $K$-group (which will be the case in practice), it will follow from the known structure of the centralizers of elements of order $p$ acting on such groups that the group $\bar{J}$ is actually quasisimple. Hence, in fact,

$$\bar{J} \text{ is a component of } L(C_{\bar{K}}(\bar{a})).$$

Now if $I$ is known at the outset of the analysis, then so likewise is $J$. But then we see that our quasisimple $K$-group $\bar{K}$, acted on by the element $\bar{a}$

of order $p$, is such that $C_{\bar{K}}(\bar{a})$ has the specified group $\bar{J}$ as a component. From the table of centralizers of elements of order $p$ in the known simple groups, we can now identify the possibilities for $\bar{K}$ (in most instances, $\bar{K}$ is uniquely determined). Thus, beginning with the $p$-component $I$ of $L_{p'}(C_a)$, Theorem 4.79 provides us with a general method for determining certain $p$-components of the centralizers in $G$ of other elements of order $p$ in $C_a$, which are closely linked to the original $p$-component $I$ of $C_a$. This pumping-up process is of fundamental importance for analyzing the centralizers of elements of order $p$ in simple groups.

A second question of a similar nature is also important for the analysis; namely, if $b$ is again an element of order $p$ in $C_a$, what can we say about the embedding of $O_{p'}(C_a)\cap C_b$ in $C_b$? When $G$ is 1-balanced for $p$, we know that $O_{p'}(C_a)\cap C_b \leqslant O_{p'}(C_b)$; however, we cannot expect such a strong conclusion in general. Nevertheless, by arguments similar to those that establish the above results, we can give a rather precise description of this embedding. We need a further definition.

DEFINITION 4.80. If $X$ is a group with $O_{p'}(X)=1$ for some prime $p$, define $L^*_{p'}(X)=L(X)O_{p'}(C_X(P))$, where $P\in Syl_p(L(X)O_p(X))$. Since all choices for $P$ are conjugate by elements of $L(X)$ (by Sylow's theorem), it is immediate that $L^*_{p'}(X)$ is determined independently of the choice of $P$. For an arbitrary group $X$, define $L^*_{p'}(X)$ to be the complete inverse image in $X$ of $L^*_{p'}(X/O_{p'}(X))$.

THEOREM 4.81. *If $X$ is a $K$-group and $P$ is a $p$-subgroup of $X$ for any prime $p$, then we have*

$$O_{p'}(C_X(P)) \leqslant L^*_{p'}(X).$$

COROLLARY 4.82. *Let $G$ be a group in which all $p$-local subgroups are $K$-groups for some prime $p$. If $a, b$ are commuting elements of $G$ of order $p$, then*

$$O_{p'}(C_a)\cap C_b \leqslant L^*_{p'}(C_b).$$

Robert Gilman was the first to prove an $L$-balance theorem for certain types of $2'$-components [107]. His result has been considerably extended by Foote [105], so that it is now a key technical tool in the analysis of groups of characteristic 2 type which possess a proper "characteristic generated core" (see Section 4.13 for the definition).

To state Foote's result, we need the notion of an "Aschbacher" block.

DEFINITION 4.83. A group $X$ is called a *weak block* provided that it satisfies the following conditions:

(1) $X = O^2(X)$ (i.e., $X$ has no normal subgroups of index 2).
(2) $F^*(X) = O_2(X)$.
(3) $X/O_2(X)$ is either quasisimple or of prime order.
(4) *If* $1 = V_0 < V_1 < \cdots < V_n = O_2(X)$ is a sequence of normal subgroups of $X$ such that $V_i/V_{i-1}$ is elementary abelian and irreducible as an $X$-module, then $X$ acts nontrivially on $V_i/V_{i-1}$ for exactly *one* value of $i$, $1 \leqslant i \leqslant n$.

Moreover, $X$ is called a *block* (or *Aschbacher block*) if $X$ is a weak block in which $X$ acts trivially on $O_2(X)/\Omega_1(Z(O_2(X)))$ [hence in the case of blocks, the distinguished member $V_i$ of the series is necessarily contained in $\Omega_1(Z(O_2(X)))$ and we have $[O_2(X), X] \leqslant \Omega_1(Z(O_2(X)))$ (the reader should be cautioned that this definition of block has no connection with Brauer's notion of a block of irreducible characters of a group)].

For the applications, the notion of block suffices; however, Foote's result holds for the wider class of weak blocks.

THEOREM 4.84. *Let $L$ be a $2'$-component of the group $X$ such that $L$ is a weak block and a $K$-group. If $a$ is an involution of $X$ and $K$ is a $2'$-component of $C_a$ such that $K$ is a weak block, then one of the following holds*:

(i) *$K$ centralizes $L$.*
(ii) *$K \leqslant L$.*
(iii) *$L^a \neq L$ and $K = \{xx^a | x \in L\}$ (i.e., $K$ is a "diagonal" of $LL^a$).*

This is a close analogue of Theorem 4.77. The reason for not stating it solely in terms of $2'$-layers is that it is difficult to describe the action of the $2'$-component $K$ of $L_2(C_a)$ on $O_2(X)$. We note also that the only $K$-group property of $L$ needed for the proof is that $L/O_2(L)$ has a solvable outer automorphism group.

The significance of Aschbacher blocks for the study of groups of characteristic 2 type will be described in Section 4.13.

## 4.6. $p$-Fusion

The term *p-fusion*, due to Brauer, refers to the conjugacy in a group $G$ of subsets of a Sylow $p$-subgroup of $G$; specifically, two such subsets are *fused* in $G$ if they are conjugate by an element of $G$. This is a topic with an extensive history, having its origins in the transfer homomorphism and classical Burnside and Frobenius normal $p$-complement theorems (Theorem 1.20). There have been many extensions of their results, which provide conditions for a group $G$ to contain a normal subgroup of index $p$, for example, Thompson's transfer lemma (Proposition 1.21). When such results are applied to a simple group $G$, their effect is to force restrictions on the structure of a Sylow $p$-subgroup of $G$.

These transfer-type results deal with the "top" of a group. There is a second, fundamental theorem of Glauberman, known as the *$Z^*$-theorem* [110], which deals with its base (Proposition 4.95 below). The $Z^*$-theorem is a beautiful generalization of a theorem of Brauer and Suzuki [48], which asserts that a core-free group with quaternion Sylow 2-subgroups necessarily has a center of order 2, and hence is not simple (Theorem 4.88 below). Together, Bender's strong embedding theorem and Glauberman's $Z^*$-theorem undoubtedly constitute the two most important tools of local analysis.

The $Z^*$-theorem has itself undergone many important generalizations in recent years. These extensions will be described in subsequent sections. But we emphasize that just as the proof of the $Z^*$-theorem makes explicit use of the Brauer–Suzuki theorem, so all these extensions invoke the $Z^*$-theorem to cover a minimal case. We should also mention that the $Z^*$-theorem represents the single result of general local analysis that requires Brauer's theory of modular characters.

However, we begin with Alperin's fundamental fusion theorem, which asserts that in any finite group, $p$-fusion is determined $p$-locally in a very precise sense [1]. (See also [130, Section 7.2].) There have been a number of refinements of Alperin's theorem; we shall state Goldschmidt's [119].

DEFINITION 4.85. Let $G$ be a group, $P$ a Sylow $p$-subgroup of $G$, and $\mathfrak{D}$ a set of subgroups of $P$. If $A$, $B$ are nonempty subsets of $P$ and $g \in G$, we say that $A$ is *$\mathfrak{D}$-conjugate* to $B$ *via* $g$ if there exist subgroups $D_1, D_2, \ldots, D_n$ in $\mathfrak{D}$ and elements $g_1, g_2, \ldots, g_n$ in $G$ such that

(1) $g_i \in N_G(D_i)$, $1 \leqslant i \leqslant n$.
(2) $A \leqslant D_1$ and $A^{g_1 \cdots g_i} \leqslant D_{i+1}$, $1 \leqslant i \leqslant n-1$.
(3) $g = g_1 g_2 \cdots g_n$ and $A^g = B$.

Furthermore, we call $\mathfrak{D}$ a *conjugation family* (for $P$ in $G$) provided that whenever $A$ and $B$ are nonempty subsets of $P$ which are conjugate in $G$, then $A$ and $B$ are necessarily $\mathfrak{D}$-conjugate via some $g \in G$. When this is the case, one can always choose the subgroups $D_i$ and the elements $g_i$ so that, in addition,

(4) $g_i$ is a $p$-element, $1 \leqslant i \leqslant n-1$, and $D_n = P$.

Thus the conjugating elements can be chosen either to be $p$-elements or to normalize $P$.

THEOREM 4.86. *Let G be a group, P a Sylow p-subgroup of G, and let $\mathfrak{D}$ be the set of subgroups D of P with the following properties*:

(a) $N_P(D) \in Syl_p(N_G(D))$.
(b) $N_G(D)$ *is p-constrained.*
(c) *D maps onto* $O_p(N_G(D)/O_{p'}(N_G(D)))$.
(d) *Either* $D = P$ *or* $N_G(D)/D$ *has a strongly p-embedded subgroup.*

*Then* $\mathfrak{D}$ *is a conjugation family for P in G.*

Thus, in effect, the theorem asserts that any fusion in $G$ of subsets of $P$ can be "factored" into a product of conjugations, each of which occurs in a $p$-local subgroup of $G$ of a very particular shape.

Note that $Z(P) \leqslant N_P(D)$ and $Z(P)$ centralizes $D$. Hence together (b) and (c) imply the additional property:

(e) $Z(P) \leqslant D$.

Now we turn to the Brauer–Suzuki theorem. To state it, we need a definition.

DEFINITION 4.87. For any group $X$, $Z^*(X)$ will denote the complete inverse image in $X$ of $Z(X/O(X))$.

Clearly $Z(X/O(X))$ is an abelian 2-group and $Z^*(X) \triangleleft X$, so by the Feit–Thompson theorem, $Z^*(X) \leqslant Sol(X)$. The $Z^*$-theorem (and the Brauer–Suzuki particular case) gives a sufficient condition for $Z^*(X)$ to contain $O(X)$ properly. We state the Brauer–Suzuki theorem in the following way:

THEOREM 4.88. *If G is a group with* $m_2(G) = 1$, *then* $Z^*(G) > O(G)$. *In particular, if* $|G| > 2$, *then G is not simple.*

The condition $m_2(G) = 1$ implies (by Proposition 1.35) that a Sylow 2-subgroup $S$ of $G$ is either cyclic or quaternion. However, in the first case, as $\Omega_1(S)\cong Z_2$, it is immediate from Proposition 4.7 that $\mathrm{Aut}(S)$ is a 2-group, which means that $N_G(S)=C_G(S)$, whence $S\leqslant Z(N_G(S))$. But then by Burnside's transfer theorem [Theorem 1.20(ii)], $G$ has a normal 2-complement. Thus $G=SO(M)$ and $S\leqslant Z^*(G)$. Hence the theorem is trivial in the cyclic case and so it is really a theorem about groups with quaternion Sylow 2-subgroups $S$.

Brauer and Suzuki's original proof required modular character theory, but subsequently when $|S|>8$, they obtained a proof using only results from ordinary character theory (their proof is reproduced in Chapter 12 of my book [130]). In 1974 Glauberman obtained an ordinary character-theoretic proof for the remaining case $|S|=8$ [115]. However, the argument is much trickier than either the $|S|>8$ case or the modular character-theoretic proof in the $|S|=8$ case. On the other hand, all known proofs of the $Z^*$-theorem itself require modular character theory. Because of this, we have chosen to follow the modular approach in the quaternion case as well, since then there are certain similarities in the arguments.

Indeed, both proofs then depend upon Brauer's so-called *second main theorem*, which relates $p$-blocks of characters of a subgroup $H$ of $G$ to $p$-blocks of characters of $G$ (for any prime $p$) by means of a well-defined correspondence known as the Brauer correspondence (see [84, 126, 182]). In the quaternion and $Z^*$-theorems, the prime $p=2$ and the subgroup $H=C_x$ for a suitable 2-element $x$ of $G$. Moreover, the theorem is applied to the so-called *principal* 2-block $B_0$ of $G$, which by definition is the block containing the trivial character $1_G$ of $G$. The importance of the principal block for Theorem 4.88 as well as the $Z^*$-theorem can be seen from the following fact.

Proposition 4.89. *In any group $X$, if $\chi_1, \chi_2, \ldots, \chi_r$ are the irreducible characters of the principal 2-block of $X$, then*

$$O(X)=\bigcap_{i=1}^{r}\ker(\chi_i).$$

We shall outline the proof of Theorem 4.88 when $S$ is quaternion of order 8 but shall omit all character-theoretic technicalities. We let $G$ be a counterexample of least order. It follows easily that $O(G)=1$ and $G$ does not have a normal 2-complement. We shall argue that the unique involution $t$ of $S$ lies in $Z(G)$, whence $G$ is not a counterexample.

Write $S=\langle x, y\rangle$ with $x^4=y^4=1$, $x^2=y^2=t$, and $y^{-1}xy=x^{-1}$. Note that all subgroups of $S$ of order 4 are cyclic.

LEMMA 4.90. *All subgroups of $S$ of order 4 are conjugate in $N_G(S)$.*

Indeed, as $G$ does not have a normal 2-complement, Frobenius's theorem [Theorem 1.20(i)] implies that $N_G(T)/C_G(T)$ is not a 2-group for some subgroup $T$ of $S$. But if $T<S$, Aut$(T)$ is a 2-group, so we must have $T=S$ and $N_G(S)$ must contain a 3-element that cyclically permutes the three (cyclic) subgroups of $S$ of order 4 under conjugation.

It is not difficult to prove

LEMMA 4.91. *The principal 2-block of $G$ consists of seven irreducible characters $1_G=\chi_0,\chi_1,\chi_2,\dots,\chi_6$.*

Brauer's second main theorem is used to establish relations among the character values of these $\chi_i$, $0\leqslant i\leqslant 6$. Ultimately Brauer and Suzuki prove

PROPOSITION 4.92. *For a suitable choice of $\chi_1$, $\chi_2$, and $\chi_3$, and suitable signs $\delta_1,\delta_2=\pm 1$, we have*

(i) $1+\delta_1\chi_1(t)-\delta_2\chi_2(t)=0$.
(ii) $1+\delta_1\chi_1(1)+\delta_2\chi_2(1)=0$.
(iii) $1+\delta_1[\chi_1(t)^2/\chi_1(1)]+\delta_2[\chi_2(t)^2/\chi_2(1)]=0$.

The proposition has the following key consequence:

PROPOSITION 4.93. *We have $\chi_1(t)=\chi_1(1)$.*

Indeed, $\chi_i(1)=\deg(\chi_i)$ is a positive integer, so $\delta_2=-\delta_1$ by (ii), whence by (i) and (ii),

$$\chi_2(1)=\delta_1+\chi_1(1)\quad\text{and}\quad\chi_2(t)=-\delta_1-\chi_1(t).$$

Substituting in (iii), we obtain

$$0=1+\delta_1\frac{\chi_1(t)^2}{\chi_1(1)}-\delta_1\frac{[-\delta_1-\chi_1(t)]^2}{[\delta_1+\chi_1(1)]}.$$

Clearing denominators, this gives

$$\begin{aligned}0=&\,\delta_1\chi_1(1)+\chi_1(1)^2+\chi_1(t)^2+\delta_1\chi_1(t)^2\chi_1(1)\\&-\delta_1\chi_1(1)-2\chi_1(1)\chi_1(t)-\delta_1\chi_1(t)^2\chi_1(1),\end{aligned}$$

whence

$$0=\chi_1(1)^2+\chi_1(t)^2-2\chi_1(1)\chi_1(t),$$

and so

$$0=[\chi_1(t)-\chi_1(1)]^2,$$

forcing $\chi_1(t)=\chi_1(1)$.

Now we can obtain a final contradiction. First, Proposition 1.4(i) yields now that $t\in\ker(\chi_1)$. Since $\chi_1\neq 1_G$, $N=\ker(\chi_1)<G$. Set $T=S\cap N$, so that $T\in Syl_2(N)$. If $T$ is cyclic, then again by Burnside's transfer theorem, $N$ has a normal 2-complement. But $O(N)\leqslant O(G)=1$, so $N=T$ and hence $t\in Z(G)$, as required. In the contrary case, $T=S$ and $N$ has quaternion Sylow 2-subgroups. Since $N<G$, the theorem holds for $N$ and as $O(N)=1$, it follows that $t\in Z(N)$. By Sylow's theorem, $t$ is thus the unique involution of $N$ and as $N\lhd G$, we conclude at once that $t\in Z(G)$ in this case as well.

If $G$ is a group with $m_2(G)=1$ and $S\in Syl_2(G)$, then $S$ contains only one involution $z$. Obviously then $S-\langle z\rangle$ has no involutions and so $z$ cannot fuse in $G$ to an element of $S-\langle z\rangle$. Glauberman realized that it was this property rather than the precise structure of the Sylow 2-subgroup which underlay the character-theoretic analysis of Brauer and Suzuki. Thus the following definition is needed to state his $Z^*$-theorem.

DEFINITION 4.94. Let $G$ be a group and $S$ a Sylow 2-subgroup of $G$. An involution $z$ of $S$ is said to be *isolated* in $S$ (with respect to $G$) if $z$ is not conjugate in $G$ to any element of $S-\langle z\rangle$.

Clearly then $z$ is not conjugate in $S$ to any element of $S-\langle z\rangle$, so an isolated involution necessarily lies in $Z(S)$.

THEOREM 4.95. ($Z^*$-THEOREM). *Let $G$ be a group and $S$ a Sylow 2-subgroup of $G$. If $z$ is an isolated involution of $S$, then $z\in Z^*(G)$.*

Again our discussion will omit modular character-theoretic technicalities. The first step is a purely group-theoretic consequence of isolation.

PROPOSITION 4.96. *If $S\in Syl_2(G)$ and $t\in\mathfrak{I}(S)$ is isolated in $S$ with respect to $G$, then $[g,t]=g^{-1}tgt$ has odd order for all $g\in G$.*

Indeed, let $w$ be the product $t^g t$. To establish the proposition, we must show that $|w|$ is odd. Since $w$ is the product of two involutions, it is inverted by both $t$ and $t^g$ by Proposition 2.42 (as $t$ and $t^g$ are involutions). Let $|w| = 2^a \cdot n$ with $n$ odd and set $r = \frac{1}{2}(n+1)$, so that $r$ is an integer. Since $t^g$ inverts $w$, and $t^g = (t^g)^{-1}$ as $t$ is an involution, we have

$$t^g w^r t^g = w^{-r}.$$

Hence

$$t^{gw^r} t = (w^{-r} t^g w^r) t = w^{-r} (t^g w^r t^g) t^g t = w^{-r} (w^{-r}) w = w^{1-2r} = w^{-n}. \quad (4.17)$$

Since $(w^{-n})^{2^a} = 1$, $w^{-n}$ is a power of 2. But $w^{-n}$ is the product of the two involutions $t^{gw^r}$ and $t$, so again by Proposition 2.42, $\langle t, t^{gw^r} \rangle$ is a 2-group.

Hence, by Sylow's theorem, $t^x$ and $t^{gw^r x}$ are in $S$ for some $x \in G$. Since each is conjugate to $t$ in $G$, it follows therefore from the isolation of $t$ that

$$t^x = t^{gw^r x} = t,$$

whence

$$t = t^{gw^r}.$$

But then by (4.17),

$$w^{-n} = t^{gw^r} t = t^2 = 1,$$

whence $w^n = 1$. Since $n$ is odd, we conclude that $|w|$ is odd, as required.

As usual, we begin with a counterexample $G$ to Theorem 4.95 of least order. In particular, a Sylow 2-subgroup of $G$ is not quaternion by the Brauer–Suzuki theorem, so we have

LEMMA 4.97. *$S - \langle t \rangle$ contains an involution.*

Using the minimality of $G$ and the isolation of $t$, one easily obtains

LEMMA 4.98.

(i) *$O(N) = 1$ for any normal subgroup $N$ of $G$.*
(ii) *$t \in Z(S)$.*
(iii) *$C_u < G$ for any $u \in \mathscr{I}(S)$.*

The first modular character-theoretic result we need is a general "orthogonality" property concerning the values of the irreducible characters in a given block. We need it only for the principal 2-block $B_0$ of $G$. Let then $1_G = \chi_0, \chi_1, \ldots, \chi_r$ be the irreducible characters in $B_0$.

PROPOSITION 4.99. *Let $x, y \in G$ with $x$ an involution and either $y=1$ or $y$ an involution not conjugate to $x$ in $G$. Then*

$$\sum_{i=0}^{r} \chi_i(x)\chi_i(y)=0.$$

Brauer's second main theorem is used in the proof of the next result (which also depends upon the preceding four lemmas and propositions).

PROPOSITION 4.100. *If $u$ is an involution of $S-\langle t\rangle$, then for any $x, y \in G$ and any $i$, $1 \leqslant i \leqslant r$, we have*

$$\chi_i(t^x u^y)=\chi_i(tu).$$

Using this result together with multiplication in the complex group ring* of $G$, Glauberman is now able to determine the value of $\chi_i$ on $t$ for suitable $i$, namely,

PROPOSITION 4.101. *Let $u$ be an involution of $S-\langle t\rangle$. If $\chi_i(u)\neq 0$, $1 \leqslant i \leqslant r$, then $\chi_i(t)=-\chi_i(1)$ [i.e., $\chi_i(t)=-\deg(\chi_i)$].*

Now Propositions 4.99 and 4.101 yield an immediate arithmetic contradiction. Indeed, fix an involution $u \in S-\langle t\rangle$. By Proposition 4.101 for each $i$, $1 \leqslant i \leqslant r$, we have

$$\chi_i(u)\left[\chi_i(t)+\chi_i(1)\right]=0, \tag{4.18}$$

since one of the two factors is 0. On the other hand, $1_G(x)=1$ for all $x \in G$, so

$$1_G(u)\left[1_G(t)+1_G(1)\right]=2. \tag{4.19}$$

Summing (4.18) over all $i$ and adding (4.19), we thus obtain

$$\begin{aligned} 2 &= \sum_{i=0}^{r} \chi_i(u)\left[\chi_i(t)+\chi_i(1)\right] \\ &= \sum_{i=0}^{r} \chi_i(u)\chi_i(t)+\sum_{i=0}^{r} \chi_i(u)\chi_i(1). \end{aligned} \tag{4.20}$$

*By definition, the *group ring* $G[\mathbb{C}]$ of the group $G$ is a vector space over $\mathbb{C}$ with the elements $g_1, g_2, \ldots, g_n$ of $G$ as a basis and the product of two elements $g=\sum_i c_i g_i$, $c_i \in \mathbb{C}$, and $g'=\sum_j c_j' g_j$, $c_j' \in \mathbb{C}$, of $G[\mathbb{C}]$ given by the rule

$$gg'=\sum_{i,j} c_i c_j'(g_i g_j).$$

But $u$ and $t$ are not conjugate in $G$ as $t$ is isolated, so each of the two sums in (4.20) is 0 by Proposition 4.99. Thus $2=0$, which is absurd and the theorem is proved.

I think the reader will agree that the Brauer–Suzuki and $Z^*$-theorems are wonderful illustrations of the power, depth, and elegance of Brauer's theory of modular characters and blocks of characters.

It is difficult to overestimate the importance of the $Z^*$-theorem in the study of simple groups. As a striking example, we mention Glauberman's theorem, establishing a special case of the celebrated Schreier conjecture, which asserts that the outer automorphism group of every simple group is solvable. The full conjecture is, in fact, a corollary of the classification of simple groups inasmuch as each known simple group has a solvable outer automorphism group. But Glauberman's theorem constitutes the only progress ever made on a direct attack on the conjecture.

THEOREM 4.102. *Let $G$ be a group with $O(G)=1$ (in particular, simple) and let $S\in Syl_2(G)$. Then $C_{\mathrm{Aut}(G)}(S)$ has abelian Sylow 2-subgroups and a normal 2-complement. In particular, $C_{\mathrm{Aut}(G)}(S)$ is solvable.*

Set $C=C_{\mathrm{Aut}(G)}(S)$ and consider the semidirect product of $G$ and $C$. By definition, $C$ centralizes $S$, so if $T\in Syl_2(C)$ and we let $z$ be an involution of $Z(T)$, it follows that $z\in Z(ST)$. Observe that $z\notin Z^*(G\langle z\rangle)$. Indeed, in the contrary case, as $O(G)=1$, $z\in Z(G\langle z\rangle)$, whence $z$ acts trivially on $G$, contrary to the fact that $z$ is an automorphism of $G$ of period 2. Hence, by the $Z^*$-theorem, there is $g\in G$ such that $z^g\in S\langle z\rangle$ with $z^g\neq z$.

This is the situation that Glauberman must analyze. If $T$ is not abelian, he can choose $z=[a,b]$ for suitable $a,b\in T$. On the other hand, if $T$ is abelian, but $C$ does not have a normal 2-complement, it follows from Burnside's transfer theorem [Theorem 1.20(ii)] that $z$ can be chosen so that for some $x\in N_C(T)$ of odd order, $z^x\neq z$. In both cases, Glauberman uses Alperin fusion to "decompose" the conjugacy $z^g\neq z$ and combines this with the fact that $a$, $b$, and $x$ centralize $S$ to reach a contradiction. The argument itself is fairly straightforward; Glauberman's brilliance here lies rather in his realization that the $Z^*$-theorem had a strong implication for the automorphism group of a finite group.

We conclude this section with an application of several of the above ideas by outlining a theorem of Alperin concerning simple groups of 2 rank 2 [4]. We remark that, in general, the analysis of 2-fusion in a simple group of low 2-rank yields very strong conclusions about the possible structures of its Sylow 2-subgroups.

THEOREM 4.103. *If $G$ is a simple group of 2-rank 2 and $S$ a Sylow 2-subgroup of $G$, then $S$ is either dihedral, quasi-dihedral, wreathed, or of type $U_3(4)$ [i.e., isomorphic to a Sylow 2-subgroup of $U_3(4)$].*

We can suppose that $S$ is not dihedral or quasi-dihedral, so that $S$ contains a normal four subgroup $V$ by Proposition 1.35(iv). Since $V \lhd S$, $V$ contains an involution $z \in Z(S)$. We fix such an element $z$.

By the $Z^*$-theorem, there is $g \in G$ and $t \in S - \langle z \rangle$ such that $t^g = z$. Now let $\mathfrak{D}$ be the conjugation family of Theorem 4.87. We conclude from the theorem that there exists $D \in \mathfrak{D}$ and $x \in N_G(D)$ such that

$$u = z^x \neq z.$$

Since $C_S(D) \leqslant D$ by definition of $\mathfrak{D}$, we have $z \in D$, whence $z \in Z = \Omega_1(Z(D))$. Hence also $u = z^x \in Z$ as $x$ normalizes $D$ and $Z \operatorname{char} D$. Thus $Z$ is noncyclic. Since $m_2(D) \leqslant m_2(S) = 2$, we conclude that

$$Z = \Omega_1(D) = \langle z, u \rangle. \tag{4.21}$$

Now let $N = N_G(V)$, $C = C_G(V)$, and $T = S \cap C$. We have $S \in Syl_2(N)$ as $V \lhd S$. Thus $T \in Syl_2(C)$ and $|S:T| \leqslant 2$ (the latter as $N/C$ is isomorphic to a subgroup of $\Sigma_3$).

We now consider two cases according as $u \in T$ or $u \notin T$. If $u \in T$, then $\langle u, V \rangle$ is elementary; and again as $m_2(S) = 2$, it follows that $u \in V$, whence $V = Z$. Thus $N_G(D)$ normalizes $V$ by (4.21) and so $N_G(D) \leqslant N$. Hence $u$ is conjugate to $z$ in $N$, so certainly $N \neq SC$. This implies that $N/C \cong Z_3$ or $\Sigma_3$. Now a Frattini argument yields that $N_N(T)$ contains a 3-element $y$ which transitively permutes the involutions of $V$. Since $V = \Omega_1(T)$, $y$ thus transitively permutes the involutions of $T$ and so G. Higman's theorem [170] is applicable. In the present case, it yields that $T$ is either homocyclic abelian or of type $U_3(4)$.

If $T = V$, then either $S = V$ or $S$ is dihedral of order 8, contrary to assumption; so if $T$ is homocyclic abelian, then necessarily $|T| \geqslant 16$. We conclude easily in this case that either $S$ is wreathed or $S = T$. In the latter case, a Sylow 2-subgroup of $G$ has the form $Z_{2^n} \times Z_{2^n}$ for some $n > 2$. However, a theorem of Brauer (which is proved by means of modular character theory) asserts that no simple group has a Sylow 2-subgroup of this form [43]. Thus the theorem holds if $T$ is homocyclic abelian. Likewise it holds if $S = T$ is of type $U_3(4)$. On the other hand, if $S > T$ and $T$ is of type $U_3(4)$, an analysis of the automorphism group of $T$ leads to a contradiction. Thus the theorem holds if $u \in T$.

Suppose finally that $u \notin T$. Then $S = T\langle u \rangle$ and $T < S$, so $D = E\langle u \rangle$, where $E = D \cap T$. Also $u \notin E$. But $u \in Z \leqslant Z(D)$, and consequently $D = E \times \langle u \rangle$. Since $m_2(D) \leqslant 2$, it follows that $m_2(E) = 1$, whence $E$ is either cyclic or quaternion. But $z \in D \cap T = E$, so $\langle z \rangle = \Omega_1(E)$. If $|E| \geqslant 4$, it is immediate that $\langle z \rangle \operatorname{char} D = E \times \langle u \rangle$. Clearly then $z$ and $u$ cannot be conjugate in $N_G(D)$. Hence we must have $|E| = 2$, whence $D = \langle z, u \rangle$ is a four group. Since $C_S(D) \leqslant D$, we conclude now from Proposition 1.16 and Theorem 1.17 that $S$ is either dihedral or quasi-dihedral, contrary to assumption.

The analysis of 2-fusion in a group $G$ usually does not use the full force of simplicity, but only the fact that $G$ has no isolated involution and no normal subgroups of index 2. Thus it applies to the wider class of "fusion-simple" groups.

Definition 4.104. A group $G$ is said to be *fusion simple* provided $G$ has no normal subgroups of index 2 and no isolated involution.

For example, Theorem 4.102 is easily extended to the fusion-simple case; the only distinction turns out to be that $Z_{2^n} \times Z_{2^n}$, $n \geqslant 2$, must be added to the list of possibilities for $S$. (But as a result, Brauer's theorem [43] is no longer needed. Of course, it is still needed to *classify* fusion simple groups with such Sylow 2-subgroups.)

## 4.7. Stability and Characteristic Subgroups for Odd Primes

In Corollary 4.18, we have described a basic result of the odd-order paper for primes $p$ for which $m_p(G) \geqslant 3$, namely, each $P \in Syl_p(G)$ has a unique maximal signalizer $W$ in $G$. Pursuing the analysis further, along the lines of the proof of Theorem 4.35, Thompson was able to establish the corresponding assertion:

$$\Gamma_{P,2}(G) \leqslant N_G(W) \leqslant G. \tag{4.22}$$

Hence, in the case $W \neq 1$, it follows (as $G$ is simple) that $G$ possesses a proper 2-generated $p$-core. (Of course, as no analogue of the Bender–Aschbacher classification theorem exists for odd $p$, this represented only the first-level analysis of the complete odd-order problem.)

On the other hand, when $W = 1$ (or as we would now say, when the signalizer functor is trivial), there is no obvious candidate for a *proper* subgroup of $G$ having the property that it contains $\Gamma_{P,2}(G)$. Thompson's first proof of the existence of such a subgroup was extremely difficult.

Shortly thereafter, he discovered a brilliant, conceptual proof (it is this proof which appears in the final paper), based on so-called *factorization lemmas*, which have had a profound impact on the development of simple group theory and which introduced for the first time what is now called the *Thompson* subgroup of a $p$-group. The proof then went through a second simplification, based on Glauberman's $ZJ$-theorem [130, Theorem 8.2.11], which showed that for odd primes, the so-called "Thompson subgroup" had certain remarkable properties, which allowed one to dispense with factorization lemmas altogether.

However, no satisfactory analogue of Glauberman's theorem is known to hold for the prime 2, whereas factorization lemmas carry over in many cases. As a result, we have the ironic twist of fate that the ideas developed for studying *odd* local subgroups have come to play a fundamental role in the analysis of the 2-local structure of groups of characteristic 2 type.

We leave the discussion of factorization lemmas until Section 4.11 and focus here on Glauberman's achievements for odd primes. The starting point is the concept of $p$-stability that Walter and I introduced to establish an extension of Thompson's odd-order uniqueness results, applicable to the study of groups with dihedral Sylow 2-subgroups.

DEFINITION 4.105. Let $p$ be an odd prime and $X$ a group with $O_p(X)=1$. A faithful representation $\phi$ of $X$ on a vector space $V$ over $GF(p)$ is *linearly* $p$-stable if no $p$ element of $\phi(X)$ has a *quadratic* minimal polynomial in its action on $V$.

We say that $X$ is *linearly p-stable* if each such faithful representation of $X$ is linearly $p$-stable.

The natural representation of $SL(2, p^n)$ is not linearly $p$-stable. (We have noted this fact in Section 4.1 in the case $p^n=3$.) This definition is useful only for odd primes; for if $p=2$, every element of $\phi(X)$ of order $p$ is an involution and so necessarily has a quadratic minimal polynomial.

Walter and I, using a result of Dickson's on the generation of the group $SL(2, p^n)$ [130, Theorem 2.8.4] proved the following result.

THEOREM 4.106. *Let p be an odd prime and X a group with* $O_p(X)=1$. *If X is not linearly p-stable, then X involves* $SL(2, p)$ [*i.e., some quotient of a subgroup of X is isomorphic to* $SL(2, p)$]. *In particular, a Sylow 2-subgroup of X is neither abelian nor dihedral.*

The second assertion is a consequence of the fact that $SL(2, p)$ has quaternion Sylow 2-subgroups.

We now extend this notion to the $p$-local situation. Let $Q$ be a nontrivial normal $p$-subgroup of the group $X$ and set $\bar{X} = X/\phi(Q)$, so that $\bar{Q} = Q/\phi(Q)$ is elementary abelian. Setting $\tilde{X} = \bar{X}/C_{\bar{X}}(\bar{Q})$, we can view $\tilde{X}$ as a group of linear transformations of $\bar{Q}$, regarded as a vector space over $GF(p)$. If $x$ is a $p$-element of $X$ such that $[Q, x, x] = 1$, then $[\bar{Q}, \tilde{x}, \tilde{x}] = 1$, which in additive notation for $\bar{Q}$ asserts that $\tilde{x}$ satisfies the equation $(X-1)^2 = 0$. Hence if $\tilde{x} \neq 1$, then $\tilde{x}$ has a quadratic minimal polynomial on $\bar{Q}$. Thus if also $O_p(\tilde{X}) = 1$, then $\tilde{X}$ is not linearly $p$-stable. These remarks serve as motivation for the following definition.

DEFINITION 4.107. Let $p$ be an odd prime and $X$ a group with $O_{p'}(X) = 1$ and $O_p(X) \neq 1$. We say that $X$ is *p-stable* provided that for every normal $p$-subgroup $Q$ of $X$ and every $p$-element $x$ of $X$ such that $[Q, x, x] = 1$, if we set $\tilde{X} = X/C_X(Q)$, then $\tilde{x} \in O_p(\tilde{X})$. More generally, if $O_{p'}(X) \neq 1$, we say that $X$ is *p-stable* if $X/O_{p'}(X)$ is $p$-stable.

The point of the definition is that if $X$ is not $p$-stable and $O_{p'p}(X) > O_{p'}(X)$, it is easily shown that some section of $X$ is not linearly $p$-stable. Hence we have

THEOREM 4.108. *If $p$ is an odd prime and $X$ is a non-$p$-stable group with $O_{p'p}(X) > O_{p'}(X)$, then $X$ involves $SL(2, p)$.*

The $ZJ$-theorem deals with $p$-constrained $p$-stable groups; its statement involves the Thompson subgroup of a $p$-group, which, in practice, has been defined in several slightly different ways. In the present context, we use the following:

DEFINITION 4.109. If $P$ is a $p$-group, $p$ any prime, let $\mathscr{A}(P)$ be the set of abelian subgroups of $P$ of maximal order and define

$$J(P) = \langle A \mid A \in \mathscr{A}(P) \rangle.$$

Then $J(P)$ is called the *Thompson* subgroup of $P$. It is clearly characteristic in $P$.

For factorization lemmas, it is better to define the Thompson subgroup to be

$$J_e(P) = \langle A \mid A \in \mathscr{A}_e(P) \rangle,$$

where $\mathcal{Q}_e(P)$ denotes the set of *elementary* abelian subgroups of $P$ of maximal order (equivalently, of maximal rank).

As noted, $J(P)$ char $P$. Furthermore, if $Q$ is any subgroup of $P$ containing $J(P)$, it is immediate from the definitions that $J(P)=J(Q)$. This conclusion is basic, for it implies that *$J(P)$ is characteristic in any subgroup of $P$ containing $J(P)$.* $J_e(P)$ has the same properties.

THEOREM 4.110 ($ZJ$-THEOREM). *Let $p$ be an odd prime and $X$ a $p$-constrained group with $O_{p'}(X)=1$. If $X$ is $p$-stable, then $Z(J(P))$ is a characteristic subgroup of $X$. In particular, $Z(J(P))\lhd X$.*

This is a remarkable conclusion since, even with the restriction of $p$-stability, $X$ can be fairly wild. Glauberman's proof of this theorem involves brilliant use of commutator relations. In his subsequent investigations Glauberman has pushed this technique to the status of a fine art.

To exhibit the power of this result, we present an application, which will include the $W=1$ case of the odd-order paper.

THEOREM 4.111. *Let $G$ be a group of characteristic $p$ type in which all $p$-locals are $p$-stable, $p$ an odd prime. If $P\in Syl_p(G)$, then we have*

$$\Gamma_{P,1}(G)\leqslant N_G(Z(J(P))).$$

Thus $N_G(Z(J(P)))$ is strongly $p$-embedded in $G$. If $G$ is simple of odd order with all proper subgroups solvable, it is immediate that all $p$-locals are $p$-constrained and $p$-stable. Moreover, in the $W=1$ situation, it is not difficult to prove that every $p$-local has a trivial $p'$-core as well, so the hypotheses of the theorem are satisfied in this case.

The proof is by contradiction. Set $M=N_G(Z(J(P)))$ and let $H$ be a $p$-local subgroup of $G$ such that $H\not\leqslant M$ with $Q=P\cap H$ of largest possible order. Assuming the theorem is false, $Q\neq 1$. We argue that $Q\in Syl_p(H)$. This is clear if $Q=P$ as $P\in Syl_p(G)$; so we may assume $Q<P$. Then $Q<N_P(Q)$ by Theorem 1.10 and so $N_G(Q)\leqslant M$ by the choice of $H$ and $Q$. In particular, if $Q\leqslant R\in Syl_p(H)$, it follows that $N_R(Q)\leqslant M\cap H$. Since $P\in Syl_p(M)$, this implies that $(N_R(Q))^x\leqslant P$ for some $x\in M$. If $N_R(Q)>Q=P\cap H$, maximality of $H$ and $Q$ then yields that $H^x\leqslant M$, whence $H\leqslant M$, contradiction. Thus $N_R(Q)=Q$ and we conclude from Corollary 1.11 that $Q=R\in Syl_p(H)$. Also, as $G$ is of characteristic $p$ type, $H$ is $p$-constrained with $O_{p'}(H)=1$.

Now Glauberman's theorem implies that $Z(J(Q))\lhd H$, so if we set $N=N_G(Z(J(Q)))$, we have $H\leqslant N$. If $Q=P$, then $N=M$, so $H\leqslant M$, contrary to assumption. Hence $Q<P$ and so $N_P(Q)>Q$, again by Theorem 1.10. But $N_P(Q)$ normalizes $Z(J(Q))$ which is characteristic in $Q$, so $N_P(Q)\leqslant N$. Thus $P\cap N>Q$ and so $N\leqslant M$ by our maximal choice of $H$. Again $H\leqslant M$, contradiction.

With a little further arguing, Glauberman is able to draw additional conclusions, which require the following definition.

DEFINITION 4.112. Let $G$ be a group, $P$ a Sylow $p$-subgroup of $G$ for some prime $p$, and $H$ a subgroup of $G$ containing $P$. Then $H$ *controls p-fusion* in $G$ if two subsets of $P$ conjugate in $G$ are conjugate in $H$; and $H$ *controls p-transfer* in $G$ if the largest abelian $p$-factor group of $G$ is isomorphic to that of $H$.

By the focal subgroup theorem (Theorem 1.19), if $H$ controls $p$-fusion in $G$, it also controls $p$-transfer. However, the converse need not be the case since less information than complete $p$-fusion is required to determine the image of $G$ in $P/P'$ under the transfer homomorphism.

First we give a local version of Glauberman's results.

THEOREM 4.113. *Let $X$ be a p-constrained, p-stable group with $O_{p'}(X)=1$ and let $P\in Syl_p(X)$, p an odd prime. Then $N_X(Z(J(P)))$ controls p-fusion in $X$ and hence also p-transfer. In particular, $X$ has a normal p-complement if and only if $N_X(Z(J(P)))$ does.*

The final assertion follows from control of $p$-transfer together with Frobenius's normal $p$-complement theorem [Theorem 1.20(i)].

Using Theorem 4.113 and induction on $G$, Glauberman obtains the following criterion for a group $G$ to have a normal $p$-complement (cf. [130, Theorem 8.3.1]).

THEOREM 4.114. *Let $P$ be a Sylow p-subgroup of the group $G$, p an odd prime. If $N_G(Z(J(P)))$ has a normal p-complement, then so also does $G$.*

The celebrated Frobenius conjecture, Thompson's proof of which began the whole show, is an immediate consequence of the theorem.

THEOREM 4.115. *If a group $G$ admits an automorphism $\phi$ of prime order q fixing only the identity element of $G$, then $G$ is nilpotent.*

Let $G$ be a minimal counterexample. By Sylow's theorem, $\phi$ leaves invariant a Sylow $q$-subgroup $Q$ of $G$. If $Q\neq 1$, Proposition 1.9 immediately yields that $C_Q(\phi)\neq 1$, contrary to the fact that $\phi$ is fixed-point-free on $G$. Thus $Q=1$ and so $G$ is a $q'$-group. Thus $G$ and $\phi$ have coprime orders.

Consider first the case in which $G$ contains no nontrivial proper characteristic subgroups. Since $G$ is not nilpotent, it is not a 2-group, so $|G|$ is divisible by some odd prime $p$. By our assumption on the nonexistence of characteristic subgroups, $O_{p'}(G)=O_p(G)=1$. Furthermore, as $C_G(\phi)=1$ and $G,\phi$ have coprime orders, Theorem 4.2 implies that $\phi$ leaves invariant some Sylow $p$-subgroup $P$ of $G$. It follows that $\phi$ also leaves $N=N_G(Z(J(P)))$ invariant [as $Z(J(P))$ char $P$]. Since $O_p(G)=1$, $N<G$. But $\phi$ is fixed-point-free on $N$ as it is on $G$, so $N$ must be nilpotent by the minimality of $G$. In particular, $N$ has a normal $p$-complement. Hence by Theorem 4.114, also $G$ has a normal $p$-complement, so $G=PO_{p'}(G)=P$ is nilpotent, contradiction.

Thus we are reduced to the case in which $G$ possesses a nontrivial proper characteristic subgroup $H$. Then $\phi$ acts on both $H$ and $G/H$ and $\phi$ is fixed-point-free on $H$. It is an easy exercise that the same is true for the action of $\phi$ on $G/H$. Hence, again by the minimality of $G$, $H$ and $G/H$ are both nilpotent, so $G$ is solvable. Thus the normal $p$-complement theorem has, in fact, reduced us to the solvable case.

Using the kind of reduction arguments we described in Section 4.1, it is not difficult to argue that $H$ and $G/H$ are elementary abelian $r$- and $s$-groups for suitable primes $r\neq s$, that $\phi$ acts irreducibly on $G/H$ and that the group $X=(G/H)\langle\phi\rangle$ acts irreducibly on $H$. Now applying Clifford's theorem (Theorem 1.5) to the action of $X$ on $H$ relative to the normal subgroup $G/H$ of $X$, one easily concludes that $H$ [regarded as a vector space over $GF(r)$] is the direct sum of subspaces $H_1, H_2,\ldots, H_q$ cyclically permuted by $\phi$. But then if $v_1$ is a nonzero vector of $H_1$, it follows that

$$v=v_1+\phi(v_1)+\cdots+\phi^{q-1}(v_1)$$

is a nontrivial vector of $H$. However, as $\phi^q=1$, we see that $\phi(v)=v$, contrary to $C_G(\phi)=1$.

If $P\in Syl_p(G)$, it is immediate from Alperin's fusion theorem that any subgroup of $G$ containing $\Gamma_{P,1}(G)$ controls $p$-fusion in $G$. Hence as a corollary of Theorem 4.111, we have

THEOREM 4.116. *Let $G$ be a group of characteristic $p$ type in which all $p$-locals are $p$-stable, $p$ an odd prime. If $P\in Syl_p(G)$, then $N_G(Z(J(P)))$ controls $p$-fusion and $p$-transfer in $G$.*

This entire discussion suggests several interesting general questions.

A. If $G$ is a group of characteristic $p$ type and $H$ is a $p$-local subgroup of $G$, under what conditions is $H$ contained in a $p$-local subgroup of $G$ that also contains some Sylow $p$-subgroup of $G$?
B. If $G$ is a group and $P \in Syl_p(G)$, $p$ a prime, is there a "functorially defined" nonidentity characteristic subgroup $K$ of $P$ for which $N_G(K)$ controls $p$-fusion or at least $p$-transfer in $G$?
C. What can be said about the structure of a group $X$ with $O_p(X)=1$, $p$ an odd prime, which is not linearly $p$-stable?

Question A is concerned with the problem of "pushing up" $p$-local subgroups, an essential part of any theorem which asserts that a group $G$ of characteristic $p$ type has a proper 2-generated $p$-core or strongly $p$-embedded subgroup. The primary interest occurs for $p=2$ and the major results will be discussed in Section 4.13.

The general answer to Question B is negative for $p$-fusion, the groups $GL(3, p^n)$ providing an easy counterexample; $p$-stability seems to be an essential requirement. Remarkably, Glauberman has shown that an affirmative answer for $p$-transfer holds under very broad circumstances [117]. In fact, the only restriction is $p \geqslant 5$. The proof of this result involves extremely subtle commutator calculations coupled with tremendous ingenuity. To state it, we require a definition that extends some of the properties of the Thompson subgroup.

DEFINITION 4.117. A *p-conjugacy functor* $K$ on a group $G$, $p$ a prime, is a mapping $K$: $P \mapsto K(P)$ defined on the set of $p$-subgroups $P$ of $G$ with the following properties:

(a) $K(P) \leqslant P$.
(b) *If* $P \neq 1$, *then* $K(P) \neq 1$.
(c) *For any* $g \in G$, $K(P^g)=(K(P))^g$.

It is immediate that the map $P \mapsto J(P)$, $P$ a $p$-subgroup of $G$, defines a $p$-conjugacy functor on $G$.

THEOREM 4.118. *Let $G$ be a group, $p \geqslant 5$ a prime, and $P \in Syl_p(G)$. Then there is a p-conjugacy functor $K$ on $G$ such that $N_G(K(P))$ controls p-transfer in $G$. Moreover,* $K(P)$ char $P$.

Glauberman defines for any $p$-group $P$ two chains of characteristic subgroups of $P$, one increasing and the other decreasing, the terminal member of either of which can be taken as $K(P)$. We shall not attempt to describe them explicitly here, nor shall we discuss the many other important results that Glauberman has established in this direction. We mention only that a corollary of the last theorem, when combined with Burnside's theorem [Theorem 4.130] that a group of order $p^a q^b$ is solvable (applied here in the case $p=2$, $q=3$), proves the following long-standing conjecture.

THEOREM 4.119. *If $G$ is a group in which every Sylow subgroup is its own normalizer, then $G$ is not simple.*

Thompson was the first to study Question C in a general setting (Theorem 4.107 can be viewed as a particular case of C) and he obtained a complete solution when $p \geqslant 5$ [293]. One may as well assume that $X$ itself is a group of linear transformations of a vector space $V$ over $GF(p)$, that $X$ acts irreducibly on $V$, and that $X$ is generated by the set of $p$-elements $D$ of $X$ with quadratic minimal polynomial on $V$. Thompson calls $(X, V)$ a *quadratic pair* under these conditions.

Note that if $x \in D$, then the quadratic condition implies that $x^p = 1$, so $x$ has order $p$.

THEOREM 4.120. *If $(X, V)$ is a quadratic pair for the prime $p \geqslant 5$, then we have*

(i) *$X$ is semisimple with quasisimple components $X_i$, $1 \leqslant i \leqslant m$, $V$ is the tensor product of subspaces $V_i$, $1 \leqslant i \leqslant m$, and $(X_i, V_i)$ is a quadratic pair for each $i$, $1 \leqslant i \leqslant m$.*
(ii) *Each $X_i \in Chev(p)$* [excluding $E_8(p^r)$], $1 \leqslant i \leqslant m$.

Near the beginning of his analysis, Thompson proves the following result, which shows that the hypothesis here is closely related to the Fischer-type situation. For simplicity, we state it only in the case that $D$ is a single conjugacy class of elements of $X$.

PROPOSITION 4.121. *If $x, y \in D$, then either $\langle x, y \rangle$ is a $p$-group, or $\langle x, y \rangle \cong SL(2, p^n)$ for some $n$.*

There are now several proofs of Thompson's quadratic pair theorem, each of which ends up constructing the underlying geometry [equivalently, the $(B, N)$-pair structure] from properties of $D$. Thompson makes explicit use of the module $V$; but Betty Stark has shown [268], using Thompson's sharpened form of Proposition 4.121, that the theorem can be reduced to a purely group-theoretic characterization of the groups of Lie type of odd characteristic by centralizers of involutions (Aschbacher's classical involution theorem [13]; see D13 at the end of Section 1.1).

The case $p=3$ is very interesting since a number of noncharacteristic 3-groups have quadratic pair representations. Thompson's former student Chat Ho has nearly completed the analysis in this case [173–177]. (Also his results include a proof of Thompson's theorem.) The essential difficulty is that the analogue of Proposition 4.121 includes the following groups:

$$\text{3-groups}, \quad SL(2,3^n), \quad SL(2,3)\times Z_3, \quad SL(2,5). \tag{4.23}$$

Thus there are a great number of possibilities for the underlying "geometry" when $p=3$.

In Thompson's analysis, the "root" subgroups play a basic role. We define them here only when $D$ is a single conjugacy class. If $x \in D$, set

$$D_x = \{y \in D \mid V(y-1) = V(x-1)\} \quad \text{and} \quad E_x = D_x \cup \{1\}.$$

Then it is easily shown that $E_x$ is an elementary abelian $p$-group. $E_x$ is called a *root subgroup* of $X$. Since $D$ is a single conjugacy class, $|E_x|$ is independent of the choice of $x \in D$.

Ho has proved

THEOREM 4.122. *If $(X,V)$ is a quadratic pair for $p=3$ and the root subgroups of $X$ have order exceeding 3, then the conclusions of Thompson's quadratic pair theorem hold with $p=3$.*

Thus it is the case of root subgroups of order 3 which leads to groups outside of Chev(3). In this case, if $E=\langle x\rangle$ and $F=\langle y\rangle$ are root subgroups of $X$, then (4.23) can be refined to the statement:

$$\langle E, F\rangle \cong Z_3, Z_3\times Z_3, SL(2,3), SL(2,5), SL(2,3)\times Z_3, \text{ or a nonabelian group of order 27.} \tag{4.24}$$

Ho has proved

THEOREM 4.123. *Let $(X,V)$ be a quadratic pair for $p=3$ with $X$ quasisimple and assume that the root groups of $X$ have order 3. If $\langle E, F\rangle \cong$*

*$SL(2,3)\times Z_3$ for every pair of root groups $E$, $F$ of $X$, then one of the following holds*:

(i) $X \in Chev(3)$ [*excluding* $E_8(3^r)$, $r \geqslant 1$].
(ii) $X \cong \hat{A}_n$, $n > 5$.
(iii) $X \cong U_n(2)$, $Sp(6,2)$, $D_4(2)$, *or* $G_2(4)$.
(iv) $X \cong J_2$, $Suz$, *or* $.1$.

Thus only the $SL(2,3)\times Z_3$ case of root groups of order 3 remains open to complete the classification of all quadratic pairs.

## 4.8. The Bender Method, Small Class Sylow 2-Subgroups, Strong Closure, and the $p^a q^b$-Theorem

The Bender method, which we have illustrated in connection with Goldschmidt's rank 4 solvable signalizer functor theorem (Theorem 4.38), has also led to an alternate approach for studying cores of centralizers of involutions in simple groups. Because the approach depends upon the $Z^*$-theorem and $p$-stability, we have deferred its discussion until those two topics were treated. As previously noted, the Bender method developed out of his simplification of Thompson's odd-order uniqueness theorems. Shortly thereafter Bender applied his ideas to the analysis of simple groups with abelian Sylow 2-subgroups and obtained a dramatic simplification of Walter's earlier classification of such groups [27]. More recently (with the aid of some character-theoretic ideas of Glauberman), he has applied the method to obtain an equally striking simplification of Walter's and my original classification of groups with dihedral Sylow 2-subgroups [32, 33].

The underlying philosophy of the Bender method differs sharply from that of the signalizer functor method. Indeed, the latter is essentially *constructive*—i.e., one constructs a proper subgroup $M$ of $G$, which, if not strongly embedded, at least controls a critical amount of 2-fusion in $G$. In contrast, Bender focuses directly on a maximal subgroup $M$ of $G$ containing $C_G(x)$ for some involution $x$ of $G$. The question of whether $M$ controls suitable 2-fusion related to the conjugacy class of $x$ is then reduced to properties of $F^*(M)$ and its embedding in $G$. The method applies when $F(M)\neq 1$ and $F^*(M)$ has order divisible by at least two primes. It is based on a general uniqueness result, whose proof is quite elementary [involving repeated use of the Thompson (A$\times$B)-lemma (Proposition 4.9)].

To state it, we need a preliminary notion.

DEFINITION 4.124. Let $H$ be a maximal subgroup of the group $G$ such that

(a) $F(H)\neq 1$.
(b) $|F^*(H)|$ is divisible by at least two primes.

For any maximal subgroup $K$ of $G$, write $H\rightsquigarrow K$ if for some nontrivial subgroup $Q$ of $F(H)$, we have $N_{F^*(H)}(Q)\leq K$.

This may look like a weak relationship between the subgroups $H$ and $K$; but that is deceptive, for the following result shows its strong consequences on the structure of $H\cap K$ and includes a criterion for the equality of $H$ and $K$.

PROPOSITION 4.125. *Let $G$ be a simple group and $H, K$ maximal subgroups of $G$ such that $H\rightsquigarrow K$ [so that $F(H)\neq 1$ and $|F^*(H)|$ is divisible by at least two primes]. If $\pi$ denotes the set of primes dividing $|F(H)|$, then*

$$O_\pi(F(K))\leq H \quad \text{and} \quad O_{\pi'}(F(K))\cap H=1.$$

*Furthermore, if in addition $K\rightsquigarrow H$, then $H=K$.*

Clearly the proposition does not cover the case in which either $F(H)=1$ or $F^*(H)=F(H)$ has prime power order. In the applications to groups of odd order and to groups with abelian or dihedral Sylow 2-subgroups, one uses the proposition to reduce to a single residual case, namely, when the maximal subgroup $M$ of $G$ containing $C_G(x)$ satisfies

$$F^*(M)=F(M)=O_p(M) \qquad \text{for some odd prime } p. \tag{4.25}$$

This case requires a special argument. In each of the above three problems, $SL(2, p)$ is not involved in $G$, so all $p$-local subgroups of $G$ are *$p$-stable* by Theorem 4.108, and hence Glauberman's $ZJ$-theorem (Theorem 4.110) can be applied. It yields the powerful conclusion that for $P\in Syl_p(G)$, $N_G(Z(J(P)))$ is the *unique* maximal $p$-local subgroup of $G$ containing $P$, thus providing an effective replacement for Proposition 4.125 under the assumptions of (4.25).

The Bender method is so powerful when it is applicable that there have been attempts (notably by Thompson) to use it in place of signalizer functors for studying cores of centralizers of involutions in arbitrary simple groups. Unfortunately the case in which $F^*(M)$ is a $p$-group for some odd prime $p$ and $M$ is *not* $p$-stable has so far constituted an insurmountable obstacle to this approach.

At the end of the section we shall give a further illustration of the Bender method, proving the Burnside $p^a q^b$-theorem by purely local group-theoretic techniques. The standard classical proof (e.g., see [130, Theorem 4.3.3]) involves character theory but is, of course, much shorter. The present proof is due to Goldschmidt [120], for the case $p$ and $q$ both odd, and Hiroshi Matsuyama [213], who by a clever argument involving involutions showed that Goldschmidt's proof could be extended to the case $p$ or $q=2$. [Subsequently S. Smith and R. Solomon (unpublished) gave a unified treatment of the Goldschmidt–Matsuyama argument. We shall follow their proof here. Bender [30] has also given a proof of the theorem by local methods, but his argument is somewhat longer.]

Note that as every minimal simple group has order divisible by at least three primes by Thompson's $N$-group theorem (see Section 1.1 for its statement), Burnside's theorem is also a corollary of that result. In fact, soon after his classification of $N$-groups, Thompson (unpublished) derived a self-contained local group-theoretic proof of the $p^a q^b$-theorem, using ideas from the $N$-group argument. However, that argument also is more elaborate than the Goldschmidt–Matsuyama proof.

We also note that the present argument does not explicitly use Proposition 4.125; in fact, a careful reading of the proof will hopefully suggest to the reader a basis for establishing the proposition.

But, before turning to the Burnside theorem, we would first like to state the principal classification results to which the Bender method has successfully been applied. (The dihedral theorem has already been stated in Section 1.1). We should emphasize, however, that Bender's method has so far been used only in that portion of the given problem in which the group under investigation is not of characteristic 2 type, other techniques being required to cover the characteristic 2 type case. Furthermore the 2-rank 2 cases of these theorems rely on the corresponding cases of the general classification of simple groups of 2-rank 2, a result which will be fully discussed in the sequel.

THEOREM 4.126. *If $G$ is a simple group with abelian Sylow 2-subgroups, then $G \cong L_2(q)$, $q \equiv 3,5 \pmod 8$, $L_2(2^n)$, $n \geqslant 2$, $J_1$, or ${}^2G_2(3^n)$, $n$ odd, $n>1$.*

The characteristic 2 type case of this problem is very easy, for if $S \in Syl_2(G)$, it is immediate under that assumption, as $S$ is abelian, that $N_G(S)$ is strongly embedded in $G$, so that $G$ is determined directly from Bender's classification of such groups (Theorem 4.24).

Goldschmidt has obtained an important extension of Bender's proof of this result, which simultaneously generalizes Glauberman's $Z^*$-theorem as well as Shult's weak closure theorem [252]. To state it, we need the following definition.

DEFINITION 4.127. Let $G$ be a group, $P$ a Sylow $p$-subgroup of $G$, $p$ any prime, and $Q$ a subgroup of $P$. We say that $Q$ is *strongly closed* in $P$ with respect to $G$ provided that whenever $x \in Q$ and $x^g \in P$ for $g \in G$, we have $x^g \in Q$.

Obviously $P$ itself is strongly closed in $P$. If $\Omega_1(P)$ is elementary abelian, then clearly it consists precisely of the set of all elements of $P$ of order $p$ together with the identity of $P$ and so is strongly closed in $P$ with respect to $G$. In particular, this is the case if $P$ is abelian or if $p=2$ and $P$ is quaternion.

Goldschmidt was interested in the case in which a Sylow 2-subgroup $S$ of a group $G$ contained a nontrivial strongly closed abelian subgroup $A$. (If $S$ is abelian, we can take $A=S$; while if $S$ has an isolated involution $x$, we can take $A=\langle x \rangle$.) Goldschmidt proved [123]

THEOREM 4.128. *Let $G$ be a simple group and $S$ a Sylow 2-subgroup of $G$. If $S$ contains a nontrivial abelian subgroup $A$ which is strongly closed in $S$ with respect to $G$, then either $S$ is abelian or $G$ has a strongly embedded subgroup. Thus $G \cong L_2(2^n)$, $n \geq 2$, $U_3(2^n)$, $n \geq 2$, $Sz(2^n)$, $n$ odd, $n > 1$, $L_2(q)$, $q \equiv 3,5 \pmod 8$, $J_1$, or ${}^2G_2(3^n)$, $n$ odd, $n > 1$.*

The primary difference between the Goldschmidt and the Bender situation is that in the abelian problem *all* proper subgroups of $G$ are $K$-groups, whereas Goldschmidt is only able to assert this inductively for certain proper subgroups of $G$. Namely, if $H < G$ and $B = H \cap A^g \neq 1$ for some $g \in G$, then $\langle B^H \rangle$ is a $K$-group. This makes for a more delicate analysis of the 2-fusion of elements of $A$ and the structure of the corresponding maximal subgroup $M$. We note that Aschbacher's criterion for strong embedding (Theorem 4.31) is invoked at a critical juncture and enables Goldschmidt to conclude that

$$G = \langle C_a | a \in A^{\#} \rangle.$$

There is one other general classification problem to which the Bender method can be applied, namely, in groups with Sylow 2-groups of nilpotency class 2, which Gilman and I have classified–using signalizer functors, however [108]. Although, *a priori*, $SL(2, p)$ can occur as a proper section of a minimal counterexample $G$ in this problem, we were able to eliminate this possibility very easily by showing in such a case that a Sylow 2-subgroup $S$ of $G$ had a nontrivial strongly closed abelian subgroup and then invoked Goldschmidt's theorem above to derive a contradiction. However, we deliberately chose to use the signalizer functor method rather than the Bender approach in the hope of gaining further insight into the general centralizer of involution core problem.

But just as the abelian Sylow 2-group problem is a special case of the strongly closed abelian 2-group problem, so the class 2 Sylow 2-group problem is a particular case of the classification of groups with a strongly closed 2-subgroup of nilpotency class 2. This problem has recently been successfully treated by Peter Rowley [242, 243], using the Bender method and generalizing Goldschmidt's abelian argument. His result is as follows:

THEOREM 4.129. *Let $G$ be a simple group and $S$ a Sylow 2-subgroup of $G$. If $S$ contains a subgroup $A$ of nilpotency class 2 which is strongly closed in $S$ with respect to $G$, (in particular, if $S$ has nilpotency class 2), then $G \cong L_2(q)$, $q \equiv \pm 1 \pmod 8$, $q \not\equiv \pm 1 \pmod{16}$, $U_3(2^n)$, $n \geqslant 2$, $Sz(2^n)$, $n$ odd, $n \geqslant 1$, $L_3(2^n)$, $n \geqslant 1$, $PSp(4, 2^n)$, $n \geqslant 2$, or $A_7$.*

The characteristic 2 type case of this problem is considerably more difficult than in the abelian Sylow 2-group (or strongly closed abelian) situation. Rowley's argument is patterned after Gilman's and mine in the class 2 Sylow 2-group problem [108]. It depends in a critical way upon Thompson "failure of factorization," a topic to be discussed in Section 4.12, and involves an analysis of the maximal 2-local subgroups in $G$ containing $S$. Assuming $S$ possesses no nontrivial strongly closed abelian subgroup, this analysis eventually yields that $S$ lies in exactly two such maximal 2-local subgroups $M$ and $N$, whose structure and embedding in $G$ are precisely determined—indeed, so precisely determined that one can argue that the subgroup $G_0 = \langle M, N \rangle$ is a split $(B, N)$-pair of rank 2. Given the structure of $M$ and $N$ and using the Fong–Seitz classification theorem (Theorem

3.14), the only possibilities are $G_0 \cong L_3(2^n)$ or $PSp(4,2^n)$. The final step is to show that $G=G_0$, thus determining the possibilities for $G$.

Now at last to the Burnside $p^aq^b$-theorem. Besides illustrating the Bender method, the proof provides an excellent example of local group-theoretic analysis in action!

THEOREM 4.130. *Every group of order $p^aq^b$, where $p$ and $q$ are primes, is solvable.*

We consider a minimal counterexample $G$. It is immediate that $G$ is simple with every proper subgroup solvable. In particular, $|G|$ is divisible by both $p$ and $q$. For definiteness, suppose that $p^a<q^b$.

We shall use the following easy result.

LEMMA 4.131. *If $G=AB$ with $A<G$, then $B$ normalizes no nonidentity subgroup of $A$. In particular, a Sylow $p$-subgroup of $G$ normalizes no nonidentity $q$-subgroup of $G$ (and vice versa).*

PROOF. Suppose $B$ normalizes $1\neq X\lhd A$. Then for $g\in G$, as $G=AB=BA$, we have $g=ba$ with $b\in B$ and $a\in A$, so $X^g=X^{ba}=X^a\leqslant A$. Hence $Y=\langle X^g \mid g\in G\rangle\leqslant A<G$. But $Y\lhd G$ and $X\leqslant Y$, thus contradicting the simplicity of $G$.

If $P\in Syl_p(G)$ and $Q\in Syl_q(G)$, then $|G|=|P||Q|$, so $G=PQ$. Thus the second assertion is a particular case of the first.

We now divide the proof into two cases.

*Case 1.* For any maximal subgroup $M$ of $G$, $F(M)$ is of prime power order.

*Case 2.* For some maximal subgroup $M$ of $G$, $F(M)$ is not of prime power order.

We first treat Case 1. (The Bender method will be used in Case 2.) Thus we assume the hypothesis of Case 1 in Lemmas 4.132–4.136. The key result we need is the following:

LEMMA 4.132. *If $P\in Syl_p(G)$, $Q\in Syl_q(G)$, $x\in Z(P)^{\#}$, and $y\in Z(Q)^{\#}$, then $\langle x, y\rangle=G$.*

PROOF. Suppose false, in which case $\langle x, y\rangle\leqslant M$ for some maximal subgroup $M$ of $G$. As we are in Case 1, we can suppose for definiteness that

$F(M)=O_p(M)$. Replacing $P$ and $Q$ by suitable $G$-conjugates, if necessary, we can suppose without loss that $O_p(M)\leqslant P$ and $y\in Y=Z(Q)\cap M$. Set $Z=Z(P)$. Since $M$ is solvable, $C_M(F(M))\leqslant F(M)$, by Proposition 1.25, so $Z\leqslant O_p(M)$. Thus $\langle Z^Y\rangle$ is a $p$-group. Let $X$ be a $Y$-invariant $p$-subgroup of $G$ chosen maximal subject to containing $\langle Z^Y\rangle$ and generated by $G$-conjugates of $Z$. Set $N=N_G(X)$ and let $R\in Syl_p(N)$, so that $X\lhd R$ and $N<G$.

It will suffice to prove that $R\in Syl_p(G)$. Indeed, if that is the case, then $G=RQ=QR$, whence [as $Y\leqslant Z(Q)$], $\langle Y^G\rangle=\langle Y^{QR}\rangle=\langle Y^R\rangle\leqslant N<G$, contradiction. To establish this fact, we need only show that $N_G(R)\leqslant N$, for then $R\in Syl_p(N_G(R))$, in which case $R\in Syl_p(G)$ by Corollary 1.11.

Let $u\in N_G(R)$. We have that $X^u\leqslant R$ and we must prove that $X^u=X$ [in which case $u\in N=N_G(X)$]. But $X$ is generated by $G$-conjugates of $Z$, so it will suffice to show, for any $g\in G$ such that $Z^g\leqslant X$, that $Z^{gu}\leqslant X$.

Set $C=C_G(Z^g)$ and $D=C_G(Y)$. Since $P^g\leqslant C$ and $Q\leqslant D$, we have $G=CD$. Thus we can write $u=cd$ with $c\in C$, $d\in D$. Now $Z^{gu}\leqslant R$ and so normalizes $X$. Hence $XZ^{gu}$ is a group and therefore so is $XZ^{gua}$ for every $a\in Y$ (as $X$ is $Y$-invariant). Thus $W=\langle XZ^{gu}\rangle^Y=\langle XZ^{gua}\,|\,a\in Y\rangle=X\langle Z^{gua}\,|\,a\in Y\rangle=X\langle Z^{guY}\rangle$. But as $c$ centralizes $Z^g$ and $d$ centralizes $Y$, $V=\langle Z^{guY}\rangle=\langle Z^{gdY}\rangle=\langle Z^{gYd}\rangle=\langle Z^{gY}\rangle^d$. Since $Z^g\leqslant X$ with $X$ invariant under $Y$, it follows that $V$ is conjugate under $d$ to a $p$-group and so is itself a $p$-group. However, $W=XV$, and $|W|=|X||V|/|X\cap V|$, so also $W$ is a $p$-group. Furthermore, by its definition $W$ is $Y$-invariant and is generated by $G$-conjugates of $Z$. Since $X\leqslant W$, we conclude now from our maximal choice of $X$ that $W=X$. Since $Z^{gu}\leqslant W$, this yields the desired inclusion $Z^{gu}\leqslant X$, and the lemma is proved.

As an immediate corollary, we obtain

LEMMA 4.133. *If $Q\in Syl_q(G)$ and $y\in Z(Q)^{\#}$, then $y$ normalizes no nontrivial $p$-subgroup of $G$. A similar statement holds with $p$ and $q$ interchanged.*

PROOF. By symmetry, it suffices to prove the first assertion. Suppose $R$ is a nontrivial $y$-invariant $p$-subgroup of $G$. Then $N=N_G(R)<G$ and $N$ contains both $y$ and $Z(P)$ for any Sylow $p$-subgroup $P$ of $G$ containing $R$. But then if $x\in Z(P)^{\#}$, $\langle x, y\rangle\leqslant N<G$, contrary to the previous lemma.

Now we can prove, using special properties of involutions:

LEMMA 4.134. *G has odd order.*

PROOF. Suppose false, and for definiteness assume $q=2$. Let $\mathcal{Y}$ be a conjugacy class of 2-central involutions of $G$. Since $G$ is simple, $O_2(G)=1$, so by Theorem 2.66, there exists $y, y' \in \mathcal{Y}$ such that the group $\langle y, y' \rangle$ is not a 2-group. But as $y, y'$ are involutions, Proposition 2.41 implies that $\langle y, y' \rangle$ is a *dihedral group* (of order $2k$, where $k=|yy'|$ and $y, y'$ each invert $yy'$). In particular, $k$ is not a power of 2. Since $|G|=p^a 2^b$, it follows that $p$ divides $k$. Hence $y$ inverts the unique subgroup $R$ of $\langle yy' \rangle$ of order $p$. However, as $y$ is 2-central, $y \in Z(Q)$ for some Sylow 2-subgroup $Q$ of $G$, and so the previous lemma is contradicted.

Now in this case, we can exploit Glauberman's $ZJ$-theorem.

LEMMA 4.135

(i) *Every maximal subgroup of $G$ contains a Sylow $p$- or $q$-subgroup of $G$.*
(ii) *A Sylow $p$- or $q$-subgroup of $G$ is contained in a unique maximal subgroup of $G$.*

PROOF. Let $M$ be a maximal subgroup of $G$ and suppose for definiteness that $F(M)$ is a $p$-group. Let $P \in Syl_p(M)$. Since $M$ is solvable of *odd* order (as $p$ and $q$ are both odd), $M$ is $p$-stable by Theorem 4.108, so by the $ZJ$-theorem $M \leqslant N_G(Z(J(P)))$. Hence by the maximality of $M$, equality must hold. Using Corollary 1.11 and the fact that $J(P)$ char $P$, we conclude, as usual, that $P \in Syl_p(G)$, so (i) holds.

Next let $N$ be any maximal subgroup of $G$ containing $P$. By Lemma 4.131, $P$ normalizes no nontrivial $q$-subgroup of $N$, so $O_q(N)=1$. Since $F(N)$ has prime power order by assumption, $F(N)$ is thus a $p$-group and so the argument of the preceding paragraph can be repeated to yield that $N=N_G(Z(J(P)))=M$. Hence (ii) holds for $p$; and by symmetry, it holds as well for $q$.

As an immediate corollary, we have

LEMMA 4.136. *If $P \in Syl_p(G)$, then $Z(P)$ is contained in a unique maximal subgroup of $G$. A similar statement holds for $q$.*

PROOF. If false, the previous lemma implies that $Z(P) \leqslant N$ for some maximal subgroup $N$ which contains a Sylow $q$-subgroup of $G$. But then $\langle Z(P), Z(Q) \rangle \leqslant N < G$, contrary to Lemma 4.132.

Now we can eliminate Case 1.

PROPOSITION 4.137. *Case* 1 *does not hold.*

PROOF. Suppose false. For definiteness, assume that $p^a > q^b$. This implies that any two Sylow $p$-subgroups $P_1$, $P_2$ have nontrivial intersection. Indeed, in the contrary case, $|G| \geqslant |P_1 P_2| \geqslant p^{2a} > p^a q^b = |G|$, a contradiction. We shall use this fact in the proof.

Let $P \in Syl_p(G)$. The preceding analysis applies to $P$, and so $Z(P)$ is contained in a unique maximal subgroup $M$ of $G$ by Lemma 4.136. Since $P$ lies in some maximal subgroup of $G$, this forces $P \leqslant M$. Since $G$ is simple, there is $R \in Syl_p(G)$ with $R \nleqslant M$. Choose $R$ so that $R_0 = R \cap M$ is of maximal order. Conjugating $R$ by an element of $M$ does not affect the size of the intersection. Hence as $P \in Syl_p(M)$, we can assume without loss, in view of Sylow's theorem, that $R_0 \leqslant P$. Then $R_0 = R \cap P \neq 1$ by the previous paragraph.

Set $N = N_G(R_0)$; since $R_0 \neq 1$, $N < G$. Since $R_0 \leqslant P$, $Z(P) \leqslant N$ and as $N < G$, the uniqueness of $M$ implies that $N \leqslant M$. In particular, $N_R(R_0) \leqslant M$, whence $N_R(R_0) = R_0$. But then $R = R_0$ by Theorem 1.10, so $R \leqslant M$, contrary to our choice of $R$.

Thus Case 2 holds and so there is a maximal subgroup $M$ of $G$ such that $F(M)$ is not of prime power order. To treat this case, we need two preliminary lemmas, the first of which is a general consequence of Thompson's (A×B)-lemma (Proposition 4.9).

LEMMA 4.138. *If $P$ is a $p$-subgroup of the $p$-constrained group $X$, then* $O_{p'}(N_X(P)) \leqslant O_{p'}(X)$.

If $\bar{X} = X/O_{p'}(X)$, one shows easily, using a Frattini-type argument, that $N_X(P)$ maps onto $N_{\bar{X}}(\bar{P})$, so it suffices to prove the result for $\bar{X}$. Thus we can assume without loss that $O_{p'}(X) = 1$, in which case we must show that $Y = O_{p'}(N_X(P)) = 1$. Set $Q = O_p(X)$ and $Q_0 = C_Q(P)$. Then $Q_0 \leqslant N_X(P)$, so $[Y, Q_0] \leqslant Y$. On the other hand, $[Y, Q_0] \leqslant Q$ as $Q \lhd X$. Thus $[Y, Q_0] \leqslant Y \cap Q = 1$ (as $Y$ is a $p'$-group and $Q$ is a $p$-group). Hence $Y$ centralizes $Q_0 = C_Q(P)$, so by the (A×B)-lemma, $Y$ centralizes $Q$. However, $C_X(Q) \leqslant Q$ as $X$ is $p$-constrained with $O_{p'}(X) = 1$. Hence $Y \leqslant Q$, forcing the desired conclusion $Y = 1$.

The second result is a well-known and easily verified property of cyclic $p$-groups [130, Lemma 5.4.1].

LEMMA 4.139. *If $X$ is a cyclic $p$-group of order $p^n$, $p$ a prime, then* Aut($X$) *is abelian of order* $(p-1)p^{n-1}$.

Set $F=F(M)=F_p\times F_q$, where $F_p$, $F_q$ are the Sylow $p$- and $q$-subgroups of $F$, and set $Z=Z_p\times Z_q$, where $Z_p=Z(F_p)$, $Z_q=Z(F_q)$. Since $F$ is not of prime power order, we have $Z_p\neq 1$ and $Z_q\neq 1$.

We first eliminate the case that $F$ is "small."

LEMMA 4.140. *Either $m_p(F_p)\geqslant 2$ or $m_q(F_q)\geqslant 2$.*

PROOF. Suppose false and, for definiteness, assume $p<q$, so that $q$ is odd. Since $m_p(F_p)=m_q(F_q)=1$, $F_q$ is cyclic and $F_p$ is either cyclic or quaternion by Proposition 1.35. Assume first that $F_p$ is quaternion (whence $p=2$) and let $P\in Syl_2(M)$. Since $F_p\triangleleft M$, $F_p\leqslant P$. In particular, $Z(P)$ centralizes $F_p$. By Lemma 4.139, $P/C_P(F_q)$ is abelian, so $P'\leqslant C_P(F_q)$. Thus $U=Z(P)\cap P'$ centralizes $F=F_p\times F_q$. However, as $M$ is solvable, $C_M(F)\leqslant F$ by Proposition 1.25, so $U\leqslant F_p$. But $F_p$ and hence $P$ are nonabelian, so $P'$ and hence $U\neq 1$. Also $U\leqslant Z(F_p)$ as $U\leqslant Z(P)$, so, in fact, $U=Z(F_p)$ $(\cong Z_2)$. Thus $U\triangleleft M$, so $M=N_G(U)$ by the maximality of $M$. But by its definition, $U$ char $P$, so the usual argument on normalizers of nilpotent subgroups implies that $P\in Syl_p(G)$. However, as $P$ normalizes $F_q\neq 1$, this contradicts Lemma 4.131.

Hence $F_p$ and $F_q$ are both cyclic. The proof is similar to the previous case. Let $Q\in Syl_q(M)$. Lemma 4.139 implies that $\mathrm{Aut}(F_p)$ has no prime divisors exceeding $p$, so as $p<q$, $Q$ necessarily centralizes $F_p$. Hence again by Proposition 1.25, $Q/F_q\leqslant\mathrm{Aut}(F_q)$ and so is abelian by Lemma 4.139. Thus $Q'\leqslant F_q$. Also, as $C_Q(F_q)=F_q$, clearly either $Q=F_q$ or $Q'\neq 1$. Correspondingly set $V=F_q$ or $Q'$, so that in either case $V$ char $Q$ and $V\neq 1$. Also as $F_q$ is cyclic, $V\triangleleft M$. Thus as with $U$ above, $M=N_G(V)$ and $Q\in Syl_q(G)$. Since $Q$ normalizes $F_p\neq 1$, Lemma 4.131 is again contradicted.

LEMMA 4.141. *$M$ is the unique maximal subgroup of $G$ containing $Z$.*

Let $M_1$ be a maximal subgroup of $G$ containing $Z$. Now $N_{M_1}(Z_q)\leqslant N_G(Z_q)=M$ and $Z_p\leqslant O_{q'}(M)$, so $Z_p\leqslant O_{q'}(N_{M_1}(Z_q))$. Hence by Lemma 4.138, $Z_p\leqslant O_{q'}(M_1)=O_p(M_1)\leqslant F(M_1)$. Similarly $Z_q\leqslant F(M_1)$. In particular, $Z_p$ centralizes $O_q(M_1)$, so $O_q(M_1)\leqslant N_G(Z_p)=M$. Similarly $Z_q\leqslant O_q(M_1)\leqslant F(M_1)$ and $O_p(M_1)\leqslant M$. But $F(M_1)=O_p(M_1)\times O_q(M_1)$ (as $M_1$ is a $\{p,q\}$-group), so $F(M_1)\leqslant M$.

Since $Z_p\times Z_q\leqslant F(M_1)$, both $p$ and $q$ divide $|F(M_1)|$, so the preceding argument applies equally well to $M_1$. Since $Z\leqslant M$, it follows that $F\leqslant M_1$. But again we have $F_p\leqslant O_{q'}(N_{M_1}(F_q))$, so $F_p\leqslant F(M_1)$ by Lemma 4.138. Similarly $F_q\leqslant F(M_1)$. Thus $F=F_p\times F_q\leqslant F(M_1)$. By symmetry, the reverse inclusion also holds, so that $F=F(M_1)$. But $M=N_G(F)$ and $M_1=N_G(F(M_1))$, so $M=M_1$, proving the lemma.

Now we use Lemma 4.140. We have that $m_r(F_r)\geqslant 2$ for $r=p$ or $q$, say $r=p$. Thus $F_p$ contains a subgroup $V=Z_p\times Z_p$. Since $Z=Z(F)$, $Z$ centralizes $V$, so $Z\leqslant C_v=C_G(v)$ for every $v\in V^{\#}$. Hence by the preceding lemma, we have

LEMMA 4.142. *$C_v\leqslant M$ for every $v\in V^{\#}$.*

Now let $R\in Syl_p(M)$ with $V\leqslant R$. Using Lemma 4.142, we prove finally

LEMMA 4.143. *$F_q$ is the unique maximal $R$-invariant $p'$-subgroup of $G$.*

PROOF. If $X$ is any $R$-invariant $p'$-subgroup of $G$, then $X$ is $V$-invariant, so as $V$ is noncyclic abelian, $X=\langle C_X(v)|v\in V^{\#}\rangle$ by Theorem 4.13. Hence $X\leqslant M$ by Lemma 4.142. Set $\overline{M}=M/F_q=M/O_q(M)$, so that $C_{\overline{M}}(O_p(\overline{M}))\leqslant O_p(\overline{M})$ by Proposition 1.25. But $O_p(\overline{M})\leqslant\overline{R}$ and so $[\overline{X},O_p(\overline{M})]\leqslant\overline{X}\cap O_p(\overline{M})=1$ (as $\overline{X}$ is a $p'$-group). Hence $\overline{X}$ centralizes $O_p(\overline{M})$ and so $\overline{X}\leqslant O_p(\overline{M})$, whence $\overline{X}=1$. We conclude that $X\leqslant F_q$ and the lemma follows.

This in turn yields the desired conclusion.

PROPOSITION 4.144. *Case 2 does not hold.*

PROOF. Indeed, continuing the preceding analysis, for any $y\in N_G(R)$, $F_q^y$ is an $R$-invariant $p'$-subgroup of $G$, so $F_q^y=F_q$ by Lemma 4.143. Thus $y\in N_G(F_q)=M$ and consequently $N_G(R)\leqslant M$. As usual, this yields that $R\in Syl_p(G)$. Since $R$ normalizes $F_q\neq 1$, Lemma 4.131 is once again contradicted.

Together Propositions 4.137 and 4.144 establish Theorem 4.130.

## 4.9. Product Fusion and Strong Closure

In the course of determining all simple groups $G$ with a given type of Sylow 2-group, one is often forced to consider a subsidiary problem involving the classification of groups with Sylow 2-subgroups $S$ of the form

$S=S_1\times S_2$ with $S_i\neq 1$, $i=1,2$. Harada and I were the first to face this situation in our study of groups with Sylow 2-groups isomorphic to those of $G_2(q)$, $q$ odd [134]. If $t$ is a 2-central involution of such a group $G$, and we set $C=C_t$, $\overline{C}=C/O(C)$, and $\tilde{C}=\overline{C}/\langle\tilde{t}\rangle$, our analysis of the 2-fusion of $G$ enabled us to show that $\tilde{C}$ has a normal subgroup $\tilde{X}$ of index 2 with Sylow 2-subgroup $\tilde{S}=\tilde{S}_1\times\tilde{S}_2$, where $\tilde{S}_1$, $\tilde{S}_2$ are (isomorphic) dihedral groups. Thus to pin down the structure of $\tilde{C}$ and hence of $\overline{C}$, we had first to determine the possibilities for $\tilde{X}$. [The existence of such an $\tilde{X}$ was expected, for if $G=G_2(q)$, then $C$ has a normal subgroup $X$ of index 2 with $X\cong SL(2,q)*SL(2,q)$. Thus $X/\langle t\rangle\cong L_2(q)\times L_2(q)$ and $L_2(q)$ has dihedral Sylow 2-subgroups for odd $q$.]

If $\tilde{S}$ is abelian, the possibilities for $\tilde{X}$ follow from the abelian Sylow 2-group classification theorem; so one can restrict to the nonabelian case. Harada and I established the following result [134].

THEOREM 4.145. *If $G$ is a group with no nontrivial normal subgroups of odd order or odd index and $G$ has a nonabelian Sylow 2-subgroup $S$ of the form $S=S_1\times S_2$ with $S_1$ and $S_2$ dihedral, then*

(i) *$G=G_1\times G_2$ for suitable subgroups $G_1$, $G_2$ of $G$.*
(ii) *For a suitable factorization of $S$, $S_i\in Syl_2(G_i)$, $i=1,2$.*

The point of (ii) is that the factorization of $S$ is by no means unique. Since $G_1$ and $G_2$ have dihedral Sylow 2-subgroups, their possible structures are determined from the dihedral theorem and hence so are those of $G$. Note also that the restriction on normal subgroups of odd order and index is clearly necessary to obtain a direct product factorization of $G$.

An important first step in the proof of Theorem 4.145 is the following fusion assertion.

PROPOSITION 4.146. *For a suitable factorization of $S$, $S_i$ is strongly closed in $S$ with respect to $G$, $i=1,2$.*

This is a natural result, since if $G=G_1\times G_2$ with $S_i\in Syl_2(G_i)$, then indeed $S_i$ is strongly closed in $S$ with respect to $G$. It also suggests the following general definition.

DEFINITION 4.147. Let $S\in Syl_2(G)$ for any group $G$. If $S=S_1\times S_2$ with $S_i\neq 1$ strongly closed in $S$ with respect to $G$, $i=1,2$, we say that $G$ has *product fusion* with respect to the given decomposition of $S$.

In the context of the $G_2(q)$ problem, as well as in other similar classification problems, the analysis of 2-fusion yields not only the existence of $\tilde{X}$ with Sylow 2-subgroup $\tilde{S}$ of the form $\tilde{S}=\tilde{S}_1 \times \tilde{S}_2$ but also the fact that $\tilde{X}$ has product fusion with respect to an appropriate factorization of $\tilde{S}$. Hence for the applications, it would have sufficed to assume $G$ has product fusion in Theorem 4.145.

Once the fusion pattern of $G$ was determined, Harada and I were able to use signalizer functors and Bender's strong embedding theorem to prove the existence of the required normal subgroups $G_1$, $G_2$. Soon after this, F. Smith [257, 258] treated the dihedral$\times$quasi-dihedral and quasi-dihedral$\times$ quasi-dihedral cases in similar fashion, followed by David Mason [208], who considered dihedral$\times$wreathed, etc. In each case, construction of the required signalizer functor used local balance properties of the elements of $\mathfrak{L}_2(G)$ (which had either dihedral, quasi-dihedral, wreathed, or quaternion Sylow 2-groups and so were $K$-groups by prior classification theorems). Each of these product fusion theorems has had corresponding application to some classification problems.

Some time later, Morton Harris and I, investigating groups with Sylow 2-subgroups isomorphic to those of $PSp(6, q)$, $q$ odd [164], were forced to treat *two* subsidiary product fusion problems—dihedral$\times$dihedral$\times$dihedral and $PSp(4, q)$-type$\times$dihedral. It was obviously time to consider the general case!

We succeeded in establishing a completely general product fusion theorem, under the *assumption* that suitable sections of 2-local subgroups of $G$ satisfied suitable local balance conditions, known to hold for all simple $K$-groups [138]. Using our $K$-group assumptions, we constructed 2- and 3-balanced signalizer functors with respect to suitable elementary 2-subgroups of $S$ (cf. Proposition 4.65), which we applied to an analysis of the 2-local structure of $G$.

However, it is unnecessary to state our result explicitly, for very shortly thereafter Goldschmidt greatly improved it by showing that the local balance assumptions we had imposed were, in fact, superfluous, if instead one carried out the analysis, using *core-separated* signalizer functors (cf. Proposition 4.67). Indeed, Goldschmidt was able to establish the conditions for core separation solely on the basis of the product fusion hypothesis on his group $G$ together with some general properties of the 2-local subgroups which that hypothesis implied. He thus obtained the following lovely result, of which the product fusion theorem is a particular case [124].

THEOREM 4.148. *If $G$ is a core-free group which contains the direct product of two 2-subgroups $S_1$ and $S_2$, each of which is strongly closed in a*

*Sylow* 2-*subgroup with respect to* $G$, *then*

$$\langle S_1^G\rangle\cdot\langle S_2^G\rangle=\langle S_1^G\rangle\times\langle S_2^G\rangle.$$

If the core-free assumption is dropped, one obtains, of course, the corresponding result in $G/O(G)$.

We shall outline Goldschmidt's proof, for it provides an excellent illustration of the signalizer functor method in operation. We let $G$ be a minimal couterexample. Then a reasonably straightforward induction argument implies that $G$ is simple. In particular, the condition of strong closure together with the $Z^*$-theorem (Theorem 4.95) yields now that each $S_i$ contains more than one involution, so $m_2(S_i)\geqslant 2$, $i=1,2$.

Core separation is obtained as a direct corollary of the general structure of the normalizers of subgroups of $S_1\times S_2$, which is derived from the strong closure hypothesis and induction.

PROPOSITION 4.149. *Let* $T$ *be a nontrivial subgroup of* $S_1\times S_2$ *and set* $H=N_G(T)$. *Then there exist* 2-*subgroups* $T_i\leqslant H$, $i=1,2$, *with the following properties*:

(i) $N_{S_i}(T)\leqslant T_i$.
(ii) $T_i$ *is* $G$-*conjugate to a subgroup of* $S_i$.
(iii) $T_i$ *is strongly closed in a Sylow* 2-*subgroup of* $H$ *with respect to* $G$.
(iv) $T_1T_2=T_1\times T_2$.
(v) $O(H)(\langle T_1^H\rangle\times\langle T_2^H\rangle)$ *is normal in* $H$.

Notice that (v) implies that either $T_1$ or $T_2$ [and hence by (i) either $N_{S_1}(T)$ or $N_{S_2}(T)$] centralizes any given component of $H/O(H)$. Thus as an immediate corollary of the proposition and the definition of core separation, we obtain

COROLLARY 4.150. *If* $A$ *is an elementary subgroup of* $S_1\times S_2$ *of order* 16 *with* $A_i=A\cap S_i\cong Z_2\times Z_2$, $i=1$ *and* 2, *then* $G$ *is core-separated with respect to the decomposition* $A=A_1\times A_2$.

(One has only to apply the proposition with $T=\langle a\rangle$ and $H=C_a$ for each $a\in A^{\#}$).

For *each* such choice of $A$, Goldschmidt now uses his core-separated $A$-signalizer functor (Proposition 4.67) and applies the solvable signalizer functor theorem (Theorem 4.38). He is then able to show that the resulting groups are essentially determined independently of $A$. The analysis ultimately yields the following key result.

PROPOSITION 4.151. *There exists a subgroup $W$ of $G$ of odd order with the following properties*:

(i) *$W$ is invariant under $S_1 \times S_2$.*
(ii) *If $T$ is any subgroup of $S_1 \times S_2$ such that $S_i \cap T \neq 1$ for both $i=1$ and 2, then $W$ is the unique maximal $T$-invariant subgroup of $G$ of odd order.*
(iii) *If $x_i \in \mathscr{I}(S_i)$, $i=1,2$, then $[O(C_{x_i}), x_j] \leqslant W$ for $j=1,2$.*

As indicated in Section 4.3, the signalizer functor method (for the prime 2) always involves some form of the Aschbacher-Bender results on strong embedding. Here those results are used to prove

PROPOSITION 4.152. *Let $X$ be a simple group and $T$ a nontrivial 2-subgroup which is strongly closed in a Sylow 2-subgroup $S$ with respect to $X$. If $\langle C_X(t) | t \in \mathscr{I}(T)\rangle < X$, then $X$ has a strongly embedded subgroup.*

Indeed, strong closure implies that $T \triangleleft N_X(S)$, so elements of $N_X(S)$ permute the subgroups $C_X(t)$ for $t \in \mathscr{I}(T)$ under conjugation and hence normalize the subgroup they generate. Since $X$ is simple, it follows that $M = \langle C_X(t) | t \in \mathscr{I}_2(X)\rangle N_X(S) < X$.

We now use Alperin fusion to prove that $M$ controls $X$-fusion in $S$ (i.e., two subsets of $S$ conjugate in $X$ are already conjugate in $M$). Indeed, let $\mathfrak{D}$ be the conjugacy family of Theorem 4.86, so that, in particular, $Z(S) \leqslant D$ for $D \in \mathfrak{D}$. Since $N_X(S) \leqslant M$, we need only show [in view of condition (4) following Definition 4.85] that for any $D \in \mathfrak{D}$ and 2-element $y$ of $N_X(D)$, then necessarily $y \in M$. But $Z(S) \leqslant D$ and $Z(S) \cap T \neq 1$ by Proposition 1.9 as $T \triangleleft S$. Thus $T \cap D \neq 1$. However, as $T$ is strongly closed in $S$, $y$ must normalize $T \cap D$. Since $(T \cap D)\langle y \rangle$ is a 2-group, another application of Proposition 1.9 yields that $y$ centralizes some involution $t$ of $T \cap D$. Since $C_X(t) \leqslant M$, the desired conclusion $y \in M$ follows.

Now we can prove that Aschbacher's strong embedding criterion holds for $X$. Let $z \in \mathscr{I}(Z(S)) \cap T$. Then $C_X(z) \leqslant M$. In addition, if $z^x \in M$ for $x \in X$, then $z^{xm} \in S$ for some $m \in M$ by Sylow's theorem, so by the previous paragraph $z^{xmy} = z$ for some $y \in M$. But then $xmy \in C_X(z) \leqslant M$ and so $x \in M$. We see then that $z^x \in M$ if and only if $x \in M$. Furthermore, if $z^x$ centralizes $z$, then without loss $z^x \in S$, whence by strong closure $z^x \in T$, so if $z^x \neq z$, it follows that $C_X(zz^x) \leqslant M$. Thus $M$ and $z$ indeed satisfy the hypothesis of Aschbacher's criterion (Theorem 4.31), and we conclude that $X$ possesses a strongly embedded subgroup.

Clearly the product fusion hypothesis on $S_1 \times S_2$ implies that our group $G$ has more than one conjugacy class of involutions, so as noted following Proposition 4.21, $G$ does not contain a strongly embedded subgroup. Hence as a corollary we have

COROLLARY 4.153. *For $i=1$ and 2, $G=\langle C_x | x \in \mathcal{I}(S_i)\rangle$.*

Goldschmidt uses this last result to establish an analogous generational assertion for the normal closure $K_i$ of $S_i$ in $N_G(S_j)$, $j \neq i$, $i, j=1,2$. The argument also depends on his prior classification theorem concerning groups $X$ which contain an abelian 2-subgroup strongly closed in a Sylow 2-subgroup with respect to $X$.

PROPOSITION 4.154. *For $i=1$ and 2, $K_i=\langle C_{K_i}(x) | x \in \mathcal{I}(S_i)\rangle$.*

Finally, using induction, he is able to sharpen Proposition 4.149 in the case that $T \leqslant S_i$ by showing that, in fact, $K_j \leqslant N_G(T)$ for $j \neq i$. Thus he is able to prove

PROPOSITION 4.155. *Let $1 \neq T \leqslant S_i$, $i=1$ or 2 and set $H=N_G(T)$. Then $O(H)K_j \lhd H$ for $j \neq i, j=1$ or 2.*

Now the stage is set for a final contradiction. As expected, there are two cases, according as the subgroup $W$ of Proposition 4.151 is trivial or nontrivial. Suppose first that $W=1$. If $x \in \mathcal{I}(S_1)$, Proposition 4.151(iii) then yields that $\Omega_1(S_2)$ centralizes $O(C_x)$. But $K_2 \leqslant C_x$ by Proposition 4.155, whence by Proposition 4.154, $K_2=\langle \Omega_1(S_2)^{K_2}\rangle$ centralizes $O(C_x)$. Hence $K_2=O^{2'}(K_2O(C_x))$ and so $K_2 \lhd C_x$ by Proposition 4.155. Since $x$ was arbitrary, we conclude that $\langle C_x | x \in \mathcal{I}(S_1)\rangle \leqslant N_G(K_2)$ and so is a proper subgroup of $G$ (as $G$ is simple), contrary to Proposition 4.152.

On the other hand, if $W \neq 1$, then $M=N_G(W)<G$ (again as $G$ is simple). Now for any $x \in \mathcal{I}(S_2)$ and any $1 \neq T \leqslant S_2$, Proposition 4.151(ii) implies that $W$ is the unique maximal element of $\mathcal{U}_G(\langle x\rangle \times T; 2')$, whence $W$ is invariant under $N_G(\langle x\rangle \times T)=N_{C_x}(T)$. Thus $\langle N_{C_x}(T) | 1 \neq T \leqslant S_2\rangle \leqslant M$. Since $K_2 \leqslant C_x$ and $K_2=\langle C_{K_2}(x_2) | x_2 \in \mathcal{I}(S_2)\rangle$ by Proposition 4.154, it follows that $K_2 \leqslant M$. But as $m_2(S_2) \geqslant 2$, $O(C_x) \leqslant \langle N_{C_x}(T) | 1 \neq T \leqslant S_2\rangle$ by Theorem 4.13, so also $O(C_x) \leqslant M$. Furthermore, as $S_2 \leqslant K_2$ and $O(C_x)K_2 \lhd C_x$, it follows easily by product fusion and the Frattini argument (Proposition 4.3) that $C_x=O(C_x)K_2N_{C_x}(S_2)$. However, also $N_{C_x}(S_2) \leqslant M$, so we conclude that $C_x \leqslant M$. Thus $\langle C_x | x \in \mathcal{I}(S_1)\rangle \leqslant M<G$, once again contradicting Proposition 4.152.

With a little additional effort, Goldschmidt established a definitive result about arbitrary strongly closed 2-groups. Here the term $X^{(\infty)}$ denotes the ultimate term of the derived series of the group $X$.

THEOREM 4.156. *Let $G$ be a core-free group, $S$ a Sylow 2-subgroup of $G$, and $T$ a subgroup of $S$ that is strongly closed in $S$ with respect to $G$. Then $(C_G(T))^{(\infty)} \lhd G$.*

## 4.10. Weak Closure and Trivial Intersection Sets

We should like now to describe the fundamental group-theoretic consequences obtained by Timmesfeld from his root involution theorem. To state them, we need the notion of weak closure.

DEFINITION 4.157. Let $G$ be a group, $H$ a subgroup of $G$, and $K$ a subgroup or subset of $H$. Set

$$V(\mathrm{ccl}_G(K); H) = \langle K^g | K^g \leqslant H, g \in G \rangle.$$

(ccl denotes "conjugacy class" here.) We call $V(\mathrm{ccl}_G(K); H)$ the *weak closure* of $K$ in $H$ with respect to $G$. If $V(\mathrm{ccl}_G(K); H) = K$, we say simply that $K$ is *weakly closed* in $H$ (with respect to $G$).

Very often one is interested in the case that $H = P$ is a Sylow $p$-subgroup of $G$ for some prime $p$. In that case, if $K$ is strongly closed in $P$, it is clearly weakly closed. The Thompson subgroup $J(P)$ [likewise $J_e(P)$] is an example of a weakly closed subgroup of $P$. There is a classical theorem of Otto Grün which asserts that if $Z(P)$ is weakly closed in $P$, then the largest abelian $p$-factor group of $G$ is isomorphic to that of $N_G(Z(P))$ [130, Theorem 7.5.2].

Shult studied groups generated by a class of involutions $D$ such that for $x \in D$, the weak closure of $x$ in its centralizer $C_x$ is an abelian 2-group [252]. Timmesfeld obtained as a consequence of his root involution theorem [301] the following partial extension of Shult's results.

THEOREM 4.158. *Let $G$ be a group with $O(G) = 1$ and $Z(G) = 1$, which is generated by a conjugacy class $D$ of involutions such that:*

(a) *For any $x \in D$, the weak closure of $x$ in $C_x$ is a 2-group of class at most 2.*
(b) *The product of any two distinct commuting elements of $D$ is an element of $D$.*

*Then* $G \cong L_2(q)$, $U_3(q)$, $Sz(q)$, $L_3(q)$, ${}^3D_4(q)$, $q=2^n$, $G_2(q)$, $q=2^n$, $n>1$, $A_6$ *of* $J_2$.

The theorem is proved by arguing that $D$ is a class of root involutions and then invoking the root involution theorem, checking afterward which groups on the list satisfy the given hypotheses. To prove that $D$ is indeed a class of root involutions, we must show that if $x, y \in D$ with $|xy|=2k$ for some $k>1$, then necessarily $k=2$. Now $\langle x, y\rangle$ is a dihedral group of order $4k$ and so $Z(\langle x, y\rangle)=\langle z\rangle$ has order 2. One checks from the structure of $\langle x, y\rangle$ that there is a conjugate $u$ of $x$ or $y$ in $\langle x, y\rangle$ such that $\langle x, u\rangle$ is a four group with $z=xu$. But then $u\in D$ and so by (b), also $z\in D$. Thus $\langle x, y\rangle \leqslant C_z$ and so by (a), $\langle x, y\rangle$ is a 2-group of class at most 2, which immediately forces $k=2$, as required.

Timmesfeld's next major result represents a basic extension of Theorem 4.158 [302]. At the same time, it generalizes Glauberman's $Z^*$-theorem, which corresponds to the case $|A|=2$. We state the theorem only for simple groups.

THEOREM 4.159. *Let G be a simple group and A a nontrivial elementary abelian 2-subgroup of G. If A is weakly closed in* $C_a$ *with respect to G for each* $a\in A^{\#}$, *then one of the following holds*:

(i) $G\cong L_n(q)$, $Sz(q)$, $U_3(q)$, $q=2^n$.
(ii) $G\cong A_n$, $6\leqslant n\leqslant 9$.
(iii) $G\cong M_{22}$, $M_{23}$, $M_{24}$, *or He*.

The first step in the proof is to pin down the possible structures of the group $\langle A, A^g\rangle$ for $g\in G$ when $A\cap N_G(A^g)\neq 1$. With this information, Timmesfeld is able to determine the normal subgroup $X$ of $N=N_G(A)$ generated by the conjugates of $A^g\cap N$ in $N$. In particular, he shows that the hypotheses of his theorem are satisfied in $X/A$ for the image of $A^g\cap X$ and so induction can be applied. In this way, he ultimately establishes the following result.

PROPOSITION 4.160. *Suppose* $A\cap N_G(A^g)\neq 1$ *for some* $g\in G$ *and let X denote the normal closure of* $A^g\cap N_G(A)$ *in* $N_G(A)$. *Then either* $X/A\cong L_r(2^m)$, $A_6$, $A_7$, $A_8$ *for some r and m or else* $O(X/A)\neq 1$.

Also the order of $A$ and the action of $X/A$ on $A$ are determined when $X/A$ is simple. In the linear case, with $X/A\not\cong L_3(2)$ or $L_4(2)$, Timmesfeld argues that the conjugacy class determined by any element of $A^{\#}$ is a class of root involutions of $G$, in which case $G \cong L_{r+1}(2^m)$ by the root involution theorem.

On the other hand, if $X/A \cong L_3(2)$, $L_4(2)$, $A_6$, $A_7$, or $A_8$, Timmesfeld proves that $X = N_G(A)$. Since $N_G(A)$ can also be shown to contain a Sylow 2-subgroup $S$ of $G$, the possible structures of $S$ are completely determined; and now the possibilities for $G$ follow from various prior classification theorems for groups with such Sylow 2-groups. These latter results all concern groups of low 2-rank, a subject we shall discuss in considerable detail in the sequel.

In the final case, Timmesfeld argues that $|A| = 4$ and hence that $S$ is dihedral of order 8. Now the dihedral classification theorem yields that $G \cong A_6$ or $A_7$.

The theorem has a corollary, which is of fundamental importance for the study of groups of characteristic 2 type.

THEOREM 4.161. *Let $G$ be a simple group and $A$ a nontrivial elementary 2-subgroup of $G$ with the following properties*:

(a) *$A$ is a T.I. set in $G$ (i.e., $A^g \cap A = 1$ or $A$ for $g \in G$).*
(b) *$A$ centralizes no distinct conjugate of itself in $G$. (In particular, this is the case if $A$ is weakly closed in some Sylow 2-subgroup of $G$ containing $A$.)*

*Then $G$ is isomorphic to one of the groups listed in Theorem* 4.159.

If $A$ is weakly closed and $A$ centralizes $A^g$ for $g \in G$, then $AA^g$ is a 2-group, whence $AA^g \leqslant S \in Syl_2(G)$ and so $A^g = A$ by the weak closure of $A$. Thus indeed (b) holds in the weakly closed case.

We shall prove that the hypotheses of Theorem 4.159 follow from those of Theorem 4.161. Suppose by way of contradiction that $A$ is not weakly closed in $C_a$ for some $a \in A^\#$, in which case there is $B = A^g \leqslant C_a$ with $B \neq A$ for some $g \in G$. For $b \in B^\#$, $a \in A \cap A^b$, so $A = A^b$ as $A$ is a T.I. set by hypothesis. We conclude that $B$ normalizes $A$. Likewise as $A$ is a T.I. set and $B \neq A$, we have $B \cap A = 1$. Set $C = A \cap N_G(B)$. Since $B$ normalizes $A$, it follows that $[C, B] \leqslant A \cap B = 1$, so $B$ centralizes $C$. This forces $C < A$, otherwise $B$ centralizes $A$ and then $B = A$ by hypothesis, contradiction. Since $BC$ normalizes $B$, we conclude that $BC < BA$.

It follows now that there exists $x \in BA - BC$ with $x$ normalizing $BC$. Thus $B^x \leqslant (BC)^x = BC$. But $|BC| < |BA| = |B|^2$, as $A$ is conjugate to $B$. This forces $B \cap B^x \neq 1$, so $B^x = B$ as $B$ is a T.I. set. Hence $x \in N_{BA}(B) = BC$, contrary to the choice of $x$.

## 4.11. Factorizations and 3′-Groups

We shall limit the discussion to the prime 2 and shall write $\mathcal{Q}(S)$, $J(S)$ for $\mathcal{Q}_e(S)$, $J_e(S)$, respectively, for any 2-group $S$.

The factorization methods derive from the following basic result of Thompson [288].

PROPOSITION 4.162. *Let $X$ be a solvable group with $O(X)=1$ and let $S\in Syl_2(X)$. If $\Sigma_3$ is not involved in $X$ (in particular, if $X$ is a 3′-group), then we have*

$$X=C_X(Z(S))N_X(J(S)).$$

We shall give the proof, which is quite elementary and very instructive, for it illustrates many ideas of local analysis—in particular, some used in the proof of Glauberman's $ZJ$-theorem. Set $Z=Z(S)$. Since $X$ is 2-constrained (as $X$ is solvable) with trivial core, $Z\leqslant Z(O_2(X))$ and hence $W\leqslant Z(O_2(X))$, where $W$ is the normal closure of $Z$ in $X$. Since $Z(O_2(X))$ is abelian, so therefore is $W$. Set $C=C_X(W)$ and $T=S\cap C$. Since $W\triangleleft X$, also $C\triangleleft X$, which implies that $T\in Syl_2(C)$. Also $C\leqslant C_X(Z)$ as $Z\leqslant W$.

It will suffice to prove that $J(S)\leqslant T$. Indeed, assume this is the case. Then $J(S)=J(T)\,\mathrm{char}\,T$ and so $N_X(T)$ normalizes $J(S)$. But $X=CN_X(T)$ by the Frattini argument as $T\in Syl_2(C)$, so $X=CN_X(J(S))=C_X(Z)N_X(J(S))$, as asserted.

We can therefore assume that $J(S)\not\leqslant T$, whence there is $A\in\mathcal{Q}(S)$ such that $A\not\leqslant T$. Hence $A\not\leqslant C$. Set $\bar{X}=X/C$, so that $\bar{A}\neq 1$. We claim next that $O_2(\bar{X})=1$. Indeed, let $D$ be the preimage in $X$ of $O_2(\bar{X})$. Then $C\leqslant D\triangleleft X$ and so $D=C(S\cap D)$ as $D/C$ is a 2-group. But $S\cap D$ centralizes $Z=Z(S)$ and hence so does $D$. Since $D\triangleleft X$, it follows that $D$ centralizes the normal closure $W$ of $Z$ in $X$. Since $C=C_X(W)$, this forces $D=C$ and so $1=\bar{D}=O_2(\bar{X})$, as claimed.

Let $\bar{F}=F(\bar{X})$, the Fitting subgroup $\bar{X}$. By the previous paragraph, $\bar{F}\leqslant O(\bar{X})$. But $C_{\bar{X}}(\bar{F})\leqslant\bar{F}$ by Proposition 1.25, as $\bar{X}$ is solvable. Thus $\bar{A}$ does not centralize $\bar{F}$ and so $\bar{A}$ does not centralize $O_q(\bar{F})=O_q(\bar{X})$ for some (odd) prime $q$. Since $\bar{A}$ is an elementary abelian 2-group, one proves easily, using reduction arguments of the type described in Section 4.1, that $\bar{A}$ normalizes but does not centralize a subgroup $\bar{Y}$ of $O_q(\bar{X})$ of order $q$. Then $\mathrm{Aut}(\bar{Y})$ is cyclic and so $\bar{B}=C_{\bar{A}}(\bar{Y})$ has index 2 in $\bar{A}$ with $\bar{Y}\bar{A}/\bar{B}$ a dihedral group of order $2q$.

Now by definition of $C$, $\bar{Y}$ acts faithfully on $W$ and does not centralize $W$. Since $\bar{B}\bar{Y}=\bar{B}\times\bar{Y}$ acts on $W$ and $\bar{B}$ is a 2-group, the Thompson

$(A\times B)$-lemma implies that $\bar{Y}$ does not centralize $W_0=C_W(\bar{B})$. As $W_0$ is abelian, $\bar{Y}$ does not centralize $W_1=\Omega_1(W_0)$ by Proposition 4.7. Now set $V_1=[W_1,\bar{Y}]$, so that $V_1\neq 1$. Viewing $W_1$ as a vector space over $GF(2)$, we see immediately that $\bar{Y}$ has no nontrivial fixed points on $V_1$. Note also that as $\bar{A}$ normalizes both $\bar{Y}$ and $\bar{B}$, $A$ leaves $W_0$, $W_1$, and hence $V_1$ invariant.

Next, let $B$ be the subgroup of $A$ of index 2 which maps on $\bar{B}$. Then $B$ centralizes $V_1$ and so $BV_1$ is an elementary abelian subgroup of $S$. But $A\in\mathcal{Q}(S)$, so by definition of $\mathcal{Q}(S)$, we must have $m_2(BV_1)\leqslant m_2(A)$. This forces $B_1=B\cap V_1$ to be of index at most 2 in $V_1$.

Finally let $\bar{a}\in\bar{A}-\bar{B}$ and let $\bar{Y}=\langle\bar{y}\rangle$, so that $a$ inverts $y$. Thus $\langle\bar{a},\bar{y}\rangle=\langle\bar{a},\bar{a}^{\bar{y}}\rangle$ is dihedral of order $2q$. On the other hand, as $A$ is abelian, $\bar{a}$ acts trivially on $B_1$, whence $\bar{a}^{\bar{y}}$ acts trivially on $B_1^{\bar{y}}$. But then $\langle\bar{a},\bar{a}^{\bar{y}}\rangle$ and hence $\bar{Y}$ acts trivially on $B_0=B_1\cap B_1^{\bar{y}}$. Since $Y$ has no nontrivial fixed points on $V_1$, this forces $B_0=1$. However, $|V_1:B_1|\leqslant 2$, so $|V_1:B_0|\leqslant 4$. We thus conclude that $|V_1|=4$. The only possibility therefore is $q=3$, whence $\langle\bar{a},\bar{y}\rangle\cong\Sigma_3$, contrary to our assumption that $\Sigma_3$ is not a section of $X$.

The assumption on $X$ is necessary, for let $X$ be the semidirect product of $SL(2,2^n)$ and its natural 2-dimensional module $V$ over $GF(2^n)$. [Note that $SL(2,2)\cong\Sigma_3$.] If $S\in Syl_2(X)$, one checks that $\mathcal{Q}(S)$ has exactly two members: namely, $V$ and a second member $A=Z(S)\times T$, where $Z(S)$ is a 1-dimensional subspace of $V$ (and hence of order $2^n$) and $T$ maps isomorphically on a Sylow 2-subgroup of $X/V\cong SL(2,2^n)$. Thus $J(S)=\langle V,A\rangle=S$ and it follows that $C_X(Z(S))N_X(J(S))\leqslant N_X(Z(S))$, [as $N_X(J(S))=N_X(S)\leqslant N_X(Z(S))$]. But clearly $Z(S)$ is not normal in $X$ (since $V$ is an irreducible $X$-module), so $X$ does not "factor."

Thompson established further factorizations of solvable groups. These depend upon the following definition.

DEFINITION 4.163. If $S$ is a 2-group, let $\mathcal{Q}_1(S)$ be the set of elementary abelian 2-subgroups $B$ of $S$ such that either $B\in\mathcal{Q}(S)$ or $|B|=\frac{1}{2}|A|$, where $A\in\mathcal{Q}(S)$, and set

$$J_1(S)=\langle B|B\in\mathcal{Q}_1(S)\rangle.$$

Arguing in the same way as in Proposition 4.162, Thompson proves [288]:

PROPOSITION 4.164. *Let $X$ be a solvable group with $O(X)=1$ and let $S\in Syl_2(X)$. If $X$ does not involve $\Sigma_3$ or a dihedral group of order* 10 (*in particular, if $X$ has order prime to* 3 *and* 5), *then we have*

$$X=C_X(Z(S))N_X(J_1(S)).$$

The semidirect product of a dihedral group of order 10 and its natural 4-dimensional module over $GF(2)$ is a counterexample to the proposition (as is $\Sigma_3$ on its natural 2-dimensional module).

Thompson's brilliance was shown here by his realization that yet a *third* factorization existed for $X$ and that these three factorizations together could be exploited [289, Lemma 5.53].

PROPOSITION 4.165. *Let $X$ be a solvable group with $O(X)=1$ and let $S \in Syl_2(X)$. If $X$ does not involve $\Sigma_3$, then we have*

$$X = N_X(J(S))N_X(Z(J_1(S))).$$

COROLLARY 4.166. *If $X$ and $S$ are as in Proposition* 4.164 *and we set $N_1 = N_X(Z(S))$, $N_2 = N_X(J(S))$, $N_3 = N_X(Z(J_1(S)))$, then for every $i, j$, $1 \leqslant i, j \leqslant 3$, $i \neq j$, we have*

$$X = N_i N_j.$$

The corollary follows at once from the three propositions together with the fact that $AB = BA$ whenever $AB$ is a group and that $N_X(J_1(S)) \leqslant N_X(Z(J_1(S))) = N_3$.

This is Thompson's "triple factorization lemma." Now comes his marvelous "three against two" argument [93, Lemma 8.6].

PROPOSITION 4.167. *Let $G$ be a group and $H_1$, $H_2$, $H_3$ subgroups of $G$ such that for every permutation $\pi$ of the set $\{1,2,3\}$, we have*

$$H_{\pi(3)} \leqslant H_{\pi(1)} H_{\pi(2)}.$$

*Then $H_i H_j$ is a subgroup of $G$ for all $i, j$, $1 \leqslant i, j \leqslant 3$, $i \neq j$.*

Indeed, take $i=1$, $j=2$ for definiteness. We need only show that $H_2 H_1 \leqslant H_1 H_2$ to conclude that $H_1 H_2$ is a group. But $H_2 \leqslant H_1 H_3$ and $H_1 \leqslant H_3 H_2$ by assumption, so $H_2 H_1 \leqslant (H_1 H_3)(H_3 H_2) = H_1 H_3 H_2$. But also $H_3 \leqslant H_1 H_2$, whence $H_1 H_3 H_2 \leqslant H_1(H_1 H_2)H_2 = H_1 H_2$. Thus $H_2 H_1 \leqslant H_1 H_2$, as required.

As remarked earlier, Thompson developed these ideas in connection with the odd-order problem. We illustrate their significance by proving the following result.

THEOREM 4.168. *If $G$ is a simple group in which every 2-local subgroup has order prime to both* 3 *and* 5, *then $G \cong L_2(2^n)$, $n$ odd, $n \geqslant 1$, $U_3(2^n)$, $n \not\equiv 0$* (mod 3), *or $Sz(2^n)$, $n$ odd, $n \geqslant 3$.*

Let $G$ be a minimum counterexample. We shall argue first that every 2-local subgroup of $G$ is solvable with trivial core, at which point Thompson's ideas will be immediately applicable. By assumption, every 2-local subgroup $H$ of $G$ has order prime to 3 and 5. If $H$ is nonsolvable, then $H$ has a nonabelian simple section $K$ and $|K|$ is prime to 3 and 5. Clearly then $K$ satisfies the hypothesis of the theorem, so by the minimality of $G$, $K \cong L_2(2^n)$, $U_3(2^n)$, or $Sz(2^m)$ for suitable $m$. But then 3 or 5 divides $|K|$, contradiction. We conclude that every 2-local subgroup of $G$ is solvable.

We claim next that a Sylow 2-subgroup of $G$ is connected and $m_2(G) \geq 3$. Assume false, whence $r_2(G) \leq 4$ by Corollary 1.39. Since $G$ is simple, Frobenius's normal complement theorem implies that some 2-local subgroup $H$ of $G$ does not have a normal 2-complement. Choose $H$ so that a Sylow 2-subgroup $S$ of $H$ has maximal order. Set $\bar{H} = H/O(H)$, $\bar{R} = O_2(\bar{H})$ and $\tilde{R} = \bar{R}/\phi(\bar{R})$. Since $H$ is solvable, $C_{\bar{H}}(\bar{R}) \leq \bar{R}$ and hence $C_{\bar{H}}(\tilde{R}) \leq \bar{R}$ by Proposition 1.12(iii). Thus $\bar{H}/\bar{R}$ is isomorphic to a subgroup of $\mathrm{Aut}(\tilde{R})$. But $\tilde{R}$ is elementary abelian by Proposition 1.12(i) and so has rank at most 4, since $r_2(G) \leq 4$. Thus $\mathrm{Aut}(\tilde{R}) \leq L_4(2) \cong A_8$. Hence $\bar{H}/\bar{R}$ is a nontrivial solvable subgroup of $A_8$ of order prime to 3 and 5 with $O_2(\bar{H}/\bar{R}) = 1$. The only possibility is that $\bar{H}/\bar{R}$ is of order 7. In particular, $\bar{R} = \bar{S}$.

Let $R$ be the inverse image of $\bar{R}$ in $H$. Then $R \lhd H$ as $\bar{R} \lhd \bar{H}$ and $R = O(H)S$. Hence, by the Frattini argument, $H = RN_H(S) = O(H)N_H(S)$. Since $H$ does not have a normal 2-complement, neither therefore does $H_1 = N_G(S)$. By maximality of $S$, $S \in Syl_2(H_1)$, whence $S \in Syl_2(G)$. On the other hand, $\bar{H}/\bar{R} = \bar{H}/\bar{S}$ induces a group of automorphisms of $\bar{S}$ of order 7; and using this condition, we can argue without much difficulty that $\bar{S}$ must be connected group and $m_2(\bar{S}) \geq 3$. Thus $S$ is connected and $m_2(G) \geq 3$, as asserted.

Now we can apply Theorem 1.40 and conclude that every 2-local subgroup of $G$ has a trivial core.

Next, let $S \in Syl_2(G)$ and set $N_1 = N_G(Z(S))$, $N_2 = N_G(J(S))$, and $N_3 = N_G(Z(J_1(S)))$. Then each $N_i$ is a 2-local subgroup of $G$ and so is solvable of order prime to 3 and 5 with trivial core, $1 \leq i \leq 3$. Since also $S \in Syl_2(N_i)$, applying Corollary 4.166 to the groups $N_i$, we obtain

$$N_{\pi(3)} \leq N_{\pi(1)} N_{\pi(2)} \tag{4.26}$$

for every permutation $\pi$ of the set $\{1,2,3\}$. Hence by Proposition 4.167, $M = N_1 N_2$ is a subgroup of $G$.

Suppose $M = G$. Then any element $g \in G$ has the form $g = g_1 g_2$ with $g_i \in N_i$, $i = 1,2$. But then

$$O_2(N_1)^g = O_2(N_1)^{g_1 g_2} = O_2(N_1)^{g_2} \leq S^{g_2} \leq N_2 \tag{4.27}$$

[as $O_2(N_1) \leqslant S \leqslant N_2$]. Thus the normal closure of $O_2(N_1)$ in $G$ is contained in $N_2$. However, as $O_2(N_1) \neq 1$ and $N_1 < G$, this contradicts the simplicity of $G$. We therefore conclude that $M < G$.

Finally we argue that $M$ is strongly embedded in $G$, which will certainly hold if $M$ contains the normalizer of every nonidentity subgroup of $S$. Suppose false and let $H$ be a 2-local subgroup of $G$ such that $H \nleqslant M$ with $Q = S \cap H$ of largest possible order. Clearly $Q \neq 1$. As in the proof of Theorem 4.111, $Q \in Syl_2(H)$. This time we use Proposition 4.162 rather than the $ZJ$-theorem and conclude that

$$H = N_H(Z(Q))N_H(J(Q)). \tag{4.28}$$

If $Q = S$, then by (4.28), $H \leqslant N_1 N_2 = M$, contrary to the choice of $H$, so $Q < S$. But then $N_S(Z(Q)) > Q$ and $N_S(J(Q)) > Q$, so by the maximality of $Q$, we have $N_G(Z(Q)) \leqslant M$ and $N_G(J(Q)) \leqslant M$. Thus again $H \leqslant M$ by (4.28), contradiction.

Thus $M$ is strongly embedded, as asserted, and so by Bender's theorem, $G \cong L_2(2^n)$, $n \geqslant 2$, $U_2(2^n)$, $n \geqslant 2$, or $Sz(2^n)$, $n$ odd, $n \geqslant 3$. The hypothesis on the orders of 2-locals now forces $n$ to satisfy the restrictions of the theorem. Thus $G$ is not a counterexample and the theorem is proved.

In attempting to generalize these factorization results to nonsolvable groups $X$ with $F^*(X)$ a 2-group, two distinct problems arise. First, in the proof of Proposition 4.162 (and likewise in that of Proposition 4.164), the given arguments again reduce to the module action of a group $\bar{A}\bar{Y}/\bar{B}$ on a vector space $V_1$ over $GF(2)$ with $\bar{B}$ centralizing $\bar{Y}$ and $V_1$ and with $V_1 = [V_1, \bar{Y}]$. However, now $\bar{Y}$ need not be a cyclic group, but $\bar{Y}$ may also be a quasisimple group, isomorphic to a homomorphic image of a component of $L(\bar{X})$. But in the latter case, $\bar{B} = C_{\bar{A}}(\bar{Y})$ need not be of index 2 in $\bar{A}$. The validity of the factorization of $\bar{X}$ depends upon the nature of the action of $\bar{Y}$ on this module $V_1$ and in turn reduces to considerations about the dimension of the fixed-point space $B_1 = A \cap V_1 \leqslant C_{V_1}(A)$.

The discussion above with $Y \cong SL(2, 2^n)$ and $V_1$ the natural module provides a counterexample to any proposed factorization for groups involving $SL(2, 2^n)$. For the general analysis of groups of characteristic 2 type, it is very important to know precisely which possibilities for the pair $\bar{Y}, V_1$ will lead to a factorization and which will not ($\bar{Y}$ being assumed to be a $K$-group). Such "failures of factorization" will be described in detail in the next section. Let me just say here that this phenomenon is very much associated with the groups of Lie type of characteristic 2 and the alternating groups, factorization essentially always holding if $Y$ is not of one of these forms, no matter what module $V_1$ one takes.

On the other hand, the groups $Sz(2^n)$ are somewhat special since they have order prime to 3 and are the only known simple groups with this property. Thompson was able to extend Proposition 4.162 to this case by analyzing the $GF(2)$-modules for $Sz(2^n)$ [294].

PROPOSITION 4.169. *Let $X$ be a K-group of order prime to 3 in which $F^*(X)$ is a 2-group. If $S \in Syl_2(X)$, then we have*

$$X = C_X(Z(S))N_X(J(S)).$$

In view of the dihedral group of order 10 counterexample, one cannot prove the corresponding $Z$, $J_1$ factorization for an arbitrary such group $X$.

The third $J$, $ZJ_1$ factorization is rare for nonsolvable groups. Indeed, in the proof of any such result, one again reduces (as in the proof of Proposition 4.162) to an $\bar{A}\bar{Y}/\bar{B}$ situation for some $A \in \mathcal{Q}(S)$ and again considers the largest subgroup $B$ of $A$ which maps on $\bar{B}$. For the ensuing argument, it is essential that $B$ be an element of $\mathcal{Q}_1(S)$. However, this will be the case if and only if $|A:B| \leqslant 2$ and hence if and only if $|\bar{A}:\bar{B}| \leqslant 2$. But $\bar{Y}$ may be a simple group, in which case $m_2(\bar{Y}) \geqslant 2$ and $\bar{A}/C_{\bar{A}}(Y)$ may be noncyclic. Hence one cannot assert, in general, that $\bar{B}$ has codimension at most 1 in $\bar{A}$; some additional restriction is necessary to reach such a conclusion.

Some time after completing the classification of $N$-groups, Thompson undertook the task of proving that the Suzuki groups $Sz(2^n)$ were, in fact, the only simple groups of order prime to 3 [294]. Using standard signalizer functor methods, Thompson showed that a minimal counterexample $G$ to such a theorem was of characteristic 2 type. This effort could not be completely routine since the Rudvalis group $Ru$ has $Sz(8)\times Z_2\times Z_2$ as the centralizer of one of its involutions. Even though $Ru$ is not a 3′-group, this fact is not "visible" from this centralizer.

Now every proper subgroup of $G$ is a $K$-group and so its nonsolvable composition factors are necessarily Suzuki groups. For some time Thompson sought a triple factorization for the 2-local subgroups of $G$, which would enable him to emulate the proof of Theorem 4.168. Unfortunately he never found one and was therefore forced to resort to the full panoply of techniques and the major subdivisions $e(G) \geqslant 3$, $e(G)=2$, and $e(G)=1$ of the $N$-group analysis to complete the classification of simple 3′-groups.

However, by introducing an ingeniously conceived subgroup $\hat{J}(S)$ to replace $J_1(S)$, Glauberman subsequently succeeded in proving a triple factorization for $K$-groups $X$ of order prime to 3 in which $F^*(X)$ is a 2-group (and for more general groups $X$ which do not involve $\Sigma_4$) [117]. To describe it, we need a preliminary definition.

DEFINITION 4.170. Let $T$ be a 2-group. An elementary abelian normal subgroup $V$ of $T$ is said to be *restricted* in $T$ provided that, for any elementary abelian 2-subgroup $R$ of $T/C_T(V)$, we have

$$|V/C_V(R)|>|R|^{3/2} \quad \text{and} \quad |[V,R]|>|R|.$$

Furthermore, $T$ is said to be an *E-group* if $T$ has no restricted subgroups.

The following lemma of Glauberman shows that the notion of $E$-group is not vacuous.

LEMMA 4.171. *For any* 2-*group*, $S$, $J(S)$ *is an* $E$-*group.*

Now we can define $\hat{J}(S)$.

DEFINITION 4.172. For any 2-group $S$, set

$$\hat{J}(S)=\langle T | J(S)\leqslant T\leqslant S \text{ and } T \text{ is an } E\text{-group}\rangle.$$

Thus $J(S)\leqslant\hat{J}(S)$. [Note that by the definitions, likewise $J(S)\leqslant J_1(S)$.]

We remark that the groups $Sz(2^n)$ and $U_3(2^n)$ are the only simple $K$-groups with nonabelian Sylow 2-groups which do not involve $\Sigma_4$. Thus Glauberman's factorizations deal with 2-constrained groups whose only nonsolvable composition factors either are of one of these types or else have abelian Sylow 2-subgroups.

THEOREM 4.173. *Let $X$ be a $K$-group with $F^*(X)$ a 2-group such that $X$ does not involve $\Sigma_4$. If $S\in Syl_2(X)$, then we have*

(i) $X=C_X(\Omega_1(Z(S)))N_X(J(S))$.
(ii) $X=C_X(\Omega_1(Z(S)))N_X(\hat{J}(S))$.
(iii) $X=C_X(\Omega_1(Z(\hat{J}(S))))N_X(J(S))$.

Glauberman's proof is a *tour de force* of delicate commutator calculations.

On the basis of the theorem, Glauberman readily obtains the following extension of Thompson's 3′-theorem [117].

THEOREM 4.174. *If $G$ is a simple group with nonabelian Sylow 2-subgroups which does not involve $\Sigma_4$, then $G \cong Sz(2^n)$ or $U_3(2^n)$, $n$ odd.*

Taking $G$ to be a minimal counterexample, Glauberman could have argued exactly as in Theorem 4.168 that $G$ has a strongly embedded subgroup, provided that he first showed that $G$ had to be of characteristic 2 type. However, he was able to avoid even that amount of detailed analysis by appealing to Goldschmidt's strongly closed abelian subgroup theorem rather than to Bender's strongly embedded theorem. Thus he obtained Theorem 4.174 from the following fusion result together with Goldschmidt's classification theorem (Theorem 4.126).

PROPOSITION 4.175. *Let the assumptions be as in Theorem 4.174, let $S \in Syl_2(G)$, and set $A = \langle \Omega_1(Z(S))^x \mid x \in N_G(J(S)) \rangle$. Then $A$ is a nontrivial elementary abelian subgroup of $S$ and $A$ is strongly closed in $S$ with respect to $G$.*

The point here is that even though $G$ need not be of characteristic 2 type, Goldschmidt's extension of Alperin's fusion theorem (Theorem 4.86) shows that 2-fusion is determined entirely by 2-constrained 2-local subgroups $H$ of $G$ such that $O_2(H)$ maps onto $O_2(H/O(H))$. Since any such 2-local $H$ is a $K$-group and does not involve $\Sigma_4$, it follows that $H$ has a triple factorization. Exploiting this fact, Glauberman establishes the strong closure of $A$.

## 4.12. Failure of Thompson Factorization

As we have seen in the preceding section, the question whether a group $X$ with $F^*(X)$ a 2-group satisfies either the $(Z, J)$- or $(Z, J_1)$-factorization with respect to a Sylow 2-subgroup $S$ of $X$ reduces to a module statement for a certain section $\bar{A}\bar{Y}/\bar{B}$ of $X$, where $\bar{Y}$ is a quasisimple group with $O_2(\bar{Y}) = 1$. Examining the picture a little more closely, one can rephrase these factorization questions in general terms.

If $V$ is a faithful $GF(2)$-module for the group $Y$ with $O_2(Y) = 1$, does $Y$ contain a nontrivial elementary abelian 2-subgroup $A$ such that

$$m_2(A) \geqslant m_2(V/C_V(A)) \tag{4.29}$$

or such that

$$m_2(A) \geqslant m_2(V/C_V(A)) - 1. \tag{4.30}$$

In practice, we are concerned primarily with the case in which $F^*(Y)$ is a quasisimple $K$-group [taking $\bar{A}\bar{Y}/\bar{B}$ as $Y$, one has $F^*(Y)$ quasisimple].

DEFINITION 4.176. Let $Y$ be a group with $F^*(Y)$ quasisimple and $O_2(Y)=1$. If $V$ is a faithful $GF(2)$-module for $Y$, we say that $(Y,V)$ is an *F-pair* if $Y$ contains a subgroup $A$ satisfying (4.29) and that $(Y,V)$ is an *$F_1$-pair* if $Y$ contains a subgroup $A$ satisfying (4.30). Any such subgroup $A$ is called a *nonfactor* of $(Y,V)$.

Obviously, if $(Y,V)$ is an $F$-pair, it is an $F_1$-pair. If $(Y,V)$ is an $F_1$-pair, but not an $F$-pair, then for any nonfactor $A$ of $(Y,V)$, we must have

$$m_2(A)=m_2(V/C_V(A))-1. \tag{4.31}$$

Note that if $(Y,V)$ is an $F$-pair with $|A|=2$, then $m_2(A/C_V(A))=1$ and so $A$ centralizes a hyperplane of $V$. Thus the involution of $A$ is a transvection on $V$.

One of the first results in this direction is due to McLaughlin, who classified all groups which are generated by their transvections [219].

THEOREM 4.177. *Let $V$ be an irreducible module of dimension $n$ over $GF(2)$ for the group $Y$. If $Y$ is generated by its transvections, then $Y\cong SL(n,2)$, $Sp(n,2)$, $SO^{\pm}(n,2)$, $\Sigma_{n+1}$, or $\Sigma_{n+2}$ with $n\geqslant 4$ except in the first case. [Also $n>4$ when $Y\cong SO^+(n,2)$.]*

To prove the theorem, McLaughlin builds up the geometries of the classical and symmetric groups from properties of the action of the transvections on $V$. Note that as $V$ is defined over $GF(2)$, the subgroup of $Y$ generated by a transvection is a root subgroup, so the argument can be viewed as being in the same spirit as proofs of quadratic pair and Fischer–Timmesfeld type theorems.

Aschbacher has proved the following general result [17].

THEOREM 4.178. *Let $(Y,V)$ be an $F_1$-pair in which $L=F^*(Y)$ is a quasisimple K-group. Then one of the following holds:*

(i) $L\in Chev(2)$.
(ii) $L/Z(L)\cong A_n$ *for some* $n$.
(iii) $|Z(L)|=3$, $L/Z(L)\cong U_4(3)$ *or* $M_{22}$ *and* $(Y,V)$ *is not an F-pair.*
(iv) $L\cong M_{22}$, $M_{23}$, $M_{24}$, *and* $(Y,V)$ *is not an F-pair.*

The theorem definitely shows that failure of factorization is a characteristic 2 and alternating group phenomenon. Detailed properties of the groups of Lie type of odd characteristic and of the sporadic groups are required to prove that the groups in (iii) and (iv) are the only such candidates for $Y$.

Aschbacher also analyzes the alternating case in considerable detail, determining the possible modules and embedding of nonfactors in $Y$ under various conditions. In particular, he establishes the following result.

PROPOSITION 4.179. *If $(Y, V)$ is an F-pair with $Y = A_n$ or $\Sigma_n$, $n \geqslant 9$, and $V$ is an irreducible Y-module, then $V$ is the natural module for $Y$.*

Here the *natural* modules for $A_n$ and for $\Sigma_n$ are defined as follows. Let $X$ be a split extension of $E \cong E_{2^n}$ by $\Sigma_n$, in which the action of an element of $\Sigma_n$ on $E$ is determined by its corresponding permutation action on a fixed basis $x_1, x_2, \dots, x_n$ of $E$. Clearly $\Sigma_n$ fixes the element $x = x_1 x_2 \cdots x_n$ of $E$ and also leaves invariant the subspace $E_0 = \langle x_i - x_j | 1 \leqslant i, j \leqslant n \rangle$. If $n$ is even, then $\langle x \rangle < E_0 < E$ (with $|E : E_0| = 2$) and $V = E_0 / \langle x \rangle$ is the *natural* module for $\Sigma_n$ (and $V$ is of dimension $n-2$). On the other hand, if $n$ is odd, then $E = E_0 \times \langle x \rangle$ and $V = E_0$ is the *natural* module for $\Sigma_n$ (and $V$ is of dimension $n-1$). Thus in either case the natural module for $\Sigma_n$ is the unique nontrivial composition factor of $E$ as a $\Sigma_n$-module. The restriction of $V$ to $A_n$ is the *natural* representation of $A_n$.

Cooperstein and G. Mason [72, 73] have determined the possible $F$-pairs $(V, Y)$ when $Y \in Chev(2)$. Their analysis depends on the general theory, worked out by Steinberg and Curtis [272, 75] of the irreducible $GF(p)$-representations of the groups of Lie type of characteristic $p$. This theory will be outlined in Section 14; however, we need not be more explicit here, since the final result of Cooperstein and Mason's investigations can be phrased in more classical terminology, not involving the Lie theory machinery. Furthermore, as in the alternating case, there are some low-dimensional exceptions, which we shall not attempt to describe here.

THEOREM 4.180. *Let $(Y, V)$ be an F-pair with $Y$ of Lie type defined over $GF(q)$, $q = 2^m$, and $V$ irreducible as a Y-module. Then we have*

(i) *If $Y = SL(n, q)$, then $V$ is either the standard module over $GF(q)$ or its exterior square or the dual of either of these modules.*
(ii) *If $Y = SU(n, q)$, $n \geqslant 5$, then $V$ is the standard module over $GF(q^2)$.*
(iii) *If $Y = SO^{\pm}(n, q)$, $n \geqslant 11$, then $V$ is the standard module over $GF(q)$.*
(iv) *If $Y = G_2(q)$, then $V$ is the 6-dimensional symplectic module over $GF(q)$.*

Their analysis shows that the other exceptional groups do not occur as possibilities for $Y$ in an $F$-pair. Neither does the group $SU(3,q)$. Note also that if $V$ is a $Y$-module over $GF(q)$, $q=2^n$, then $V$ is also a $Y$-module over $GF(2)$ (as required by the definition of an $F$-pair).

Exterior products and dual representations have been defined in Section 1.4. Furthermore, the "standard module" for $SU(n,q)$ is simply the restriction to $SU(n,q)$ of the standard module for $SL(n,q^2)$ over $GF(q^2)$. The "standard module" $V$ for $SO^{\pm}(n,q)$ is the restriction of the standard module $V$ for $SL(n,q)$ to the subgroup leaving invariant an appropriate nondegenerate bilinear form on $V$ (see Section 4.14). Finally the group $G_2(q)$ is always a subgroup of $B_3(q)$, which in characteristic 2 is isomorphic to $C_3(q)$ and so $G_2(q)$ is a subgroup of $Sp(6,q)$. Thus when $q=2^m$, $G_2(q)$ acts irreducibly on a 6-dimensional symplectic space over $GF(q)$ and this is the module referred to in (iv) of Theorem 4.180.

With sufficient effort, one could undoubtedly determine the possible irreducible modules for $F_1$-pairs $(Y,V)$ with $F^*(Y)$ a quasisimple $K$-group by the same general methods as were used in establishing the preceding results.

If $X$ is a group with $F^*(X)$ a 2-group and the $(Z,J)$-factorization fails with respect to $S\in Syl_2(X)$, the natural question to ask is the following: What is the structure of the normal closure $N$ of $J(S)$ in $X$?

Note that if $T=S\cap N$, $T\in Syl_2(N)$, so that $X=NN_X(T)$ by the Frattini argument. But $N_X(T)$ normalizes $J(S)$ as $J(S)\operatorname{char} T$; so, in fact,

$$X=NN_X(J(S)).$$

Thus $N$ is a measure of the "obstruction" to the normality of $J(S)$ in $X$, which indicates the significance of its structure for local analysis.

As examples, if $X$ is the semidirect product of $SL(2,2^n)$ by its standard 2-dimensional $GF(2^n)$-module, it is immediate that $N$ is $X$ itself. More generally, if $X=X_1\times X_2\times\cdots\times X_r$, where $X_i\cong SL(2,2^{n_i})$, $n_i\geqslant 1$, on its standard module $V_i$, $1\leqslant i\leqslant r$, one again checks that $N=X$.

Remarkably in the general situation the structure of $N$ appears to resemble that of the preceding direct product case. However, we must allow for certain expected indeterminancies. Indeed, let $V=\langle\Omega_1(Z(S))^X\rangle$ and set $C=C_X(V)$. Then $C$ is a normal subgroup of $X$ which centralizes $\Omega_1(Z(S))$. We have no structural information concerning $C$ and hence none about $N\cap C$. Thus all we can hope to obtain is information about the structure of the group $N/N\cap C$ $(\cong NC/C)$. Of course, this is all we need, since the group $C$ obviously factors as it centralizes $\Omega_1(Z(S))$. In particular, if $N\leqslant C$, then $J(S) \leqslant C$, whence again by a Frattini argument $X = CN_X(J(S))$ and consequently $(Z, J)$-factorization holds in $X$. Hence the only case of interest is $N \nleqslant C$.

The structure of $N/N \cap C$ is obtained by an analysis of the action of the group $X/C$ on $V$ regarded as a vector space over $GF(2)$. By definition of $C$, this action is faithful. Furthermore, as in Proposition 4.162, we check that $X/C$ has no nontrivial normal 2-subgroups, a necessary prerequisite for deriving precise structural conclusions from this action.

The first result of this type is due to Glauberman, who treated the solvable case [114].

THEOREM 4.181. *Let $X$ be a solvable group with $F(X) = O_2(X)$. Let $S \in Syl_2(X)$, set $V = \langle \Omega_1(Z(S))^X \rangle$, $C = C_X(V)$, $\bar{X} = X/C$, and $N = \langle J(S)^X \rangle$. If $N \nleqslant C$, then we have*

(i) $\bar{N} = \bar{N}_1 \times \bar{N}_2 \times \cdots \times \bar{N}_r$, *where each* $\bar{N}_i \cong \Sigma_3$, $1 \leqslant i \leqslant r$.
(ii) If $U = [V, N]$ and $V_i = [V, \bar{N}_i]$, $1 \leqslant i \leqslant r$, then
  (1) $U = V_1 \times V_2 \times \cdots \times V_r$.
  (2) $V_i \cong Z_2 \times Z_2$, $1 \leqslant i \leqslant r$.

In particular, each $V_i$ is the standard module for $L_2(2)$ $(\cong \Sigma_3)$. If by chance $N \cap C = V$ (which will be the case if $C = V$), then indeed $N$ will be a direct product of subgroups $X_i$ isomorphic to the semidirect product of $V_i$ by $\bar{N}_i$, $1 \leqslant i \leqslant r$. However, the equality $N \cap O_2(X) = V$ implies that $\bar{N}$ acts *trivially* on $O_2(X)/V$. Hence, in the contrary case, $N$ will certainly not be such a direct product. On the other hand, the semidirect product of $V$ by $\bar{N}$ will still be such a direct product, and this group clearly captures many of the structural features of $N$.

In view of Theorem 4.178, the most we could hope for in general is that $\bar{N}$ is the direct product of subgroups $\bar{N}_i$ of Lie type of characteristic 2 or alternating or symmetric groups, $1 \leqslant i \leqslant r$. We would like to be able to assert that each $(V_i, \bar{N}_i)$ is an "irreducible" $F$-pair, where $V_i = [V, \bar{N}_i]$. However, easy examples show that $V_i$ need not be irreducible; in fact, $\bar{N}_i$ may have nontrivial fixed points on $V_i$. Thus the general formulation is that the pair $(V_i/C_{V_i}(N_i), \bar{N}_i)$ be an irreducible $F$-pair.

Aschbacher's work indeed leads to just such a theorem for $K$-groups. However, the proof requires the following important property of $J(S)$.

THEOREM 4.182. *Let $X$ be a $K$-group with $F^*(X) = O_2(X)$. If $S \in Syl_2(X)$, $V = \langle \Omega_1(Z(S))^X \rangle$, and $C = C_X(V)$, then under conjugation $J(S)$ leaves* *invariant and induces inner automorphisms on each component of $X/C$ (including "solvable" components).*

The proof involves the same kind of analysis as do Theorems 4.178 and 4.180. Combining these various results, Aschbacher can prove

THEOREM 4.183. *Let $X$ be a K-group with $F^*(X)=O_2(X)$. Let $S\in Syl_2(X)$, set $V=\langle\Omega_1(Z(S))^X\rangle$, $C=C_X(V)$, $\bar{X}=X/C$, and $N=\langle J(S)^X\rangle$. If $N\not\leqslant C$, then we have*

(i) *$\bar{N}$ is the direct product of subgroups $\bar{N}_i$, $1\leqslant i\leqslant r$, where each $\bar{N}_i$ is either a classical group over $GF(2^{m_i})$, $G_2(2^{m_i})$, or $\Sigma_{n_i}$, $n_i$ odd.*
(ii) *If $U=[V,\bar{N}]$, $\tilde{U}=U/C_U(\bar{N})$, $V_i=[V,\bar{N}_i]$, and $\tilde{V}_i=V_i/C_{V_i}(\bar{N})$, then*
 (1) *Each $(V_i,\bar{N})$ is an F-pair, $1\leqslant i\leqslant r$.*
 (2) *$\tilde{U}=\tilde{V}_1\times\tilde{V}_2\times\cdots\times\tilde{V}_r$.*

We indicate the procedure. For brevity, the term "component" here will include solvable components (i.e., isomorphic to $\Sigma_3$). Let $Y$ be the (normal) subgroup of $X$ consisting of those elements which under conjugation leave invariant the components of $X$, so that by the previous theorem $J(S)\leqslant T=S\cap Y$. Since $J(S)$ char $T$ and $T\in Syl_2(Y)$, the Frattini argument, as usual, yields $X=N_X(J(S))Y$. Hence $N=\langle J(S)^X\rangle=\langle J(S)^Y\rangle$, so it suffices to prove the theorem for $Y$. But clearly $C\leqslant Y$, whence $C=C_Y(V)$. Hence our hypotheses and notation carry over to $Y$, so without loss we can suppose that $Y=X$. Thus each component of $\bar{X}$ is normal in $\bar{X}$.

Let $\bar{K}$ be the product of those components $\bar{K}_i$ of $\bar{X}$ not centralized by $J(S)$ (for definiteness, numbered $1\leqslant i\leqslant r$), with $\bar{K}=1$ if no such $\bar{K}_i$ exist. We shall argue that $\bar{N}=\bar{K}$. Set $\bar{W}=C_{\bar{X}}(\bar{K})$. Since $\bar{K}\lhd\bar{X}$, also $\bar{W}\lhd\bar{X}$, and we see that $L(\bar{W})$ is a product of components of $L(\bar{X})$, and $F^*(\bar{X})=\bar{K}F^*(\bar{W})$. Now $J(S)$ centralizes $L(\bar{W})$ by definition of $\bar{K}$. Furthermore, $J(S)$ must centralize $F(\bar{W})$, since otherwise it would follow from Theorem 4.181 that $\bar{W}$ and hence $\bar{X}$ possessed a $\Sigma_3$-component $\bar{K}_0$ not centralized by $J(S)$. But then $\bar{K}_0\leqslant\bar{K}$, contrary to the fact that $\bar{W}$ centralizes $\bar{K}$. Thus $J(S)$ centralizes $F^*(\bar{W})$. Since $\overline{J(S)}$ acts faithfully on $F^*(\bar{X})$ by Proposition 1.27 and induces inner automorphisms on $\bar{K}$, we conclude that $\overline{J(S)}\leqslant\bar{K}$. But $\overline{J(S)}$ centralizes no component of $\bar{K}$, so, in fact, $\bar{K}=\langle\overline{J(S)}^{\bar{K}}\rangle$, whence $\bar{K}\leqslant\bar{N}$. On the other hand, if $K$ denotes the preimage of $\bar{K}$ in $X$, then $J(S)\leqslant K\lhd X$ (as $\bar{K}\lhd\bar{X}$), so $N=\langle J(S)^X\rangle\leqslant\langle K^X\rangle=K$. Hence $\bar{K}=\bar{N}$, as claimed.

Finally as $\bar{N}=\bar{K}$ is a central product of its components $\bar{N}_i=\bar{K}_i$, $1\leqslant i\leqslant r$, the action of $\bar{N}$ on $V$ can be analyzed quite easily. Indeed, using the definition of $J(S)$ [together with the fact that $J(S)$ centralizes no $\bar{N}_i$], one forces $U=[V, \bar{N}]$ to be essentially a direct sum of $\bar{N}_i$-modules—the precise statement is (ii)(2), where $V_i, \tilde{V}_i$, and $\tilde{U}$ are as in (ii)—and, in addition, each $(V_i, \bar{N}_i)$ to be an $F$-pair. But then as $\bar{N}_i$ is a $K$-group, the results of Cooperstein–Mason and Aschbacher described above imply that $\bar{N}_i$ has one of the forms listed in (i), $1 \leqslant i \leqslant r$, so all parts of the theorem hold.

## 4.13. Pushing-up, Aschbacher Blocks, and the Local $C(G;T)$-Theorem.

We come now to one of the most remarkable (and difficult) chapters in the entire theory of local analysis. It represents the culmination of all questions about factorization and generation of $p$-constrained groups and forms the basis for much of the work on the classification of simple groups of characteristic 2 type. The principal architects of the theory are Aschbacher, Bernd Baumann (a student of Fischer), and Glauberman, although several others were involved in the effort, notably Neville Campbell (a student of Aschbacher), Richard Niles (a student of Glauberman), and Sims.

The first encounter with pushing-up occurred in Sims's study of primitive permutation groups in which a one-point stabilizer has an orbit of length *three*. Subsequently Glauberman attempted to put Sims's work in a general framework. Baumann came upon a pushing-up problem from a different direction, in the course of classifying simple groups $G$ that possess a 2-central involution $x$ such that $C_G(x)$ has a normal Sylow 2-subgroup. On the other hand, Aschbacher reached the problem in the course of his analysis of thin simple groups.

Furthermore, the story is by no means ended, for Goldschmidt and subsequently Andrew Chermak, Niles, and Bernd Stellmacher are in the process of extending Sims's work to a geometric theory of the generation of a group in terms of 2-local subgroups possessing the same Sylow 2-group—a theory that one can view as a more primitive form of the generation of a $(B, N)$-pair by its parabolics.

This should clearly indicate the fundamental significance of pushing-up for the study of simple groups. But more on all this below; it is time to describe the various facets of the problem. We limit ourselves here to the prime 2, even though all questions can be formulated for $p$-constrained groups with $p$ arbitrary.

Consider then a group $G$ of characteristic 2 type; thus every 2-local subgroup of $G$ is 2-constrained with trivial core. The general pushing-up question can be expressed as follows:

*If $H$ is a 2-local subgroup of $G$ which does not contain a Sylow 2-subgroup of $G$, under what conditions is $H$ contained in a 2-local subgroup $H^*$ of $G$ whose Sylow 2-subgroups have greater order than those of $H$?* (4.30)

Let $T \in Syl_2(H)$ and expand $T$ to $S \in Syl_2(G)$, so that $T < S$ by assumption. We are thus essentially asking:

*Does there exist a nontrivial subgroup $R$ of $T$ such that*
(a) $R \lhd H$.
(b) $N_S(R) > T$? (4.31)

Indeed, we may then take $H^* = N_G(R)$. Note that $N_S(T) > T$ by Theorem 1.10, so that if $R$ char $T$, then $N_S(R) \geq N_S(T) > T$. Hence for characteristic subgroups, (4.31) reduces to the following question:

*Does there exist a nontrivial characteristic subgroup $R$ of $T$ which is normal in $H$?* (4.32)

The pushing-up question is often rephrased in negative terms: what restrictions are imposed on the structure of $H$ if $T$ possesses no subgroup $R$ satisfying (4.31) or (4.32)?

Here is a variation of (4.30):

*If $H$ and $K$ are two 2-local subgroups of $G$ having a common Sylow 2-subgroup $T$, under what conditions is there a single 2-local subgroup of $G$ containing both $H$ and $K$? In particular, when does there exist a nontrivial subgroup $R$ of $T$ which is normal in both $H$ and $K$?* (4.33)

One also encounters the pushing-up question in the context of establishing uniqueness results. For example, let $S \in Syl_2(G)$ and suppose we have somehow shown that $S$ is contained in a *unique* maximal 2-local subgroup $M$ of $G$. Can we then assert that $N_G(R) \leq M$ for every $1 \neq R \leq S$? Equivalently, is it the case that

$$\Gamma_{S,1}(G) \leq M? \qquad (4.34)$$

Assuming (4.34) to be false and arguing as we did in Theorem 4.111, we quickly produce a 2-local subgroup $X$ of $G$ with the following properties:

(a) $X \not\leqslant M$.
(b) $T = S \cap X \in Syl_2(X)$. (4.35)
(c) *If $H$ is a 2-local subgroup of $G$ such that $|S \cap H| > |T|$, then $H \leqslant M$.*

To answer (4.34) in the affirmative, we must, of course, derive a contradiction from the existence of a 2-local subgroup $X$ satisfying (4.35).

Fortunately these conditions severely restrict the structure of $X$. Indeed, by the uniqueness of $M$, $T \neq S$, otherwise $X \leqslant M$. Hence $T < S$. Furthermore, by (4.35c), if $R$ is a nontrivial normal subgroup of $T$ such that $N_S(R) > T$, then $N_G(R) \leqslant M$. Since $X \nleqslant M$, it follows that

$$\langle N_X(R) | 1 \neq R \lhd T, N_S(R) > T \rangle < X. \tag{4.36}$$

In particular,

$$\langle N_X(R) | 1 \neq R \operatorname{char} T \rangle < X. \tag{4.37}$$

Since $\Omega_1(Z(T))$ and $J(T)$ are characteristic in $T$, these conditions imply, in particular, that $\langle C_X(\Omega_1(Z(T))), N_X(J(T)) \rangle < X$, so that $(Z, J)$-*generation fails for* $X$. Thus Aschbacher's Theorem 4.183 applies to $X$ and gives an initial description of the normal closure $N = \langle J(T)^X \rangle$. Furthermore, as $J(T) \leqslant T \cap N$, it is immediate from properties of the $J$-subgroup that $J(T) = J(T \cap N)$, so $J(T) \operatorname{char} T \cap N$. But $T \cap N \in Syl_2(N)$ and $N \lhd X$, so by a Frattini argument $X = NN_X(J(T))$. Since $N_G(J(T)) \leqslant M$ by (4.35c), we conclude that

$$X = N(X \cap M). \tag{4.38}$$

Thus we see that $N$ is the obstruction to the assertion $X \leqslant M$. However, the picture of $N$ given by Theorem 4.183 is determined solely from the *two* characteristic subgroups $\Omega_1(Z(T))$ and $J(T)$; whereas, in fact, (4.37) asserts that $N_X(R) \leqslant M$ for *every* nontrivial characteristic subgroup of $T$. How much greater restriction does this impose on the structure of $N$? To formalize the question, we make the following definition, due to Aschbacher.

DEFINITION 4.184. For any group $X$ and Sylow 2-subgroup $T$ of $X$, set

$$C(X; T) = \langle N_X(R) | 1 \neq R \operatorname{char} T \rangle.$$

We call $C(X;T)$ the *characteristic generated core* of $X$; clearly it is determined up to conjugacy by $T$.

In the context of (4.36) and (4.37), we have $C(X;T) \leqslant M$, and we are interested in a description of the obstruction to the assertion $C(X;T)=X$, that is, a description of the smallest normal subgroup $Y$ of $X$ such that

$$X = YC(X;T). \tag{4.39}$$

This is the precise content of Aschbacher's "local $C(G;T)$-theorem" (Theorem 4.199), which asserts that $Y$ is a product of very special kinds of blocks, which Aschbacher calls *$\chi$-blocks* (Definition 4.189; the general definition of a block has been given in Section 4.5, Definition 4.83).

Essentially the same question as just considered can arise in the course of *proving* that a Sylow 2-subgroup $S$ of $G$ lies in a unique maximal 2-local subgroup. Indeed, suppose one has managed to show that $M=C(G;S)$ is a proper subgroup of $G$. Clearly the next step would be to prove that every maximal 2-local subgroup of $G$ containing $S$ lies in $M$. If not, there will then exist a maximal 2-local subgroup $X$ of $G$ with $S \leqslant X$ and $X \nleqslant M$. Thus $C(X;S)<X$ and so this time we are interested in the structure of $Y$ in (4.39) for the case $T=S$. Again this will be determined by the local $C(G;T)$-theorem.

The local $C(G;T)$-theorem is of fundamental importance in the study of simple groups of characteristic 2 type; indeed, it represents the principal tool in the proof of the "global" $C(G;T)$-theorem, which gives a complete classification of all simple groups $G$ of characteristic 2 type with a proper characteristic generated core. As the previous discussion indicates, the global $C(G;T)$-theorem reduces to the determination of such simple groups $G$ in which some maximal 2-local subgroup of $G$ possesses a $\chi$-block. Thus it is, in fact, derived as a consequence of a second major result in the subject: the so-called "block theorem," which completely classifies simple groups of characteristic 2 type in which some maximal 2-local subgroup has a $\chi$-block.

A full discussion of the global $C(G;T)$ and block theorems, including their utilization in the analysis of simple groups of characteristic 2 type, will be given in the sequel. Here we shall be concerned only with the local $C(G;T)$-theorem.

The various questions we have raised above have such a general flavor that it may be difficult for the reader to believe that a complete analysis of most of them reduces to a single extremely special question:

*Let $X$ be a group with $F^*(X)=O_2(X)$ and $X/O_2(X)\cong L_2(2^n)$ for some $n\geq 1$ and let $T\in Syl_2(X)$. What can be said about the structure of $X$ if no nontrivial characteristic subgroup of $T$ is normal in $X$?* (4.40)

Note that in this case, the given question is equivalent to the assertion that $C(X;T)<X$. Indeed, the groups $L_2(2^n)$ have Sylow 2-subgroups which are T.I. sets. Hence if we set $\bar{X}=X/O_2(X)$, it is immediate that $N_{\bar{X}}(\bar{T})$ is the unique maximal 2-local subgroup of $\bar{X}$ containing $\bar{T}$. This clearly implies that for $1\neq R$ char $T$, either $R\triangleleft X$ or else $N_{\bar{X}}(\bar{R})=N_{\bar{X}}(\bar{T})$, whence $N_S(R)=N_X(T)$. Hence if no such $R$ is normal in $X$, we conclude that $C(X;T)=\langle N_X(R)|1\neq R \text{ char } T\rangle=N_X(T)<X$.

Therefore (4.40) is, in fact, a very special case of the local $C(G;T)$-theorem; its answer is thus the assertion that $X$ is necessarily a block under the given assumptions. Moreover, the solution in this special case leads (as we shall see) rather directly to the full local $C(G;T)$-theorem.

It was this question (4.40) in the case $n=1$ ($L_2(2)\cong\Sigma_3$) that Glauberman reached in the course of generalizing Sims's permutation group results. The general case of (4.40) first arose in Baumann's analysis of groups in which the centralizer of a 2-central involution has a normal Sylow 2-subgroup, and its solution in that situation represented the key to his classification theorem concerning such groups.

This then is the background of the theory of pushing-up; now for some of its details. We begin with Sims [254]. Thus let $G$ be a primitive permutation group in which a one-point stabilizer $X$ has an orbit of length 3. (In particular, $X$ is a maximal subgroup of $G$.) It is not difficult to translate Sims's orbit hypothesis into group-theoretic terms. Indeed, he proves

PROPOSITION 4.185. *If $G$ is a primitive permutation group in which a one-point stabilizer $X$ has an orbit of length 3, then we have*

(i) $G=\langle X, X^g\rangle$ *for some* $g\in G$.
(ii) $|X: X\cap X^g|=3$.
(iii) $X/O_2(X)\cong\Sigma_3$.
(iv) *No nontrivial normal subgroup of $G$ is contained in* $X\cap X^g$.

Observe that by (iii), $|X|=2^a\cdot 3$ for some $a$, whence by (ii), $T=X\cap X^g$ has order $2^a$. Since $X^g\cong X$, it follows that $T$ is a Sylow 2-subgroup of both $X$

and $X^g$. Sims's goal is to obtain an upper bound for the value of $a$, so without loss we can assume that $a \geq 2$, in which case $O_2(X) \cong O_2(X^g) \neq 1$. But as $G$ is primitive, $X$ (and likewise $X^g$) is a maximal subgroup of $G$, so, in fact, each is a maximal 2-local subgroup of $G$. Thus we have here an example of the situation described in (4.33).

We claim that in the present case, no nontrivial subgroup $R$ of $T$ is normal in both $X$ and $X^g$. Indeed, if such an $R$ existed, then $\langle X, X^g \rangle \leq N_G(R)$, whence $G = N_G(R)$ by (ii). But then $R \leq O_2(G)$, so $O_2(G) \neq 1$, contrary to (iv).

This then is the situation which Sims was forced to analyze. His main result is the following:

THEOREM 4.186. *If $G$ is a primitive permutation group on a set in which a one-point stabilizer $X$ has an orbit of length* 3, *then $X$ has order dividing* $3 \cdot 2^4$.

In Section 2.6, we have described Sim's procedure for associating a directed graph $\Gamma$ with any transitive permutation group $G$ on a set $\Omega$ and an orbit $\Delta = \Delta(a)$ for $a \in \Omega$ of the one-point stabilizer $G_a$ of $G$ on $\Omega - \{a\}$. [For $b \in \Omega$, $b = a^g$ for some $g \in G$ as $G$ is transitive on $\Omega$, and one defines $\Delta(b) = (\Delta(a))^g$. Moveover, by definition, the vertices of $\Gamma$ are the elements of $\Omega$, and each $b \in \Omega$ is connected by an edge precisely to the points of $\Delta(b)$.] In particular, each vertex $b$ of $\Gamma$ is "adjacent" to exactly $|\Delta|$ vertices of $\Gamma$.

In the special case that the orbit $\Delta$ has length 3, which is the situation in Theorem 4.186, the resulting graph $\Gamma$ is called *cubic* or *trivalent*, as then each vertex of $\Gamma$ is connected to exactly three other vertices of $\Gamma$. W. T. Tutte was the first to study such (finite, connected) cubic graphs [308]. In [309], he considered those which possess a vertex-transitive group of automorphisms $G$ and showed that the stabilizer in $G$ of a vertex of $\Gamma$ has order dividing $3 \cdot 2^4$. Thus Sims was able to derive Theorem 4.186 by direct application of Tutte's result.

We note that although Sims defined $\Gamma$ in terms of the orbit $\Delta$, it can equally well be defined from the cosets of the subgroups $X$ and $X^g$ of Proposition 4.185. It is the latter point of view that provides the basis for Goldschmidt's extension of Theorem 4.186 (see Theorem 4.200).

Using Sims's result, Wong went on to completely classify primitive permutation groups in which a one-point stabilizer $X$ has an orbit of length 3 [322]. Indeed, using Sims's bound on $|X|$, he first showed quite directly that

$$X \cong \Sigma_3, \quad \Sigma_3 \times Z_2, \quad \Sigma_4, \quad \text{or} \quad \Sigma_4 \times Z_2. \tag{4.41}$$

In the first case, $G$ has a self-centralizing subgroup of order 3 and a theorem of Feit and Thompson [94] applies to yield the possibilities for $G$. In the second and third cases, $G$ has dihedral or quasi-dihedral Sylow 2-subgroups and $G$ is determined from the classification theorems for such groups. In the final case, Wong first pins down the possible structures of a Sylow 2-subgroup of $G$ (of order $2^4$ or $2^5$). In particular, $G$ has a normal subgroup of index 2 and again $G$ is determined from prior classification theorems.

The final result is as follows.

THEOREM 4.187. *If $G$ is a primitive permutation group in which a one-point stabilizer has an orbit of length 3, then $G \cong A_5$, $\Sigma_5$, $PGL(2,7)$, $L_2(11)$, $L_2(q)$, $q \equiv \pm 1 \pmod{16}$, $L_3(3)$, or $\mathrm{Aut}(L_3(3))$.*

Examining Sims's proof, Glauberman was led directly to a variation of question (4.30) in the case of a 2-local subgroup $X$ of the form $X/O_2(X) \cong \Sigma_3$ [more generally, with $X/O_2(X)$ an arbitrary dihedral group, $\Sigma_3$ being dihedral of order 6]. The following is a particular case of his first effort [113].

THEOREM 4.188. *Let $X$ be a 2-local subgroup of the group $G$ such that $F^*(X) = O_2(X)$ and $X/O_2(X)$ is a dihedral group. If no 2-local subgroup of $G$ contains $X$ properly, then either $X$ contains a Sylow 2-subgroup of $G$ or else $X \cong \Sigma_4$ or $\Sigma_4 \times Z_2$. In particular, $X$ is a block in the latter two cases.*

Glauberman's result (as well as Sims's) depends not only on the embedding of $T$ in $X$, but also on the embedding of $X$ in the ambient group $G$. To make the analysis completely local, it is desirable to study the structure of $X$ as an independent problem, *free* from its embedding in the larger group $G$. The most direct way to accomplish this and still retain the essential features of the given situation is to restrict oneself to *characteristic* subgroups of $T$ in place of *normal* subgroups. This is the underlying motivation for focusing on the characteristic generated core $C(X; T)$. Fortunately it has turned out that this restriction to characteristic subgroups has sufficed for most applications to groups of characteristic 2 type. (The notable exception is Goldschmidt's geometry of "amalgams," which is more directly related to the original Sims hypothesis.)

Initially Glauberman treated the case $X/O_2(X)$ dihedral. Subsequently Baumann studied the more general $X/O_2(X) \cong L_2(2^n)$ case as part of his 2-central involution problem [23]; independently, Niles obtained a proof of the same result [221]. Shortly thereafter, Glauberman and Niles [118] sharpened the theorem by showing that it sufficed to consider a specific *pair* of characteristic subgroups of a Sylow 2-subgroup of $X$ rather than the set of all characteristic subgroups.

It will be convenient to state the result in terms of $\chi$-blocks, so we define this term first.

Let then $X$ be a block and set $U=[\Omega_1(Z(O_2(X))), X]$ and $\bar{U}=U/C_U(X)$. By the definition, $X/O_2(X)$ acts trivially on $O_2(X)/U$ and irreducibly and nontrivially on $\bar{U}$.

DEFINITION 4.189. $X$ is called a *$\chi$-block* if $X/O_2(X)\cong L_2(2^n)$, $n\geq 1$, or $A_n$, $n$ odd, $n\geq 5$, and correspondingly $\bar{U}$ is the standard or natural module for $X/O_2(X)$.

Thus in the first case $\bar{U}$ can be identified with a 2-dimensional vector space over $GF(2^n)$ and $X/O_2(X)$ with $SL(\bar{U})$. In the second case, the action of $X/O_2(X)$ is determined from the natural permutation representation of $A_n$ (cf. Theorem 4.178). Note that $L_2(4)\cong A_5$ and the standard 4-dimensional $GF(2)$-module for $L_2(4)$ is not isomorphic to the natural 4-dimensional module for $A_5$. Hence there are two distinct types of $\chi$-blocks in the case $X/O_2(X)\cong L_2(4)$.

Now we can state the Baumann–Niles theorem.

THEOREM 4.190. *Let $X$ be a group with $F^*(X)=O_2(X)$ and $X/O_2(X)\cong L_2(2^n)$ for some $n$ and let $T\in Syl_2(X)$. Then either some nontrivial characteristic subgroup of $T$ is normal in $X$ or $X$ is a $\chi$-block (of type $L_2(2^n)$).*

Because it is important for the applications, I prefer to discuss the sharpened form of the theorem, due to Glauberman and Niles. The applications, in fact, require a slight extension of the result (see Theorem 4.199), namely, with the condition $X/O_2(X)\cong L_2(2^n)$ replaced by the weaker assumptions (a) $X/Y\cong L_2(2^n)$ for some normal subgroup $Y$ of $X$ containing $O_2(X)$, and (b) $T$ is contained in a unique maximal subgroup of $X$. [Note that as a Sylow 2-normalizer in $L_2(2^n)$ is the unique maximal subgroup containing that Sylow 2-subgroup, (b) is automatically satisfied in the case $Y=O_2(X)$ of Theorem 4.190.]

I shall briefly outline the Glauberman–Niles argument, for the way the "characteristic" hypothesis is used is quite marvelous (and similar in spirit to Glauberman's definition of $\hat{J}(T)$ in Definition 4.172). As already indicated, they prove a stronger assertion, namely, that there exist *two* nontrivial characteristic subgroups $T_1$ and $T_2$ of $T$, *depending only upon the structure of* $T$, such that either

$$T_1 \leqslant Z(X) \quad \text{or} \quad T_2 \triangleleft X \tag{4.42}$$

(assuming $X$ is not a block). Moreover, by their definitions,

$$T_1 \leqslant Z(T) \quad \text{and} \quad T_2 \text{ char } C_T(Z(J(T))). \tag{4.43}$$

As we shall see, it is this sharpened form of Theorem 4.190 that allows for a quick proof of the local $C(G;T)$-theorem. Moreover, as Baumann was the first one to realize the importance of the characteristic subgroup $C_T(Z(J(T)))$ of $T$ for the study of pushing-up, it has come to be called the *Baumann* subgroup of $T$ and is denoted by $\tilde{J}(T)$.

We argue by contradiction. First of all, by Thompson factorization, one immediately obtains:

PROPOSITION 4.191. *One of the following holds*:

(i) $Z(T)=Z(X)$.
(ii) $J(T) \triangleleft X$.
(iii) *If* $V=\Omega_1(Z(O_2(X)))$, $U=[V, X]$, *and* $\bar{U}=U/C_U(X)$, *then* $\bar{U}$ *is the standard module for* $X/O_2(X)$.

If (i) holds, then $T_1 \triangleleft X$ ($T_1$ and $T_2$ will be defined shortly), contrary to our assumption that Theorem 4.190 is false. Furthermore, if (ii) holds, it is not difficult to show [as (i) fails] that, in fact, $\tilde{J}(T)=\tilde{J}(O_2(X))$, whence $\tilde{J}(T) \triangleleft X$. Since $T_2$ char $\tilde{J}(T)$, also $T_2 \triangleleft X$, again a contradiction. Thus we have:

LEMMA 4.192. (i) $Z(T) \neq Z(X)$.
(ii) $J(T) \nleqslant O_2(X)$.

Hence Proposition 4.191 (iii) must hold. Since $X$ is not a block, it follows from the definition that $X/O_2(X)$ must act nontrivially on $O_2(X)/V$. This easily yields:

LEMMA 4.193. (i) *$T$ has class at least 3 (in particular,* $[T, \phi(T)] \neq 1$).
(ii) $\Omega_1(Z_2(T)) \leqslant O_2(X)$.

[Here $Z_2(T)$ denotes the "second center" of $T$, i.e., the preimage in $T$ of $Z(T/Z(T))$.]

Now they define $T_1$ as follows:

$$T_1=\Omega_1(Z(T))\cap T_0,\qquad \text{where } T_0=[T,\phi(T)]\phi(\phi(T)).\qquad (4.44)$$

By Lemma 4.193(i), $T_0$ and hence also $T_1$ are *nontrivial* characteristic subgroups of $T$ [and $T_1\leqslant Z(T)$]. (Notice how delicate is the choice of $T_1$.)

The definition of $T_2$ is even more subtle. We need a key preliminary result, due to Baumann.

PROPOSITION 4.194. *If we set* $Y=\langle \tilde{J}(T)^X\rangle$, *then* $\tilde{J}(T)\in Syl_2(Y)$.

In effect, the proposition can be viewed as a reduction to the case $T=\tilde{J}(T)$. Note that as $\tilde{J}(T)\not\leqslant O_2(X)$, we conclude that $Y$ covers $X/O_2(X)$ [i,e., $Y/O_2(Y)\cong X/O_2(X)$]. This in turn implies that $F^*(Y)=O_2(Y)$. Moreover, the same argument that yields Proposition 4.191 and Lemmas 4.192 and 4.193 shows that the conclusions of these lemmas hold with $Y$ and $\tilde{J}(T)$ in place of $X$ and $T$, so that all the properties established for $X$ indeed carry over to $Y$.

Now comes the really exotic twist. As Glauberman has philosophically phrased it, consider the "Platonic family $\mathfrak{F}$ of imaginary groups $Y^*$" with the following properties:

(a) $\tilde{J}(T)\in Syl_2(Y^*)$.
(b) $F^*(Y^*)=O_2(Y^*)$.
(c) $Y^*/O_2(Y^*)\cong L_2(2^{n^*})$ for some $n^*$.
(d) $Z(\tilde{J}(T))\neq Z(Y^*)$ and $J(\tilde{J}(T))\not\leqslant O_2(Y^*)$.
(e) $\Omega_1(Z_2(\tilde{J}(T)))\leqslant O_2(Y^*)$.

As we have seen, $Y$ itself is an element of $\mathfrak{F}$, so $\mathfrak{F}$ is nonempty.

Glauberman and Niles now define $T_2$ as follows:

$$T_2=\langle Z(O_2(Y^*))\,|\,Y^*\in\mathfrak{F}\rangle.\qquad (4.45)$$

It is essentially obvious that $\mathrm{Aut}(\tilde{J}(T))$ permutes the elements of $\mathfrak{F}$ among themselves and hence also permutes the subgroups $Z(O_2(Y^*))$ for $Y^*\in\mathfrak{F}$. Therefore $\mathrm{Aut}(\tilde{J}(T))$ leaves invariant the subgroup $T_2$ which they generate. Thus we have

PROPOSITION 4.195. *$T_2$ is a nontrivial characteristic subgroup of $\tilde{J}(T)$.*

Roughly speaking, $\mathfrak{F}$ is the family of "potential" minimal counterexamples to Theorem 4.190 and $T_2$ is the group generated by the normal closures of $Z(\tilde{J}(T))$ in each such group. Despite the "idealized" nature of its definition, it is nevertheless possible to prove things about $T_2$! The key result is the following:

PROPOSITION 4.196. *For any* $Y^* \in \mathfrak{F}$, $Z(O_2(Y))Z(O_2(Y^*)) \lhd Y^*$.

Hence as an immediate consequence, we obtain

PROPOSITION 4.197. *We have* $T_2 = Z(O_2(Y))T_2 \lhd Y$.

This completes the proof of Theorem 4.193. We introduce the following terminology.

DEFINITION 4.198. The subgroups $T_1, T_2$ of the 2-group $T$ will be called the Glauberman–Niles *characteristic pair* of $T$.

By combining Theorem 4.190 with his results from the previous section (specifically Theorems 4.182 and 4.183), Aschbacher is able to establish the following more precise form of the local $C(G; T)$-theorem. [We note that his published version of the theorem (which predates the Glauberman–Niles theorem) does not require $X$ to be a $K$-group. Indeed, the critical components of $X/O_2(X)$ are in the end identified by means of Timmesfeld's root involution theorem, Theorem 2.71. However, the present argument is simpler than the original and suffices for application to the classification theorem.]

THEOREM 4.199. *Let $X$ be a $K$-group with $F^*(X) = O_2(X)$, let $T \in Syl_2(X)$, let $T_1, T_2$ be the Glauberman–Niles characteristic pair of $T$, and set $Y = \langle C_X(T_1), N_X(T_2)\rangle$. If $Y < X$, then $X$ contains a nontrivial subgroup $L$ with the following properties*:

(i) $X = LY$.
(ii) $L = L_1 L_2 \cdots L_d$, *where each $L_i$ is a $\chi$-block of $X$*, $1 \leqslant i \leqslant d$.
(iii) $[L_i, L_j] = 1$ *for all* $i \neq j$, $1 \leqslant i, j \leqslant d$.

Since $Y \leqslant C(X; T)$, the theorem equally well gives a precise description of the obstruction to the assertion $X = C(X; T)$. Thus it yields as a corollary the local $C(G; T)$-theorem described in (4.39).

We outline the proof. Once again let $V = \langle \Omega_1(Z(X))^T \rangle$, $C = C_X(V)$ and $\bar{X} = X/C$. Since $C$ centralizes $\Omega_1(Z(T)) \leqslant V$, clearly $C \leqslant Y$. Also set $W = \langle C_X(\Omega_1(Z(T))), N_X(J(T))\rangle$. Since $T_1 \leqslant \Omega_1(Z(T))$ and $T_2$ char $J(T)$

char $J(T)$, we see that $W \leqslant Y$. Hence we can assume that $W < X$. Thus Thompson factorization fails in $X$, so we can apply Theorem 4.183 (together with the usual Frattini argument) to obtain that

$$X = NW = NY,$$

where $N = N_1 N_2 \cdots N_r$ with $C \leqslant N_i$ and the images $\bar{N}_i$ of $N_i$ in $\bar{X}$ classical groups over $GF(2^{m_i})$, $G_2(2^{m_i})$, or $A_{n_i}$, $n_i$ odd, $1 \leqslant i \leqslant r$. Moreover, the action of each $\bar{N}_i$ on $V$ is determined in accordance with Theorem 4.183(ii). In addition, $N \lhd X$, and $[\bar{N}_i, \bar{N}_j] = 1$ for $i \neq j$.

Suppose some $N_i$ contains a $\chi$-block $L_i$ which covers $\bar{L}_i$. In particular, $L_i$ is subnormal in $N_i$, which implies that $\bar{L}_i = \bar{N}_i$ [whence $N_i = L_i(T \cap N_i)$]. Also as $N_i$ is subnormal in $X$, so also is $L_i$ and hence $L_i$ is a $\chi$-block of $X$. Furthermore, if also $N_j$ contains a $\chi$-block $L_j$, then likewise $\bar{L}_j = \bar{N}_j$, so $[\bar{L}_i, \bar{L}_j] = [\bar{N}_i, \bar{N}_j] = 1$. But now using the three-subgroups lemma together with the specified action of $L_i$ on $O_2(L_i)$, we easily conclude that $[L_i, L_j] = 1$.

We see then that to prove Theorem 4.199, it will suffice to show for each $i$ that either $N_i \leqslant Y$ or $N_i$ contains a $\chi$-block $L_i$. Indeed, for definiteness, assume $N_i \nleqslant Y$, $1 \leqslant i \leqslant d$, and $N_i \leqslant Y$ for $i > d$, and set $L = L_1 L_2 \ldots L_d$. Then as $X = NY$ and $N_i = L_i C$, $1 \leqslant i \leqslant d$, with $C \leqslant Y$ it is immediate that $X = LY$. Also, as we have seen, distinct $L_i$ centralize each other, so the desired conclusions will follow.

The above argument allows us to analyze each $N_i$ separately, $1 \leqslant i \leqslant d$; and the aim is to use Theorem 4.190 to show that $N_i$ contains a $\chi$-block covering $L_i$. Since the argument is identical for all $i$, we take $i = 1$. Let $S$ be the subgroup of $T$ consisting of those elements which under conjugation leave $N_1$ invariant and induce inner automorphisms on $\bar{N}_1$. By Theorem 4.182,

$$J(T) \leqslant S.$$

Also clearly $T \cap N_1 \leqslant S$. Hence if we set $X_1 = N_1 S$, then we have

$$S \in Syl_2(X_1),$$

$$F^*(X_1) = O_2(X_1) \qquad (\geqslant O_2(X)),$$

and

$$\bar{X}_1 / \bar{C} O_2(\bar{X}_1) \cong \bar{N}_1.$$

Furthermore, $X_1 \nleqslant Y$.

Of course, in applying Theorem 4.190 to $X_1$, *we must work with the Glauberman–Niles characteristic pair* $S_1, S_2$ *of* $S$ *rather than with* $T_1, T_2$. But for this to be effective in proving Theorem 4.199, the pair $S_1, S_2$ must first be related to $T_1, T_2$. Since $J(T) \leqslant S$, we have, in any event, $J(S)=J(T)$. Also $\Omega_1(Z(T)) \leqslant \Omega_1(Z(S)) \leqslant \Omega_1(Z(O_2(X_1)))$. Using these facts together with the known action of each $\bar{N}_i$ on $V$, one can prove

PROPOSITION 4.200. (i) $\tilde{J}(S)=\tilde{J}(T)$, *whence* $S_2=T_2$;
(ii) $S_1 \geqslant T_1$.

Thus if we set $Y_1=\langle C_{X_1}(S_1), N_{X_1}(S_2)\rangle$, the proposition implies that $Y_1 \leqslant Y$. Because of this, it will suffice to prove Theorem 4.199 for $X_1$ and $S$. Indeed, then $X_1=JY_1$, where $J=J_1J_2\cdots J_e$ with each $J_i$ a $\chi$-block of $X_1$ and $[J_i, J_j]=1$ for $i\neq j$. Since $Y_1 \leqslant Y$ and $X_1 \not\leqslant Y$, it follows that $J \not\leqslant Y$. But the preimage $A$ of $O_2(\bar{X}_1)$ in $X_1$ is contained in $Y$ (as $A \leqslant CS$), so some $J_i \not\leqslant A$. Since $X_1/A \cong \bar{N}_1$, $J_i$ covers $\bar{N}_1$, which as we have seen above, suffices to complete the proof of Theorem 4.199. Hence without loss we can assume that $X_1=X$, in which case $X=N$ and $\bar{N}=\bar{N}_1$.

Consider first the case that $\bar{N}\in Chev(2)$ with $\bar{N}$ of Lie rank at least 2. We shall argue that $X=Y$, which will be a contradiction. Proposition 2.17 implies in this case that $\bar{N}$ is generated by its minimal parabolics $\bar{P}$ containing $\bar{T}$. Hence we need only show that the preimage $P$ in $X$ of any such $\bar{P}$ is contained in $Y$. As $\bar{N}$ is either classical or of type $G_2$, it is easy to check that the Levi factor of $\bar{P}$ is either a single linear or unitary group (of Lie rank 1). Hence by Proposition 2.17 and the discussion following, $\bar{P}=\bar{H}\bar{K}$, where $\bar{H}$ is a Cartan subgroup of $\bar{P}$, $O_2(\bar{P}) \leqslant \bar{K} \triangleleft \bar{P}$, and $\bar{K}/O_2(\bar{P}) \cong SL_2(2^m)=L_2(2^m)$ or $SU_3(2^m)$ for some $m$. Since $\bar{N}$ is defined over some $GF(2^n)$, $|\bar{H}|$ is odd, so if $K$ denotes the preimage of $\bar{K}$ in $X$, then $T \leqslant K$ and by a Frattini argument $P=KN_P(T)$. Since $N_P(T) \leqslant Y$, it thus suffices to show that $K \leqslant Y$.

We shall reduce to the extended form of the Glauberman–Niles theorem. Let $D$ be the preimage of $O_2(\bar{P})$ in $X$ and set $R=D\cap T$, so that $R\in Syl_2(D)$ and $D=CR$. In particular, $D \leqslant Y$ and $K/D \cong \bar{K}/O_2(\bar{P})$. Set $D_0=N_D(R)$ and $K_0=N_K(R)$. Then again by a Frattini argument, $K_0/D_0 \cong K/D$, so we need only show that $K_0 \leqslant Y$. [Note that as $R\in Syl_2(D)$, we also have that $D_0/R$ is of odd order.] Next let $K_1$ be minimal in $K_0$, subject to $T \leqslant K_1$ and $K_1$ covering $K_0/D_0$. Then we need only show that $K_1 \leqslant Y$. Setting $D_1=D_0\cap K$, we have likewise $K_1/D_1 \cong K/D$, $R \triangleleft D_1$, and $|D_1/R|$ is odd.

Now if $K_1/D_1 \cong SU_3(2^m)$, then as noted after Theorem 4.180, $(Z, J)$-factorization holds in $K_1$, in which case $K_1 \leqslant Y$. Thus we can assume that $K_1/D_1 \cong L_2(2^m)$. Furthermore, it follows from the minimality of $K_1$ and the structure of $L_2(2^n)$ that $D_1N_{K_1}(T)$ is the unique maximal subgroup of $K_1$

containing $T$. Hence the hypotheses of the full Glauberman–Niles theorem are indeed satisfied by $K_1$ and so either $T_1$ or $T_2$ is normal in $K_1$ (in which case $K_1 \leqslant Y$) or else $K_1$ is a $\chi$-block.

However, we claim that the latter possibility cannot occur here. Indeed, $\bar{K}/O_2(\bar{P})$ acts nontrivially on $\overline{O_2(P)} = \bar{R}$ by the Borel–Tits theorem, so $\bar{K}_1/\bar{R}$ acts nontrivially on $\bar{R} = R/O_2(X)$. On the other hand, as $F^*(X) = O_2(X)$, $K_1$ does not centralize $O_2(X)$. Hence $K_1$ has at least *two* nontrivial composition factors within $R$. But then it is immediate from the definition that $K_1$ cannot be a block.

Hence if $\bar{N} \in Chev(2)$, then $\bar{N}$ is necessarily of Lie rank 1 [whence $\bar{N}/Z(\bar{N}) \cong L_2(2^n)$, $Sz(2^n)$, or $U_3(2^n)$]. However, as $Y < X$, $(Z, J)$-factorization fails in $X$, so again the only possibility is that $\bar{N} \cong L_2(2^m)$. We now repeat the entire preceding argument with $\bar{N}$ itself in the role of $\bar{P}$. Again the Glauberman–Niles theorem yields that the (corresponding) subgroup $K_1$ of $X$ is either contained in $Y$ (contradiction) or else $K_1$ is a $\chi$-block. However, this time the latter alternative does not lead to a contradiction, but rather to the conclusion that $X$ itself contains a $\chi$-block covering $\bar{N}$, in which case the theorem holds.

The argument in the alternating case proceeds in essentially the same fashion, using an appropriate generation of these groups.

Aschbacher has also established an important variation of the local $C(G;T)$-theorem, in which $T$ is not required to be a Sylow 2-subgroup of $X$. To motivate the result, return to the situation with $M$ and $X$ at the beginning of the section. In investigating the relationship between these two subgroups of the group $G$, the group $Y = X \cap M$ will certainly be important. In some circumstances, one can show that $N_G(R) \leqslant X$ for every $1 \neq R$ char $O_2(Y)$. It follows for any such $R$ that $N_M(R) \leqslant X \cap M = Y$. This turns out to be a powerful restriction on the group $M$ as the following theorem of Aschbacher clearly illustrates [20]. Here we have replaced $M$ by $X$ and put $T = O_2(Y)$ to conform to the notation of the preceding theorems. Furthermore, to state it, we broaden the definition of $C(X;T)$ to cover arbitrary 2-subgroups of a group $X$. Thus for any 2-subgroup $T$ of the group $X$, we set

$$C(X;T) = \langle N_X(R) \mid 1 \neq R \text{ char } T \rangle.$$

Here then is the statement of Aschbacher's *weak* local $C(G;T)$-theorem.

THEOREM 4.201. *Let $X$ be a $K$-group with $F^*(X) = O_2(X)$ and let $T$ be a 2-subgroup of $S$ such that $Y = N_X(T)$ has the following properties*:

(a) $T = O_2(Y)$.
(b) $Y = C(X;T)$.

*Then either $Y = X$ (and $T = O_2(X)$) or one of the following holds:*

(i) *$X$ contains a $T$-invariant $\chi$-block $L$ with $LT/O_2(LT) \cong L_2(2^n)$ for some $n \geqslant 1$.*

(ii) *If we set $\bar{X} = X/O_2(X)$, then $L(\bar{X})$ possesses a $\bar{T}$-invariant component $\bar{L}$ such that one of the following holds:*

  (1) *$\bar{L} \in Chev(2)$, $\bar{T}$ induces inner automorphisms on $\bar{L}$, and if $\tilde{L} = \bar{L}\bar{T}/C_{\bar{L}\bar{T}}(\bar{L})$, then $N_{\tilde{L}}(\tilde{T})$ is a proper parabolic of $\tilde{L}$ with $\tilde{T} = O_2(N_{\tilde{L}}(\tilde{T}))$.*

  (2) *$\bar{L} \cong A_n$ or $\Omega_n^{\pm}(2)$, $\bar{T} \cap \bar{L} = 1$, and $\bar{T}/C_{\bar{T}}(\bar{L})$ has order 2 and induces a transposition or transvection, respectively, on $\bar{L}$.*

Aschbacher argues first that $T$ leaves invariant all $2'$-components of $X$ (including solvable ones). If all such $2'$-components are contained in $Y$, it is immediate that $T = O_2(X)$ and $Y = X$, so he can assume that there is a $2'$-component $L$ with $L \not\leqslant Y$. If $T \cap L \in Syl_2(L)$, he applies the local $C(G;T)$-theorem to $LT$ to conclude that $L$ is $\chi$-block. The assumptions in the present case then force $L$ to be of $L_2(2^n)$-type. The possibilities under (ii) arise when $T \cap L \notin Syl_2(L)$ by a fairly straightforward argument.

To see that the conclusion suggests further types of pushing-up questions, suppose, for example, that $\bar{L}$ satisfies (ii)(1) and let $X_0$ be the preimage of $\bar{L}\bar{T}$ in $X$. Then our conditions immediately imply:

(a) $F^*(X_0) = O_2(X_0)$.

(b) $X_0^* = X_0/O_2(X_0) \in Chev(2)$.

(c) $O_2(X_0) \leqslant T$ and $T^* = O_2(P^*)$ for some nontrivial proper parabolic subgroup $P^*$ of $X_0^*$.

(d) No nontrivial characteristic subgroup of $T$ is normal in $X_0$.

These conditions are a variation of those studied by Glauberman–Niles (but with $T$ not a Sylow 2-subgroup of the given group $X_0$). Under their assumptions, they were able to prove that the corresponding group $X$ was necessarily a block (Theorem 4.190). Thus the natural question to ask here is whether $X_0$ must be a block under the present conditions. Campbell has obtained an affirmative answer when $X_0^* \cong L_3(2^n)$ and $P^*$ is a particular parabolic subgroup [56]:

THEOREM 4.202. *Let $X$ be a group with $F^*(X) = O_2(X)$ and $\bar{X} = X/O_2(X) \cong L_3(2^n)$, $n > 1$. Let $P$ be a subgroup of $X$ containing $O_2(X)$ such that $\bar{P}$ is a maximal parabolic of $\bar{X}$ and set $T = O_2(P)$. If $S$ is a Sylow 2-subgroup of $P$ (and hence of $X$) containing $T$, assume that $C_X(Z(S)) \leqslant P$. Then either some nontrivial characteristic subgroup of $T$ is normal in $X$ or $X$ is a block.*

We remark that $\bar{X}$ possesses exactly two nontrivial proper (and hence maximal) parabolic subgroups $\bar{P}_1$, $\bar{P}_2$ containing $\bar{S}$. If the first alternative of the theorem fails, then the preimage in $X$ of precisely one of these will satisfy the given assumption on $P$. [This follows easily from the fact that $(Z, J)$-factorization must then fail for $X$ with respect to $T$.]

We conclude now with a few remarks about Goldschmidt's fundamental extension of the Sims–Tutte theorem [Theorem 4.186]. Goldschmidt considered the following situation:

(a) $G$ is a finite group generated by a pair of subgroups $X_1$, $X_2$.
(b) $|X_i: X_1 \cap X_2| = 3$ for $i = 1, 2$. (4.46)
(c) No nontrivial normal subgroup of $G$ is contained in $X_1 \cap X_2$.

Thus Sims's conditions (Proposition 4.185) represent the special case of (4.46) in which $X_2$ is $G$-conjugate to $X_1$. Motivation for studying groups of this form stems from the fact that $(B, N)$-pairs arising from Chevalley groups $G$ of Lie rank 2 defined over $GF(2)$ afford examples of (4.46). In those cases, the corresponding cubic graph of $G$ is precisely the Tits building associated with its $(B, N)$-pair presentation.

Indeed, if $G$ is any group satisfying (4.46), one can associate with $G$ a "bipartite" graph $\Gamma$ whose vertices are the set of right cosets of both $X_1$ and $X_2$ in $G$, with two vertices $X_1 g_1$, $X_2 g_2$, for $g_1, g_2 \in G$, connected by an edge if and only if

$$X_1 g_1 \cap X_2 g_2 \neq \varnothing .$$

(Cosets of the form $X_i g_1$, $X_i g_2$ are not connected by an edge, $i = 1, 2$.)

Clearly the group $G$ preserves incidence and so acts *transitively on the edges* of $\Gamma$. Thus you see the precise distinction between the Sims–Tutte and Goldschmidt situations: in the first case $G$ is both vertex- and edge-transitive, while in the second it is only necessarily edge-transitive.

For brevity, we shall call a group $G$ satisfying (4.46) a *weak* $(B, N)$-*pair* of rank 2 over $GF(2)$ and call $X_1$, $X_2$ *associated* (minimal) *parabolics* of $G$.

Goldschmidt completely determines the possible structures of the associated parabolics of such weak $(B, N)$-pairs. The analysis is very elegant, although quite elementary, and involves a delicate interplay of the underlying geometry with the group action on $\Gamma$.

We state his result as follows:

THEOREM 4.203. *If $G$ is a weak $(B, N)$-pair of rank 2 over $GF(2)$ with associated parabolics $X_1$, $X_2$, there are exactly 15 possibilities for the pair $(X_1, X_2)$. In particular, $X_i$ has order dividing $3 \cdot 2^7$ for both $i = 1, 2$.*

Goldschmidt pins down the precise structure of $X_1$ and $X_2$ in each of the 15 cases and gives examples of $G$ which realize each possibility.

The weak $(B, N)$-pair notion has a natural generalization: to groups of arbitrary rank defined over arbitrary finite fields. Chermak has studied certain difficult rank 3 cases over $GF(2)$, arising from some pushing-up questions in groups of characteristic 2 type; Niles has given a lovely characterization of the groups in $Chev(p)$ as weak $(B, N)$-pairs of rank $n$ in terms of conditions on the structure of the group $\langle X_i, X_j \rangle$ generated by each pair of the associated set of minimal parabolics $X_1, X_2, \ldots, X_n$ of $G$; and Stellemacher has extended Goldschmidt's result to weak $(B, N)$-pairs of rank 2 over $GF(2^m)$ [274].

Clearly this is a rich subject for further investigation. However, it is difficult to predict its future impact on the existing analysis of simple groups of characteristic 2 type.

## 4.14. Properties of $K$-Groups: Generalities

As repeatedly emphasized, local analysis requires for its success specific properties of the $K$-groups involved in the simple group under investigation. A great number of these appear only as proverbial "preliminary lemmas" within the many long classification papers, while others have been derived as independent efforts. It is fully evident to the practitioners that this important chapter of simple group theory is at present in a very unsatisfactory state. Ideally one would like to have a "general theory" of $K$-groups (and hence by the classification theorem, of all finite groups) analogous to the beautiful theory of solvable groups that has evolved since the original work of P. Hall in the 1920s and 1930s [156–159]. Such a theory would then provide the basis for the analysis of a minimal counterexample to the classification of simple groups in much the same way that the solvable theory did for groups of odd order and $N$-groups. However, the development of such a systematic theory of $K$-groups remains for the future.

The standard procedure for establishing a general property of $K$-groups $X$ is to reduce the problem to a question about suitable simple (and quasisimple) sections of $X$ (as, for example, in treating global balance or failure of factorization in 2-constrained groups). Thus our attention is soon focused on specific questions about the known simple groups. The major properties of simple $K$-groups which are needed for local analysis all seem to fall within (at least) one of the following categories:

A. Schur multipliers
B. Automorphisms
C. Centralizers of elements of prime order

D. Balance
E. Generation
F. Subgroup structure
G. Fusion
H. Signalizers
I. Modular representations
J. Small groups

To do much beyond simply listing the principal results within each category would extend this section unreasonably. However, because a complete list of all quasisimple $K$-groups is an absolute prerequisite for carrying out general local analysis, we shall make an exception in the case of Schur multipliers and discuss this one topic in some detail, in the next section.

Most properties of simple $K$-groups require separate analyses for groups of Lie type, alternating groups, and sporadic groups. In fact, the sporadic groups must usually be treated one at a time, with calculations and arguments depending upon the precise definition of the group. For the alternating groups, most properties are verified by direct calculation, using the natural permutation representation. On the other hand, for the groups of Lie type, there are often three subcases:

(1) The classical groups.
(2) The exceptional groups.
(3) The Suzuki–Ree groups $Sz$ $(={}^2B_2)$, ${}^2G_2$, ${}^2F_4$.

For the classical groups, analysis is based on properties of their underlying geometry (first systematically studied by Dieudonné [81, 82]; while for the exceptional groups the arguments usually stem from the theory of "algebraic linear groups" and the realization of these exceptional groups as the set of "rational" points of a suitable "endomorphism" of the corresponding algebraic group. Sometimes the latter approach can be used for the classical groups as well. On the other hand, some properties of the groups of Lie type (for example, the determination of their Schur multipliers) depends upon their Chevalley–Steinberg presentations by generators and relations. Finally the Suzuki–Ree groups often require special arguments, based on their $(B, N)$-pair definitions.

Besides this case division, there is a further complication: many of the basic properties of simple $K$-groups hold "in general," but their precise formulation includes a number of "exceptions"; and pinning these down exactly often requires considerable effort. Clearly, then, complete verification of the needed properties of simple $K$-groups is a very elaborate endeavor.

In this section we make some general comments about the groups of Lie type: the geometry of the classical groups, the relationship of the groups of Lie type to linear algebraic groups, and their modular representations.

We hope these remarks will give the reader some clue of what the analysis of groups of Lie type entails.

## A. The Geometry of the Classical Groups

Let $V$ be an $n$-dimensional vector space over $GF(q)$. The general linear group $GL(n,q)$ and the related groups $SL(n,q)$, $PGL(n,q)$, and $PSL(n,q)$ can all be studied from the natural action of $GL(n,q)$ on $V$. For example, in Section 1.5, we have described the centralizers of some involutions in $SL(n,q)$, using their matrix representations.

Let $f$ be a bilinear form on $V$ with values in $GF(q)$, [i.e., $f(au+bv,w)=af(u,w)+bf(v,w)$ for $u,v,w\in V$ and $a,b\in GF(q)$, with a similar identity holding for the second variable]. If $f(v,v)=0$ for all $v\in V$, $f$ is *skew-symmetric* or *alternating*; and if $f(u,v)=f(v,u)$ for all $u,v\in V$, $f$ is *symmetric*. $f$ is said to be *nonsingular* if $f(u,v)=0$ for all $v\in V$ implies that $u=0$. A subspace $U$ of $V$ is *nonsingular* (relative to $f$) if the restriction of $f$ to $U$ is a nonsingular bilinear form on $U$.

Note that if $q=2^n$, then $-v=v$ for $v\in V$ and so skew-symmetric implies symmetric. As a result, a little care must be taken with the definition of symplectic and orthogonal groups in this case; however, we shall ignore the point here and assume throughout that $q$ is odd, noting only that the various results stated here have analogues for $q$ even.

For any basis $(v)=\{v_1,v_2,\ldots,v_n\}$ of $V$, the matrix $(f(v_i,v_j))$ is called the *matrix* of $f$ *relative* to $v$. Associated with this matrix is a quadratic form

$$Q(x)=\sum_{i,j} f(v_i,v_j)x_ix_j.$$

The quadratic form $Q$ in turn determines $f$ (as $q$ is assumed odd). A change of basis of $V$ gives rise to an *equivalent* bilinear form $f'$ and corresponding *equivalent* quadratic form $Q'(x)$.

Two vectors $u,v\in V$ are called *orthogonal*, written $u\perp v$, if $f(u,v)=0$; and for any subspace $U$ of $V$, the *orthogonal complement* $U^{\perp}$ of $U$ is the set of $u'\in V$ such that $f(u,u')=0$ for all $u\in U$. We say that $V$ is the *orthogonal sum* of subspaces $V_1,V_2,\ldots,V_r$ if $V$ is the direct sum of the $V_i$ and the $V_i$ are mutually orthogonal.

A two-dimensional subspace $U$ of $V$ is called a *hyperbolic plane* provided $U$ has a basis $u,u'$ satisfying $f(u,u)=f(u',u')=0$, $f(u,u')=1$, and $f(u',u)\neq 0$. In particular, $U$ is nonsingular. Note that if $f$ is skew-symmetric or symmetric, then correspondingly $f(u', u) = -1$ or 1. Thus correspondingly the matrix of $f$ restricted to $U$ is

$$\begin{pmatrix} 0 & 1 \\ -1 & 0 \end{pmatrix} \quad \text{or} \quad \begin{pmatrix} 0 & 1 \\ 1 & 0 \end{pmatrix}.$$

A hyperbolic plane contains *isotropic* vectors, i.e., nonzero vectors $v$ such that $f(v, v) = 0$. It can be shown that any nonsingular plane which contains an isotropic vector is necessarily hyperbolic.

The principal structure theorems are as follows.

THEOREM 4.204. *If $f$ is a nonsingular skew-symmetric, then $n$ is even and $V$ is the orthogonal sum of hyperbolic planes.*

Hence, up to equivalence, there is only one nonsingular skew-symmetric bilinear form, and with respect to a suitable basis every such form can be written as

$$\begin{bmatrix} A & & & 0 \\ & A & & \\ & & \ddots & \\ 0 & & & A \end{bmatrix}, \qquad \text{where } A = \begin{bmatrix} 0 & 1 \\ -1 & 0 \end{bmatrix}.$$

In the orthogonal case, the situation is slightly more complicated. Let $a$ be a fixed element of $GF(q)$ which is not a square in the multiplicative group $GF(q)^{\times}$ of $GF(q)$.

THEOREM 4.205. *If $f$ is nonsingular symmetric, then $V = U \oplus W$, where $U$ is the direct sum of hyperbolic planes,* $\dim W \leqslant 2$, *and one of the following holds*:

(i) $W = 0$.
(ii) *$W$ is generated by a vector $w$ with $f(w, w) = 1$.*
(iii) *$W$ is generated by a vector $w$ with $f(w, w) = a$.*
(iv) *$W$ is generated by orthogonal vectors $w, w'$ with $f(w, w) = 1$ and $f(w', w') = -a$.*

In particular, $n = \dim V$ is even in cases (i) and (iv), and odd in cases (ii) and (iii). In case (iv), the plane $W$ contains no isotropic vectors and so is not hyperbolic. As a result, the geometries of cases (i) and (iv) are quite distinct. On the other, the distinctions between (ii) and (iii) are inessential since the quadratic form of (ii), when multiplied by $a$, becomes equivalent to that of (iii). Thus in even dimensions there are two geometries [case (i) is called of $+$*type*, case (iv) of $-$*type*], while in odd dimension there is essentially only one (for convenience, we call $V$ and $f$ of $+$*type* in the latter case). Correspondingly, $f$ is equivalent to a bilinear form having the matrix

$$\begin{bmatrix} A & & & & \\ & A & & 0 & \\ & & \ddots & & \\ & 0 & & A & \\ & & & & B \end{bmatrix},$$

where

$$A=\begin{bmatrix} 0 & 1 \\ 1 & 0 \end{bmatrix} \quad \text{and} \quad B=(0),(1),(a), \text{ or } \begin{bmatrix} 1 & 0 \\ 0 & -a \end{bmatrix}.$$

For brevity, one refers to $V$ as a *nonsingular symplectic* or *orthogonal* space according as $f$ is nonsingular and, respectively, skew-symmetric or symmetric.

We are interested in the group of nonsingular linear transformations of such a nonsingular space $V$ which *preserve* the bilinear form. Such linear transformations are called *isometries* of $V$, and correspondingly the group of isometries is called either the *symplectic* group on $V$ and denoted by $Sp(n,q)$ or the *orthogonal* group and denoted by $O^{\varepsilon}(n,q)$, where $\varepsilon=+1$ or $-1$ according as the form $f$ determining the geometry on $V$ is of $+$type or $-$type.

More generally if $V$ and $V'$ are any two nonsingular spaces of the same dimension over $GF(q)$ (with respective forms $f$, $f'$), a (nonsingular) linear transformation $T$ of $V$ onto $V'$ such that $f'(T(u),T(v))=f(u,v)$ for all $u,v\in V$ is called an *isometry*. In particular, the notion is applicable to two subspaces of $V$ of the same dimension which are nonsingular with respect to the restriction of the given bilinear form on $V$.

Clearly many properties (such as automorphisms, conjugacy classes, and centralizers of elements) of the groups $Sp(n,q)$ and $O^{\varepsilon}(n,q)$ will be deducible from their actions on $V$ and their matrix representations with respect to appropriately chosen bases in much the same way as in the linear case. Crucial to the arguments in the linear case is the fact that any nonsingular linear transformation between subspaces $U,U'$ of $V$ can be extended to a nonsingular linear transformation of $V$. The analogue of this result, due to Witt, is equally fundamental for studying the symplectic and orthogonal groups.

THEOREM 4.206. *Let $V$ be a nonsingular symplectic or orthogonal space over $GF(q)$. Then any isometry between two subspaces of $V$ can be extended to an isometry of $V$.*

The unitary groups have a similar description relative to a nonsingular "Hermitian" form. In this case, $V$ is defined over $GF(q^2)$, and for $a \in GF(q^2)$ one writes $\bar{a}$ for $a^q$. Thus the bar map is an automorphism of $GF(q^2)$ of period 2 analogous to complex conjugation in the field $\mathbb{C}$. A form on $V$ linear in the first argument is said to be *Hermitian* provided that

$$f(u,v)=\overline{f(v,u)}$$

for all $u, v \in V$. (Thus $f$ is "conjugate" linear in the second argument.) Nonsingularity and orthogonality are defined as before. One obtains a decomposition for $V$ in the nonsingular hermitian case, similar to that in the symplectic and orthogonal cases. However, this time there is exactly one geometry in each dimension, even or odd. Likewise an analogue of Witt's theorem exists, so that the unitary group $GU(n,q)$, $SU(n,q)$, $PGU(n,q)$, $PSU(n,q)=U_n(q)$ can be studied by the same methods as the linear, symplectic, and orthogonal groups. Details can be found in [7, 81, 82].

## B. Algebraic Linear Groups and the Finite Groups of Lie Type

The defining relations of the Chevalley groups over arbitrary algebraically closed fields $K$ make them strikingly similar to the complex Lie groups—all that is missing to make the analogy complete is an underlying *topology*. This is provided by the *Zariski* topology in which the open sets are complements of "subvarieties." (Since the defining relations for the Chevalley groups are polynomials, they are indeed algebraic varieties over $K$.)

From such considerations the general concept of an algebraic linear group was introduced (actually prior to Chevalley's construction of his groups). The various notions from the complex Lie theory could now be carried over to arbitrary algebraic linear groups: "connectedness"; "simple connectivity"; "connected component of the identity"; the "radical" as the maximal connected solvable normal subgroup; "semisimple" as a group with trivial radical; "torus" as an algebraic linear group algebraically (continuously) isomorphic to the direct product of copies of one-dimensional groups, each isomorphic to the multiplicative group of the field $K$; "reductive" group as one whose radical is a torus and central; "algebraic" homomorphisms and endomorphisms of algebraic linear groups; etc.

In the 1950s, Chevalley [67], using a beautiful mixture of Lie theory and algebraic geometry, completely classified the simple (and semisimple) linear algebraic groups; they turned out to be nothing else than his original Chevalley groups in disguise!

THEOREM 4.207. *If $G$ is a simple algebraic linear group over the algebraically closed field $K$, then $G$ is isomorphic to an (adjoint) simple Chevalley group over $K$.*

Thus the whole apparatus of "roots," "Euclidean vector space $E$ over $\mathbb{Q}$ of roots," "root subgroups," "Cartan subgroups," "Weyl" groups, "Bruhat" decomposition, etc., carries over to algebraic linear groups and forms part of their basic underlying theory.

If $K$ is of prime characteristic $p$, it turns out that there is a fundamental relationship between simple algebraic linear groups over $K$ and the finite Chevalley groups over $GF(p^m)$, which is extremely important for the verification of certain properties of these finite groups (and the related Steinberg twisted groups). The precise connection is too technical to describe except in schematic terms, which we proceed to do.

Just as a *real* Lie group can be defined to be the set of fixed points of the endomorphism $\alpha$ of the corresponding complex group induced from complex conjugation and the *unitary* groups to be the set of fixed points of the endomorphism $\alpha\tau$ of $GL(n, \mathbb{C})$, where $\tau$ is the transpose-inverse map, so the subgroup $G_\sigma$ of fixed points of an (algebraic) endomorphism $\sigma$ of an algebraic linear group $G$ is of equal importance in the development of the desired relationship. In particular, when $K$ is of prime characteristic $p$, the powers $\phi^m$ of the Frobenius automorphism $\phi$: $a \mapsto a^p$ for $a \in K$ induce such an endomorphism of $G$. The elements of $G_\sigma$ consist of those points of $G$ whose "coordinates" (in a suitable algebraic representation of $G$) satisfy the condition $x^{p^n} = x$, so that, in fact, the groups $G_{\phi^n}$ are precisely the corresponding finite Chevalley groups defined over $GF(p^n)$. [Of course, the version of $G_{\phi^n}$ depends upon that of $G$, e.g., $SL(m, K)_{\phi^n} = SL(m, p^n)$, while $PSL(m, K)_{\phi^n} = PSL(m, p^n)$.] Similarly by composing these endomorphisms with appropriate "graph automorphisms" of the underlying "root system" of $G$, one obtains the Steinberg twisted groups $U(n, p^m)$, $O^-(n, p^m)$, ${}^2E_6(p^m)$, and the triality groups ${}^3D_4(p^m)$. Also the Suzuki and Ree groups are described in terms of the fixed points of an algebraic endomorphism of a suitable simple linear algebraic group. (The full development appears in Steinberg [272, 273] and T. A. Springer [38].)

Summarizing we have

THEOREM 4.208. *Every finite Chevalley and Steinberg or Suzuki–Ree twisted group is the group of fixed points of an algebraic endomorphism of a simple algebraic group over the algebraic closure of $GF(p)$.*

For the balance of the discussion, we thus assume that $K$ is an algebraic closure of $GF(p)$ for some prime $p$.

Using Theorem 4.208, one can reduce questions about finite groups $G_\sigma$ of Lie type to questions about the corresponding algebraic linear group $G$. After solving the problem in $G$, one passes the resulting information back to $G_\sigma$ and obtains the desired solution in $G_\sigma$. Underlying this process is a fundamental result of Serge Lang, which we state in a form derived by Steinberg.

THEOREM 4.209. *Let $G$ be a connected linear algebraic group over $K$ and $\sigma$ an algebraic endomorphism of $G$ onto $G$ such that $G_\sigma$ is finite. Then every element of $G$ can be expressed in the form $x\sigma(x)^{-1}$ for some $x \in G$.*

On the basis of Lang's theorem, it is possible to describe precisely how a conjugacy class $X$ of elements of $G$ splits into conjugacy classes in $G_\sigma$ (i.e., the conjugacy classes of $G_\sigma$ contained in $X \cap G_\sigma$). To state the result, we need a definition.

DEFINITION 4.210. If $\sigma$ is an endomorphism of an algebraic linear group $G$, set $x \sim y$ for $x, y \in G$ provided there is $g \in G$ such that $x = gy\sigma(g)^{-1}$. Then $\sim$ is an equivalence relation on $G$, and we denote by $H^1(\sigma, G)$ the set of equivalence classes of $G$ under $\sim$.

Note that $\sigma$ leaves invariant the connected component $G^o$ of the identity element of $G$ and $G^o$ is an (algebraic) normal subgroup of $G$. Thus $\sigma$ induces an (algebraic) endomorphism of $G/G^o$ (also denoted by $\sigma$) and so $H^1(\sigma, G/G^o)$ is well defined. Furthermore, for any $x \in G$, $C_G(x)$ is an algebraic subgroup of $G$ and if $\sigma$ fixes $x$, then $\sigma$ leaves $C_G(x)$ invariant. Hence, in this case, $H^1(\sigma, C_G(x)/C_G(x)^o)$ is also well defined, where again $C_G(x)^o$ is the connected component of the identity element of $C_G(x)$.

THEOREM 4.211. *Let $G$ and $\sigma$ be as in Theorem 4.209. If $X$ is a $\sigma$-invariant conjugacy class of elements of $G$, then*

(i) *$\sigma$ fixes some element $x$ of $X$.*
(ii) *The conjugacy classes of $G_\sigma$ into which $X \cap G_\sigma$ splits are in one-to-one correspondence with the elements of $H^1(\sigma, C_G(x)/C_G(x)^o)$.*

COROLLARY 4.212. *Let $G, \sigma, X$, and $x$ be as in Theorem 4.211. If $C_G(x)$ is connected, then no splitting takes place. In other words, two elements of $X \cap G_\sigma$ which are conjugate in $G$ are also conjugate in $G_\sigma$.*

Theorem 4.211 and its corollary are powerful tools for studying the conjugacy classes of elements of $G_\sigma$ as well as the structure of the centralizers in $G_\sigma$ of "semisimple" elements (elements of order prime to the characteristic $p$ of $K$).

Lang's theorem is also useful for studying outer automorphisms of $G_\sigma$. Indeed, it has the following key consequence.

THEOREM 4.213. *Let $G$ be a connected linear algebraic group over $K$ and $\sigma$ an algebraic endomorphism of $G$ onto $G$ such that $G_\sigma$ is finite. Let $G^*$ be the semidirect product $G\langle\sigma\rangle$ of $G$ by $\sigma$. Then for any $x \in G$ the elements $x\sigma$ and $\sigma$ of $G^*$ are conjugate by an element of $G$.*

Steinberg has completely classified algebraic endomorphisms $\sigma$ of simple algebraic groups such that $\sigma$ is onto and $G_\sigma$ is finite [273, Section 11], showing that any such endomorphism $\sigma$ is, up to a permutation of the "roots" of the given group, essentially induced from a field automorphism (or from the special automorphisms of $B_2$, $G_2$, and $F_4$ which exist when $p=2$, 3, and 2, respectively).

Computation of centralizers of elements (and fixed points of automorphisms) of the finite groups of Lie type is based on another fundamental result of Steinberg about algebraic linear groups $G$. With $G$ viewed as a matrix group, an element $x$ of $G$ is called *semisimple* if it is diagonalizable. The inner automorphism of $G$ induced by conjugation by $x$ is an example of a "semisimple automorphism." The notion can be extended to outer automorphisms as well (we omit the precise definition).

Steinberg's result is the following:

THEOREM 4.214. *If $G$ is a simply connected linear algebraic group and $\sigma$ a semisimple automorphism of $G$, then the fixed subgroup $G_\sigma$ of $\sigma$ is a connected reductive algebraic group.*

Thus to compute the centralizer of a $\sigma$-fixed semisimple element (or automorphism) $x$ of a finite group $X=G_\sigma$ of Lie type, we pass to the corresponding algebraic group $G$ (or its simply connected cover) and from the general theory determine the semisimple automorphism $t$ of $G$ corresponding to $x$, compute $G_t$, using the preceding theorem to determine its structure, and finally intersect with $X$. In practice, the process is quite delicate, of course, but this describes the procedure.

We remark that a Cartan subgroup of an algebraic group $G$ is known to contain a representative of every conjugacy class of semisimple elements, a fact that greatly facilitates the computation of centralizers. On the other hand, centralizers of arbitrary elements and, in particular, "unipotent" elements are, in general, very difficult to calculate.

## C. Modular Representations of Groups of Lie Type

In previous sections we have seen how questions concerning the local structure of a simple group reduce to questions about $GF(p)$-representations of $K$-groups. Usually (often with considerable effort) these questions are further reduced to the quasisimple case. In most instances the subsequent analysis splits into two cases according as the given quasisimple $K$-group $X$ is or is not of Lie type of characteristic $p$. The arguments for the latter case tend to be special, depending upon particular properties of $X$. For example, the existence of a suitable *solvable* subgroup of $X$ might be used to reduce the problem to representations of solvable groups. On the other hand, when $X\in Chev(p)$, the arguments fall back on the general theory of $p$-modular representations of such groups developed by Steinberg and Curtis [271, 272, 74].

Before describing some of its general results, we wish first to emphasize that the theory is far from complete. Indeed, for most of the groups, one does not even know the *degrees*, let alone the actual representations of the modular irreducibles. What happens in practice—say, in the analysis of factorization failure—is this: the general theory enables one to show that only certain very special representations play a critical role, such as the "natural" representation or its symmetric square, representations that are well understood.

We also remark that the theory of *complex* representations of the groups of Lie type is also presently in a state of rapid development (but likewise with many basic questions still open), based on the key work of George Lusztig and Pierre Deligne [205], which greatly extended the earlier pioneering results of J. A. Green [145] on the general linear groups.

However, this theory does not seem to enter into local analytic group theory in a significant way.

The theory of modular representations begins with the representation theory of simple Lie algebras. Just as there is a $\mathbb{Z}$-lattice associated with the adjoint representation of $L$ (cf. Section 2.1), the same holds for every finite-dimensional complex representation of $L$. The entire process then lifts to an arbitrary field $K$ and ultimately to the corresponding Chevalley group over $K$. If $K$ is algebraically closed, we thus obtain, in view of Chevalley's classification theorem, a picture of the representation theory of simple algebraic linear groups over $K$ and more generally of reductive groups.

To keep the discussion brief, we pass directly to the representation theory of the groups. Likewise we restrict $K$ to be an algebraic closure of $GF(p)$ for some prime $p$, even though some of our assertions hold more generally. In addition, we limit ourselves to the Chevalley groups themselves, noting, however, that Steinberg has shown that the entire theory carries over to the twisted groups.

We first describe the situation when $K$ is algebraically closed. Let then $G$ be a semisimple algebraic linear group over $K$. Borel and Tits [37] have shown that all irreducible representations of $G$ are determined in a precise way from their so-called "rational" representations. To define this term, let $\phi$ be a representation of $G$ on a vector space $V$ over $K$ (all representations to be considered are finite dimensional) and let $(v)$ be a basis for $V$, so that the map: $g \mapsto (\phi(g)_{(v)})$ determines a matrix representation of $G$. As $G$ is an algebraic linear group, the elements $g \in G$ are $n \times n$ matrices $(g_{ij})$ with suitably determined coordinates $g_{ij}$, $1 \leqslant i, j \leqslant n$.

Definition 4.215. $\phi$ is said to be a *rational* representation of $G$ provided the entries of the matrices $(\phi(g)_{(v)})$ are *rational functions* of the variables $g_{ij}$.

In view of the formula for change of basis, the rationality of a representation of $G$ is independent of the choice of basis of $V$.

Let $\Sigma$ be the root system of $G$, $\chi_\alpha = \langle X_\alpha(t) \rangle$ its root subgroups, $H = \langle H_\alpha \rangle = \langle h_\alpha(t) \rangle$ a Cartan subgroup, $U = \langle \chi_\alpha | \alpha \text{ a positive root} \rangle$, and $B = HU$ a Borel subgroup. As we know, $G$ is described in terms of its Bruhat decomposition with respect to $B$ and $N = N_G(H)$, where $N/H = W$ is the Weyl group of $G$. Furthermore, $\Sigma$ spans a Euclidean vector space $E$ over $\mathbb{Q}$ with inner product $\langle\ ,\ \rangle$ such that $\langle \alpha, \beta \rangle \in \mathbb{Z}$ for all $\alpha, \beta \in \Sigma$.

Fundamental to the representation theory of $G$ are the notions of "weight" and "weight space."

DEFINITION 4.216. An element $\lambda \in E$ is called a *weight* (of $\Sigma$) provided $\langle\lambda, \alpha\rangle \in \mathbb{Z}$ for all $\alpha \in \Sigma$.

Thus the elements of $\Sigma$ are themselves examples of weights.

DEFINITION 4.217. Let $\phi$ be a rational representation of $G$ on a vector space $V$ over $K$. For any weight $\lambda$ of $\Sigma$, let $V_\lambda$ be the subset of vectors $v \in V$ such that for each element $h_\alpha(t) \in \mathrm{H}$, we have

$$\phi(h_\alpha(t))(v) = t^{\langle\lambda, \alpha\rangle} v.$$

[Clearly $V_\lambda$ is a subspace and each $\phi(h_\alpha(t))$ acts "diagonally" on $V_\lambda$.] We call $V_\lambda$ the *weight space* of $\phi$ (associated with $\lambda$).

Note that as $H$ is abelian, $V_\lambda$ is $H$-invariant for any weight $\lambda$. The basic fact here is that the restriction of $\phi$ to $H$ gives rise to a decomposition of $V$ as a direct sum of its ($H$-invariant) weight spaces, entirely analogous to the decomposition of a semisimple Lie algebra into root spaces with respect to a Cartan subalgebra.

PROPOSITION 4.218. *If $\phi$ is a rational representation of $G$ on a vector space $V$ over $K$, then $V$ is the direct sum of its weight spaces.*

In particular, $\phi$ has only a finite number of nontrivial weight spaces. Now we can state the main result concerning the representations of $G$.

THEOREM 4.219. *Let $\phi$ be an irreducible rational representation of $G$ on a vector space $V$ over $K$. Then there exists a uniquely determined weight $\lambda$ with the following properties*:

(i) $\dim_K(V_\lambda) = 1$.
(ii) *$U$ acts trivially on $V_\lambda$.*
(iii) *If $\mu$ is a weight such that $V_\mu \neq 0$, then $\mu = \lambda - \beta$, where $\beta$ is a suitable sum of positive roots of $\Sigma$.*
(iv) *$\langle\lambda, \alpha\rangle$ is a nonnegative integer for every positive root $\alpha \in \Sigma$.*

*Conversely, for any $\lambda \in E$ satisfying* (iv), *there is a unique (up to equivalence) irreducible rational representation $\phi$ of $G$ such that* (i), (ii), (iii) *hold for $\lambda$.*

DEFINITION 4.220. The weight $\lambda$ of Theorem 4.219 is called the *highest weight* of $\phi$.

Steinberg has proved a lovely result, which allows one to describe all irreducible rational representations of $G$ as tensor products of a certain restricted set of $p^l$ such representations, where $l$=rank of $G$=rank of Weyl group $W$ of $G$. (Although tensor products of representations have been defined in Section 1.4 only for finite groups, the definition carries over to arbitrary groups.)

To state the theorem, we first describe these "basic" representations. Let $\alpha_1, \alpha_2, \ldots, \alpha_l$ be a simple subsystem of roots of $\Sigma$. Note that for a given weight $\lambda$, the values $\langle \lambda, \alpha \rangle$ for $\alpha \in \Sigma$ are determined from the subset $\langle \lambda, \alpha_i \rangle$, as the $\alpha_i$ span $\Sigma$, $1 \leqslant i \leqslant l$.

Definition 4.221. A rational irreducible representation $\phi$ of $G$ is called *basic* if its highest weight $\lambda$ satisfies the condition

$$\langle \lambda, \alpha_i \rangle \leqslant p-1$$

for all $i$, $1 \leqslant i \leqslant l$.

In view of Theorem 4.219, there exist precisely $p^l$ such distinct basic representations.

Next the Frobenius automorphism; $t \mapsto t^p$ for $t \in K$ induces in a natural way an automorphism of $G$ (as an abstract group), which we denote by $Fr$, in which the element $g=(g_{ij})$ is transformed into $g^p=(g_{ij}{}^p)$ for every $g \in G$. If $\phi$ is any (rational) representation of $G$, the composition $\phi \circ Fr^j g \mapsto \phi(g^{p^j})$ is also a (rational) representation of $G$. Moreover, if $\phi$ is irreducible, so is $\phi \circ Fr^j$ for all $j$.

Theorem 4.222. *If $\phi$ is a rational irreducible representation of $G$, then for suitable uniquely determined positive integers $j_1, j_2, \ldots, j_m$ and basic representations $\phi_{j_1}, \phi_{j_2}, \ldots, \phi_{j_m}$, $\phi$ can be written (up to equivalence) in the form*

$$\left(\phi_{j_1} \circ Fr^{j_1}\right) \otimes \left(\phi_{j_2} \circ Fr^{j_2}\right) \otimes \cdots \otimes \left(\phi_{j_m} \circ Fr^{j_m}\right).$$

Finally we consider the finite Chevalley groups. Let then $G^*$ be such a group defined over $GF(p^m)$, $p$ a prime, and let $G$ be the corresponding linear algebraic group over an algebraically closed field $K$ of characteristic $p$, so that as we have asserted earlier, $G^*$ is the fixed-point subgroup $G_\sigma$ of a suitable endomorphism $\sigma$ of $G$. In particular, $G^* \leqslant G$, so every representation $\phi$ of $G$ on a vector space $V$ over $K$ gives rise by *restriction* to a representation of $G^*$ over $K$.

The complete analysis of this restriction process ultimately yields a general theory of the irreducible representations of $G^*$ over the *defining field* $GF(p^m)$ of $G^*$, the statements of which are quite analogous to the results for $G$ itself. We content ourselves with the following three assertions:

1. *Every irreducible representation of $G^*$ on a vector space over $GF(p^m)$ is realizable through this process of restriction.*
2. *There is a one-to-one correspondence between inequivalent such irreducible representations of $G^*$ and highest weights.*
3. *There exists a tensor product formula involving the $p^l$ basic representations (i.e., the irreducibles of $G^*$ arising by restriction from the basic representations of $G$) and the $m$ automorphisms of $G^*$ induced from the Galois group of $GF(p^m)$ over the prime field $GF(p)$.*

For emphasis, we repeat our earlier statement that the basic representations of most of the (finite) groups of Lie type (as well as of linear algebraic groups of prime characteristic) have not been determined.

## 4.15. Properties of $K$-Groups: Specifics

We now consider in succession the ten types of properties of quasisimple $K$-groups, listed at the beginning of the preceding section. As we have stipulated, only the first of these will be treated in any detail.

### A. Schur Multipliers

A thorough treatment of the general theory of Schur multipliers appears in Huppert [181, Section 5.23]. We first summarize the key definitions and general results. We fix a finite group $X$. The Schur multiplier of $X$ is given in terms of "second cohomology" groups, so we must define that term first.

Let $A$ be any abelian group and denote by $C^2(X; A)$ the set of all maps of $X\times X$ into $A$, which we make into an abelian group by pointwise addition [i.e., for $f, g\in C^2(X; A)$, $(f+g)(x, y)=f(x, y)+g(x, y)$ for all $(x, y)\in X\times X$]. Let $B^2(X; A)$ be the subgroup of $C^2(X; A)$ generated by those $f\in C^2(X; A)$ of the form

$$f(x, y)=g(x)-g(xy)+g(y)$$

for some mapping $g$ from $X$ to $A$. Let $Z^2(X, A)$ be the subgroup of

$C^2(X; A)$ generated by those $f \in C^2(X; A)$ such that

$$f(y, z)+f(x, yz)=f(xy, z)+f(x, y)$$

for all $x, y, z \in X$. Clearly $B^2(X; A) \leqslant Z^2(X; A)$.

DEFINITION 4.222. We set

$$H^2(X; A)=Z^2(X; A)/B^2(X; A)$$

and call $H^2(X; A)$ the *second cohomology group* of $X$ with *coefficients* in $A$.

[In the usual definition, $A$ is not only an abelian group but also a right $X$-module. Our definition corresponds to the "trivial" case (i.e., $ax=a$ for all $a \in A$, $x \in X$), which is all we need for the discussion.]

Now let $K$ be an algebraically closed field of characteristic 0 (e.g., $K=\mathbb{C}$) and let $K^\times$ be the multiplicative group of nonzero elements of $K$.

DEFINITION 4.223. The group $H^2(X; K^\times)$ is called the *Schur multiplier* of $X$.

Schur (who incidentally was Brauer's teacher) established the basic properties of these multipliers and computed those of the alternating groups and some of the classical groups [245, 246]. His primary result is an expression for $H^2$ in terms of a representation of $X$ as a factor group of an infinite "free" group, which, in particular, shows that the definition is independent of the field $K$. The simplest way to define free group is by means of its universal mapping property.

DEFINITION 4.224. A group $F$ is said to be *free* (of *rank* $n$) provided:

(a) $F$ is generated by $n$ elements $f_1, f_2, \ldots, f_n$.
(b) If $G$ is any group generated by $n$ elements $g_1, g_2, \ldots, g_n$, there exists a homomorphism $\phi$ of $F$ onto $G$ such that $\phi(f_i)=g_i$, $1 \leqslant i \leqslant n$.

Free groups exist in every rank and any two of the same rank are isomorphic (Huppert [181, Section 1.19]).

Thus any finite group $G$ can be represented in the form $F/R$ for some free group $F$ and normal subgroup $R$. Of course, this can be done in many ways; the only restriction on $F$ is that its rank be at least the minimal number of generators of $G$.

To state Schur's result, we need a well-known property of "finitely generated" abelian groups—i.e., abelian groups generated by a finite number of elements. In any abelian group $A$, the set $T$ of elements of $A$ of finite order forms a subgroup, called the *torsion* subgroup of $A$. Furthermore, $A$ is said to be *free abelian* (of *rank* $n$) if $A$ is isomorphic to the direct product of $n$ copies of $\mathbb{Z}$. We note also that every finite abelian group can be written as the direct product of cyclic groups $A_1, A_2, \ldots, A_n$ with $|A_i|$ dividing $|A_{i-1}|$, $2 \leqslant i \leqslant n$. Moreover, the integer $n$ is independent of the decomposition and is also called the *rank* of $A$. (Proposition 1.8 concerning abelian $p$-groups is a particular case of this result.)

PROPOSITION 4.225. *If $A$ is a finitely generated abelian group, then we have*

(i) *$A = D \times T$, where $T$ is the torsion subgroup of $A$ and $D$ is free abelian.*
(ii) *$T$ is finite.*
(iii) *If $A = D^* \times T$ with $D^*$ free abelian, then $D^* \cong D$; in particular, $D^*$ and $D$ have the same rank.*

The proposition implies that $A$ is the direct product of cyclic groups, the number of factors being rank $(D)$+rank $(T)$ and independent of the decomposition. This integer is called the *rank* of $A$. Moreover, the rank of $D$ is called the *torsion-free* rank of $A$.

Now we can state Schur's theorem.

THEOREM 4.226. *Represent $X = F/R$, where $F$ is free of rank $n$ with $R \lhd F$ and set $Y = F/[R, F]$ and $A = R/[R, F]$. Then we have*

(i) *$A$ is a finitely generated abelian group of torsion-free rank $n$.*
(ii) *$A \leqslant Z(Y)$.*
(iii) *$T = (R \cap F')/[R, F]$ is the torsion subgroup of $A$.*
(iv) *$H^2(X; K^\times) \cong T$.*
(v) *If $A = D \times T$, where $D$ is free abelian and we set $\bar{Y} = Y/D$, then*

   (1) *$\bar{Y}$ is finite and $\bar{Y}/\bar{T} \cong X$.*
   (2) *$\bar{T} \leqslant \bar{Y}' \cap Z(\bar{Y})$.*
   (3) *$\bar{T} \cong T \cong H^2(X; K^\times)$.*

Thus (iv) shows that the Schur multiplier is determined independently of $K$ and is finite, while (v) shows that there exists a covering group of $X$ whose center is isomorphic to the Schur multiplier of $X$ (namely, the group $\bar{Y}$). We call such an extension group a *universal covering* group of $X$. This term is justified by the following fact.

PROPOSITION 4.227. *If $\pi$ is a homomorphism of any group $Y$ onto $X$ such that* $\ker(\pi) \leq Z(Y) \cap Y'$, *then* $\ker(\pi)$ *is a homomorphic image of the Schur multiplier of $X$, and $Y$ is a homomorphic image of some universal covering group of $X$.*

On the other hand, it is not necessarily the case that universal covers of $X$ are uniquely determined up to isomorphism. However, this does hold in an important special case.

PROPOSITION 4.228. *If $X$ is perfect (in particular, simple), then any two universal covering groups of $X$ are isomorphic.*

The Schur multiplier, being abelian, is the direct product of its Sylow $p$-subgroups, called the *p-parts* of the multiplier. In view of Proposition 4.227, each $p$-part is determined from the largest covering group of $X$ by an abelian $p$-group. As might be expected, the latter extension is directly tied to the structure and embedding of a Sylow $p$-subgroup of $X$. Indeed, one has

PROPOSITION 4.229. *If $P \in Syl_p(X)$ and $P \leq N \leq X$, then the p-part of the Schur multiplier of $X$ is isomorphic to a subgroup of $H^2(N; K^{\times})$.*

One is therefore interested in which elements of $H^2(N; K^{\times})$ "lift" to $X$. In Cartan–Eilenberg [60] conditions are given, on any subgroup $Y$ of $X$, to lift elements of $H^2(Y; K^{\times})$ to $H^2(X; K^{\times})$. They involve the subgroups $Y \cap Y^x$ for $x \in X$. We mention only a special case of their result, to give an idea of its nature.

PROPOSITION 4.230. *Let $P \in Syl_p(X)$ and suppose that $P$ is a T.I. set in $X$. Then the p-part of the Schur multiplier of $X$ is isomorphic to that of $H^2(N_X(P); K^{\times})$.*

This describes the underlying theory of Schur multipliers; it remains to discuss their actual calculation. In view of the above results, one would anticipate two possible approaches: one depending upon a presentation of $X$ by generators and relations, and the other, by restriction to the various subgroups of $X$ (especially Sylow subgroups as well as large simple subgroups whose multipliers have already been determined); indeed, in practice, both have been used.

The simplest result of the first type is due to Schur [245].

PROPOSITION 4.231. (i) *If $X$ is cyclic or quaternion, then $X$ has trivial Schur multiplier.*

(ii) *If $X$ is a dihedral 2-group, then $X$ has $Z_2$ as Schur multiplier and the universal covering groups of $X$ are quaternion, dihedral, or quasi-dihedral.*

Combining Propositions 4.227 and 4.231, one immediately obtains

PROPOSITION 4.232. (i) *$SL_2(p)$, $p$ a prime, has trivial Schur multiplier.*
(ii) *$L_2(p)$, $p$ an odd prime, has $Z_2$ as Schur multiplier.*

Indeed, the point is that the Sylow $q$-subgroups of $SL_2(p^n)$ are known to be cyclic for all odd $q \neq p$ and to be quaternion for $q=2$ and $p$ odd. Moreover, they are isomorphic to $E_{p^n}$ for $q=p$, so that for $n=1$, also the Sylow $p$-subgroups are cyclic.

The discussion also shows that the Schur multiplier of $SL_2(p^n)$ is a $p$-group, isomorphic to the $p$-part of the Schur multiplier of $L_2(p^n)$. Moreover, in both cases a Sylow $p$-subgroup is a T.I. set, so that Proposition 4.230 can be used. In the latter case the Sylow $p$-normalizer is a Frobenius group with kernel $P \cong E_{p^n}$ and cyclic complement $H$ of order $\varepsilon(p^n-1)$, where $\varepsilon=\frac{1}{2}$ or 1 according as $p$ is odd or even. Because the action of $H$ is irreducible, the problem is reduced to the following:

Determine the largest $p$-group $Q$ of class 2 such that

(1) $Q/Z(Q) \cong P$ and $Z(Q) \leqslant Q'$.
(2) $Q$ admits an automorphism $\alpha$ of order $|H|$ which acts trivially on $Z(Q)$.

This problem is easily resolved, using "Lie ring" methods, a standard technique for investigating automorphisms of $p$-groups. As noted, this is the basic method that Higman used in his classification of "Suzuki 2-groups"

[170]. We refer the reader to [130, Section 5.6]. It turns out that $Z(Q)$ is forced to be trivial in all but the two cases $p^n=4$ or 9; correspondingly $Z(Q)\cong Z_2$ or $Z_3$ (with $Q$ quaternion of order 8 or extra-special of order 27). [Since $L_2(4)\cong L_2(5)$, the first case is, in fact, covered by Proposition 4.232.] Thus we have

PROPOSITION 4.233. *$SL_2(p^n)$ has trivial Schur multiplier if $p^n\neq 4$ or 9 and has Schur multiplier $Z_p$ if $p^n=4$ or 9.*

For the generator and relation approach, it is best to work directly with the cocycle condition in the definition of Schur multiplier. Let then $Y$ be a universal covering group of $X$, so that $Y/A\cong X$ for suitable $A\leqslant Y'\cap Z(Y)$ with $A\cong H^2(X;K^\times)$. Let $y_1, y_2,\ldots, y_n$ be a complete set of coset representatives of $A$ in $Y$, so that their images are the distinct elements of $X$. If $I$ denotes the set $\{1,2,\ldots,n\}$, there is a well-defined function $\pi$ from $I\times I\mapsto I$, determined by the condition that $y_i y_j$ and $y_{\pi(i,j)}$ lie in the same coset of $A$. Then for all $i$, $j\in I$,

$$y_i y_j=f(y_i, y_j)y_{\pi(i,j)}.$$

Thus $f$ is a function from $I\times I$ to $A$ (which we can view as a function from $X\times X$ to $A$). $f$ is called a *factor set* (or *factor system*) relative to the given coset representatives $y_1, y_2,\ldots, y_n$. Using the associative law, one obtains the basic relation

$$f(y_i, y_j)f(y_i y_j, y_k)=f(y_i, y_j y_k)f(y_j, y_k)$$

for all $i, j, k\in I$. That is, $f\in Z^2(X, A)$.

If $y_i'$, $i\in I$, is a second set of coset representatives, then $y_i'=a(y_i)y_i$ for suitable $a(y_i)\in A$, and we see that the corresponding factor system $f'$ is related to $f$ by the condition

$$f'(y_i', y_j')=a(y_i)a(y_j)a(y_i y_j)^{-1}f(y_i, y_j),$$

so that the images of $f$ and $f'$ in $H^2(X;A)$ are equal.

This should give the reader some idea of the basis for the computation of Schur multipliers.

These calculations were principally the work of three individuals: Schur, Steinberg, and Griess, but others lent a hand with specific groups or families of groups. Schur computed the multipliers of the alternating groups using their standard presentation [246]. In the Lie case, Steinberg proved two general results, based on their standard presentations [271].

THEOREM 4.234. *If $X$ is a universal Chevalley or Steinberg twisted group over $GF(p^m)$, $p$ a prime, then the $p'$-part of the Schur multiplier of the simple group $X/Z(X)$ is isomorphic to $O_{p'}(Z(X))$.*

Thus universal groups of Lie type capture all but possibly the $p$-part of the Schur multipliers of corresponding simple groups of Lie type, where $p$ denotes the characteristic of the group.

Steinberg also proved

THEOREM 4.235. *If $X$ is a simple Chevalley or Steinberg twisted group over $GF(p^n)$, then the $p$-part of the Schur multiplier of $X$ is trivial for $p^n$ sufficiently large.*

Subsequently Alperin and I treated the Suzuki groups and the Ree groups of characteristic 3 [5], while H. N. Ward [317] computed the multipliers of the Ree groups of characteristic 2 with the exception of the Tits group.

This left the computation of the $p$-part of the multiplier for certain groups of Lie type over $GF(p^n)$ with $p^n$ small plus the determination of the Schur multipliers of the sporadic groups, all very painful cases. The bulk of these were handled by Griess, largely in his doctoral thesis under Thompson [146, 147, 151]. However, Burgoyne, Feit, Fong, J. Grover, Janko, W. Lempken, Lyons, McKay, P. Mazet, Rudvalis, Thompson, and Wales were involved in one or more of the sporadic cases.

We conclude with Table 4.1, prepared by Griess, of the Schur multipliers of all finite simple groups.

In the case of the alternating and sporadic groups, we list the group together with its Schur multiplier as a pair.

In particular there are, apart from even-dimensional orthogonal groups of odd charactertistic, precisely six simple groups having noncyclic Schur multipliers:

THEOREM 4.236. *If $X$ is a known simple group, then either the Schur multiplier of $X$ is cyclic or $X$ is isomorphic to one of the following groups*: $D_{2l}(q)$, $q$ odd, $A_2(4)(\cong L_3(4))$, $D_4(2)(\cong \Omega_8^+(2))$, ${}^2A_3(3)(\cong U_4(3))$, ${}^2A_5(2)$ $(\cong U_6(2))$, ${}^2B_2(8)(\cong Sz(8))$, ${}^2E_6(2)$.

**Table 4.1.** Schur Multipliers of the Known Simple Groups

**Finite Groups of Lie Type**
The multiplier of
*$G(q)$, $q=p^n$, is $R\times P$ where $R$ is a $p'$-group and $P$ is a $p$-group.*

| *Group of Lie type* | $R$ | $R\times P$ (*when* $P\neq 1$) | $(l,q)$ |
|---|---|---|---|
| $A_l(q)$ | $\mathbb{Z}_{(l+1,q-1)}$ | $\mathbb{Z}_2$ | (1,4) |
| | | $\mathbb{Z}_2\times\mathbb{Z}_3$ | (1,9) |
| | | $\mathbb{Z}_2$ | (2,2) |
| | | $\mathbb{Z}_3\times\mathbb{Z}_4\times\mathbb{Z}_4$ | (2,4) |
| | | $\mathbb{Z}_2$ | (3,2) |
| $B_l(q)$, $l\geqslant 2$ | $\mathbb{Z}_{(2,q-1)}$ | $\mathbb{Z}_2$ | (2,2) |
| | | $\mathbb{Z}_2$ | (3,2) |
| | | $\mathbb{Z}_3\times\mathbb{Z}_2$ | (3,3) |
| $C_l(q)$, $l\geqslant 2$ | $\mathbb{Z}_{(2,q-1)}$ | $\mathbb{Z}_2$ | (2,2) |
| | | $\mathbb{Z}_2$ | (3,2) |
| $D_l(q)$, $l\geqslant 4$ | $\mathbb{Z}_{(4,q^l-1)}$, $l$ odd<br>$\mathbb{Z}_{(2,q^l-1)}\times\mathbb{Z}_{(2,q^l-1)}$, $l$ even | $\mathbb{Z}_2\times\mathbb{Z}_2$ | (4,2) |
| $E_6(q)$ | $\mathbb{Z}_{(3,q-1)}$ | | |
| $E_7(q)$ | $\mathbb{Z}_{(2,q-1)}$ | | |
| $E_8(q)$ | 1 | | |
| $F_4(q)$ | 1 | $\mathbb{Z}_2$ | (4,2) |
| $G_2(q)$ | 1 | $\mathbb{Z}_3$ | (2,3) |
| | | $\mathbb{Z}_2$ | (2,4) |
| ${}^2A_l(q)$, $l\geqslant 2$ | $\mathbb{Z}_{(l+1,q+1)}$ | $\mathbb{Z}_2$ | (3,2) |
| | | $\mathbb{Z}_4\times\mathbb{Z}_3\times\mathbb{Z}_3$ | (3,3) |
| | | $\mathbb{Z}_3\times\mathbb{Z}_2\times\mathbb{Z}_2$ | (5,2) |
| ${}^2B_2(q)\cong Sz(q)$ | 1 | $\mathbb{Z}_2\times\mathbb{Z}_2$ | (2,8) |
| ${}^2D_l(q)$, $l\geqslant 4$ | $\mathbb{Z}_{(4,q^l+1)}$ | | |
| ${}^2E_6(q)$ | $\mathbb{Z}_{(3,q+1)}$ | $\mathbb{Z}_3\times\mathbb{Z}_2\times\mathbb{Z}_2$ | (6,2) |
| ${}^2F_4(q)$ | 1 | | |
| ${}^2F_4(2)'$ | 1 | | |
| ${}^2G_2(q)$ | 1 | | |
| ${}^3D_4(q)$ | 1 | | |

**Alternating groups**

$(A_n,\mathbb{Z}_2)$, $n=5$, $n\geqslant 8$ $\qquad$ $(A_n,\mathbb{Z}_6)$, $n=6,7$

Table **4.1**—*continued*

| | Sporadic Groups | |
|---|---|---|
| $(M_{11}, 1)$ | $(M_{12}, \mathbb{Z}_2)$ | $(M_{22}, \mathbb{Z}_{12})$ |
| $(M_{23}, 1)$ | $(M_{24}, 1)$ | $(J_1, 1)$ |
| $(J_2, \mathbb{Z}_2)$ | $(J_3, \mathbb{Z}_3)$ | $(J_4, 1)$ |
| $(HS, \mathbb{Z}_2)$ | $(Mc, \mathbb{Z}_3)$ | $(\mathrm{Suz}, \mathbb{Z}_6)$ |
| $(.1, \mathbb{Z}_2)$ | $(.2, 1)$ | $(.3, 1)$ |
| $(M(22), \mathbb{Z}_6)$ | $(M(23), 1)$ | $(M(24)', \mathbb{Z}_3)$ |
| $(Ly, 1)$ | $(Ru, \mathbb{Z}_2)$ | $(ON, \mathbb{Z}_3)$ |
| $(He, 1)$ | $(F_2, \mathbb{Z}_2)$ | $(F_3, 1)$ |
| $(F_5, 1)$ | (type $F_1$, 1) | |

## B. Automorphisms

If $X$ is a group of Lie type defined over $GF(p^m)$, every element of $\mathrm{Aut}(X)$ is known to be a product of an inner, a "diagonal," a "field," and a "graph" automorphism [272].

For example, diagonal matrices in $GL(n, p^m)$ of determinant unequal to 1 determine by conjugation diagonal automorphisms of $SL(n, p^m)$. Field automorphisms are just the powers of the Frobenius map $Fr$, described in Section 4.14. Thus for $A=(a_{ij})\in SL(n, p^m)$, $a_{ij}\in GF(p^m)$, $1\leq i, j\leq n$,

$$Fr^k\colon A=(a_{ij})\mapsto A^{p^k}=\left(a_{ij}^{p^k}\right), \tag{4.47}$$

and so these automorphisms form a cyclic group of order $m$.

Likewise graph automorphisms are induced from symmetries of the Dynkin diagram of the associated Lie algebra of $X$. Thus the diagram associated with $SL(n, p^m)$

$$\circ\!-\!\!-\!\circ\!-\!\!-\!\circ \quad \cdots \quad \circ\!-\!\!-\!\circ \tag{4.48}$$

with $n-1$ nodes has a reflection of order 2, which induces an automorphism of $SL(n, p^m)$ [in this case, if we take the usual identification of the Chevalley group with $SL(n, p^m)$, the graph automorphism is the transpose-inverse map].

Diagonal automorphisms determine an abelian group of outer automorphisms [in fact, cyclic except for $D_{2n}(q)$, $n$ even, and $q$ odd], while field automorphisms always determine cyclic groups. On the other hand, graph automorphisms determine a group of order 1 or 2, except for $D_4$, in which case they determine a group isomorphic to $\Sigma_3$. Moreover, there is a general formula for the order of the diagonal group for each family of groups of Lie type. Also the group of diagonal automorphisms is normal in $\mathrm{Aut}(X)/X$ and field automorphisms commute with graph automorphisms. Summarizing, we have

THEOREM 4.237. *If $X$ is a group of Lie type, then $A=\mathrm{Aut}(X)/X$ is a solvable group with normal subgroups $D$ and $DF$, where $D$ is abelian and $F$ is cyclic; and, moreover, $A/DF \cong 1$, $Z_2$, or $\Sigma_3$.*

In the case of the alternating groups, one has

THEOREM 4.240. (i) *If $X=A_n$, $n\neq 6$, $n\geqslant 3$, then* $\mathrm{Aut}(X)\cong\Sigma_n$.
(ii) *If $X=A_6$ ($\cong L_2(9)$), then* $\mathrm{Aut}(X)\cong\mathrm{Aut}(L_2(9))$ [*$L_2(9)$ has index* 4 *in $\mathrm{Aut}(L_2(9))$*].

The automorphism group of every sporadic group has been computed (in most cases by the individual who first determined its internal structure). In the course of their work on standard components, Aschbacher and Seitz [21] completed the proof of the following result (except for Janko's group $J_4$, for which the desired assertion can be easily checked).

THEOREM 4.239. *If $X$ is a sporadic group, then*

$$|\mathrm{Aut}(X)/X|\leqslant 2.$$

Combining these results, one has

THEOREM 4.240. *For any known simple group $X$,* $\mathrm{Aut}(X)/X$ *is solvable.*

In view of Theorem 4.240, the Schreier conjecture of the solvability of the outer automorphism group of every finite simple group follows as a corollary of the classification of all finite simple groups.

### C. Centralizers of Elements of Prime Order *p*

To keep the discussion within bounds, we shall limit our statements to centralizers of involutions. Analogous results hold in many cases for centralizers of elements of odd prime order $p$, in particular, for groups $X$ of Lie type over $GF(q)$ when either $p$ divides the order of a Cartan subgroup of $X$ or $p=3$ and $q=2$ (two cases especially important for the applications). Note that we have already made some comments about centralizers of involutions in Sections 4.14 and 1.5. Again we refer the reader to [54, 55, 180, 240].

We let $X$ be a group such that $Y=F^*(X)$ is a simple $K$-group, we let $t\in\mathcal{I}(X)$, and we set $C=C_X(t)$.

THEOREM 4.241. *If $t\notin Y$, then one of the following holds*:

(i) *$C$ is solvable.*
(ii) *$F^*(C)$ is quasisimple.*
(iii) *$Y\in Chev(p)$, $p$ odd, $t$ is either a diagonal·inner or graph automorphism and $L(C)$ has two or more components.*
(iv) *$Y\in Chev(2)$, $t$ is a graph automorphism and $C$ is 2-constrained and nonsolvable.*
(v) *$Y\in Spor$ and $C$ is 2-constrained and nonsolvable* [e.g., $Y\cong M(22)$].
(vi) *$Y\in Spor$ and $C\cong Z_2\times K$, where $K$ is simple* [e.g., $Y\cong M_{12}$].

THEOREM 4.242. *If $t\in Y$, then one of the following holds*:

(i) *$L(C)$ is trivial or quasisimple.*
(ii) *$Y\in Chev(q)$, $q$ odd, and $L(C)$ has two components.*
(iii) *$Y\cong P\Omega^{\pm}(n,q)$, $q$ odd, and $L(C)$ has three or four components.*

*Furthermore, if $L(C)$ is nontrivial, then either $C/L(C)$ is solvable or $Y\cong A_n$.*

Again the possibilities for $L(C)$ are known for each choice of $Y$ and $t$. In particular, we have

THEOREM 4.243. *If $t\in Y$ and $L(C)$ is trivial, then one of the following holds*:

(i) *$C$ is solvable.*
(ii) *$F^*(C)=O_2(C)$.*
(iii) *$Y\cong A_n$, $n\equiv 1$ or 3 (mod 4), and $F^*(C)=O_2(C)\times O_3(C)$ with $O_3(C)\cong 1$ or $Z_3$.*

Combining these last three results [together with the known structure of $C$ in cases (iii) and (v) of Theorem 4.241], we obtain as a consequence the fundamental so-called "$B$-property" of the layers of the centralizers of involutions in the known simple groups.

THEOREM 4.244. *For any choice of $X$ and $t$, we have*

$$L_{2'}(C)=L(C).$$

A central chapter in the classification of simple groups is the verification of the $B$-property for centralizers of involutions in *arbitrary* core-free finite groups $G$. On the basis of Theorem 4.244, it is not difficult to show that it holds for arbitrary *$K$-groups*, thus enabling one to attack the problem inductively. A detailed outline of the proof of this fundamental, general property of finite groups will be given in the sequel.

Clearly the $B$-property holds trivially in $G$ if $O(C_G(t))=1$ for every involution $t$ of $G$. Hence, in view of Theorem 1.40 and the classification of nonconnected groups, a minimal counterexample $G$ will necessarily be *nonbalanced*. Proposition 4.58 shows then that some element of $\mathfrak{L}_2(G)$ must be *nonlocally balanced*. Clearly then to attack the "$B$-conjecture", we shall need a list of the known simple groups that are not locally balanced. The result can be stated as follows (the statement is essentially a complement to Proposition 4.56):

THEOREM 4.245. *Let $X$ be a group such that $Y=F^*(X)$ is simple. If $O(C_X(t))\neq 1$ for some $t\in\mathfrak{I}(X)$, then one of the following holds*:

(i) $Y\in Chev(p)$ *for some odd prime* $p$.
(ii) $Y\cong A_n$, $n$ *odd*.
(iii) $Y\cong L_3(4)$ *or* $He$ *and* $t\notin Y$.

The $L_3(4)$ and $He$ examples have already been described, following Proposition 4.56.

Thompson reduced a major case of the $B$-conjecture to the verification of four specific properties of the centralizers of involutions in groups of Lie type [291]. Burgoyne subsequently verified each of the desired properties [51]. To indicate their nature, we list the principal two here.

We consider a group $X$ in which $Y=F^*(X)\in Chev(p)$, $p$ odd, with $Y\not\cong L_2(p^n)$; and if $p=3$, also $Y\not\cong {}^2G_2(3^n)$. [The excluded groups are the only members of $Chev(p)$ that do not contain subgroups isomorphic to $SL(2,p)$.] Note that by definition of $Chev(p)$, $Y$ is quasisimple but not necessarily simple. Again we let $t\in\mathfrak{I}(X)$ and set $C=C_X(t)$.

THEOREM 4.246. *Assume $Y$ is simple and $O(C)\neq 1$. Then there exists an involution $u\in C$ and a subnormal subgroup $L$ of $C_X(u)$ with the following properties*:

(i) $L\cong SL(2,q)$ *for some odd* $q$ *and* $\langle u\rangle=Z(L)$.
(ii) $\langle O(C),t\rangle$ *normalizes* $L$ *and* $C_{\langle O(C),t\rangle}(L)=1$.

THEOREM 4.247. *Assume $L(C)$ has a component $L$ with the following properties*:

(a) $L \cong SL(2, q)$, *for some odd* $q$.
(b) $L$ *is normal in* $C$.
(c) $Z(L) \leqslant Z(X)$ $(\leqslant Z(Y))$.

*Then $Y/Z(Y)$ is isomorphic to $P\Omega^{\pm}(n, q)$ for some $n$ and prime power $q$.*

Further properties of centralizers of involutions are needed for many of the results listed in Sections D–J below.

## D. Balance

Proposition 4.56, Theorem 4.61, and Proposition 4.62 describe local $k$-balance for simple $K$-groups. These are the principal balance properties needed for local analysis. However, as indicated in the discussion of Goldschmidt-type signalizer functors in Section 4.4, variations of local $k$-balance are also used in the applications.

## E. Generation

We have remarked several times on the importance of generational statements for local analysis, in particular, for the construction of signalizer functors (cf. Proposition 4.65). Generational questions also arise in the stage following this construction. For example, suppose $G$ is a simple group which is 2-balanced for the prime $p$, $A$ an elementary abelian $p$-subgroup of $G$ with $m_p(A) \geqslant 4$, and the associated 2-balanced $A$-signalizer functor $\theta$ is complete. If we set $M = N_G(\theta(G; A))$, it is immediate that

$$\Gamma_{A,3}(G) = \langle N_G(E) | E \leqslant A, m_p(E) \geqslant 3 \rangle \leqslant M. \tag{4.49}$$

Assuming $\theta(G; A) \neq 1$, the ultimate goal of the analysis is to prove that $M$ is strongly $p$-embedded in $G$. In particular, we must argue that $C_a \leqslant M$ for each $a \in A^{\#}$. We cannot expect to reach such a strong conclusion solely from (4.49); however, (4.49) does yield some partial information. Indeed, as $m_p(A) \geqslant 4$, it follows from Theorem 4.13 that

$$O_{p'}(C_a) \leqslant \Gamma_{A,3}(G) \leqslant M. \tag{4.50}$$

Hence if we set $\overline{C}_a = C_a/O_{p'}(C_a)$, we can work with $\overline{C}_a$ and $\overline{A}$ inasmuch as $N_{C_a}(E)$ covers $N_{\overline{C}_a}(\overline{E})$ for every subgroup $E$ of $A$. The natural first question to ask is whether

$$L(\overline{C}_a) \leqslant \Gamma_{\overline{A},3}(\overline{C}_a) = \langle N_{\overline{C}_a}(\overline{E}) | \overline{E} \leqslant \overline{A}, m_p(\overline{E}) \rangle \geqslant 3. \qquad (4.51)$$

If (4.51) holds, then $M$ will cover $L(\overline{C}_a)$ and it will follow that $L_{p'}(C_a) \leqslant M$, which would be an important conclusion for any subsequent analysis.

One can easily reduce this problem to the corresponding generational question for $A$-invariant quasisimple sections $\overline{X}$ of $L(\overline{C}_a)$ such that $\overline{X}$ is a homomorphic image of some component of $L(\overline{C}_a)$. Clearly, for generational purposes, only the group $\tilde{A} = \overline{A}/C_{\overline{A}}(\overline{X})$ can present an obstruction to generation. If $|\tilde{A}| \leqslant p$, then $m_p(C_{\overline{A}}(\overline{X})) \geqslant 3$ and the desired generation holds trivially. Hence one can assume that $\tilde{A}$ is noncyclic. Also a suitable generational statement for $\tilde{X} = \overline{X}/Z(\overline{X})$ relative to $\tilde{A}$ implies one for $\overline{X}$ relative to $\overline{A}$. Since, in particular, $\tilde{X}$ will be a $K$-group, we are thus reduced to studying the faithful action of noncyclic elementary abelian $p$-groups on simple $K$-groups.

Seitz has obtained definitive results for odd $p$ for groups of Lie type of characteristic not $p$ [250]. The corresponding results for many sporadic groups have been computed by Lyons, using O'Nan's information concerning their local subgroups [231]. Generation typically fails for the groups in $Chev(p)$ and in the case of alternating groups, it depends strongly on the choice of elementary abelian $p$-group. We limit ourselves to the following two statements.

THEOREM 4.248. *Let $X \in Chev(q)$ and $A$ a noncyclic elementary abelian $p$-subgroup of* $\mathrm{Aut}(X)$, *$p$ odd with $p \neq q$, and assume*

$$\langle N_X(E) | E \leqslant A \text{ and } |A:E| \leqslant p \rangle < X.$$

*Then one of the following holds*:

(i) $p = 5$, $A \leqslant X$, *and* $X \cong {}^2F_4(2)'$.
(ii) $p = 3$, $A \leqslant X$, $X \cong Psp(n,2)$, $\Omega_n^{\pm}(2)$, $F_4(2)$, $F_4(4)$, ${}^2F_4(2)'$, $E_6(2)$, $E_7(2)$, $E_8(2)$, *or* $E_8(4)$.
(iii) $p = 3$, *$A$ induces inner · diagonal automorphisms on $X$, and* $X \cong L_{3^k}(4)$, $U_n(2)$, $E_6(4)$, *or* ${}^2E_6(2)$.
(iv) $p = 5$, $X \cong Sz(2^5)$ *or* ${}^2F_4(2^5)$ *and* $AX \cong \mathrm{Aut}(Sz(2^5))$ *or* $\mathrm{Aut}({}^2F_4(2^5))$.
(v) $p = 3$, $X \cong L_2(2^3)$, *and* $AX \cong \mathrm{Aut}(L_2(2^3))$.

(vi) $p=3$, $X \cong D_4(2)$, ${}^3D_4(2)$, *or* $D_4(4)$, *and* $AX = \langle a \rangle X$, *where* $a \in A^{\#}$ *induces a graph automorphism on* $X$.

(vii) $p=3$, $X \cong Psp(n, 2^3)$, $U_n(2^3)$, $\Omega_n^{\pm}(2^3)$, $G_2(2^3)$, $F_4(2^3)$, ${}^2E_6(2^3)$, $E_6(2^3)$, $E_7(2^3)$, *or* $E_8(2^3)$, *and* $AX = \langle a \rangle X$, *where* $a \in A^{\#}$ *induces a field automorphism on* $X$.

(viii) $X \cong D_4(8)$ *and* $AX = \langle a, b \rangle X$, *where* $a$ *induces a graph and* $b$ *induces a field automorphism on* $X$.

THEOREM 4.249. *Let* $X$ *be a group such that* $Y = F^*(X)$ *is a simple* $K$*-group. If* $X$ *has a strongly* $p$*-embedded subgroup for some odd prime* $p$, *then one of the following holds*:

(i) $X$ *has cyclic Sylow* $p$*-subgroups.*

(ii) $Y \cong L_2(p^n)$, $U_3(p^n)$, ${}^2G_2(3^n)$, *or* $A_{2p}$.

(iii) $p = 3$ *and* $Y \cong L_3(4)$, $\mathrm{Aut}(L_2(2^3))$, *or* $M_{11}$.

(iv) $p = 5$ *and* $Y \cong {}^2F_4(2)'$, $\mathrm{Aut}(Sz(2^5))$, $Mc$, *or* $M(22)$.

(v) $p = 11$ *and* $Y \cong J_4$.

The long list of exceptions in these two theorems leads to specific configurations in the course of the analysis of simple groups of characteristic 2 type whose elimination often requires laborious special treatments. The technical difficulties involved are often so great as to completely camouflage what had been a conceptually straightforward argument designed to cover the nonexceptional cases. Likewise, elimination of these exceptional cases substantially lengthens the overall analysis.

Similar generational statements can be established for noncyclic elementary abelian 2-groups acting on groups of Lie type of odd characteristic, alternating groups, and most likely on sporadic groups as well. [In this case fewer exceptions occur, as Bender's list of simple groups with a strongly (2-)embedded subgroup clearly indicates.] Such results are important in the study of groups of component type and, in particular, in the proof of the $B$-conjecture.

## F. Subgroup Structure

Principally, but not exclusively, local analysis requires information about the local subgroup and Sylow subgroup structure of simple $K$-groups. The Borel–Tits theorem (Theorem 1.41) shows that every maximal $p$-local in a group of $Chev(p)$ is a maximal parabolic and conversely. Thus the structure of the parabolic subgroups of such a group $X$ is very important.

On the other hand, if $q$ is prime distinct from $p$, it is difficult to formulate a general description of the $q$-local subgroups of $X$, but properties of these subgroups can be deduced from the general structure of $X$ as a $(B, N)$-pair, sometimes with much effort. A similar remark applies to the alternating and sporadic groups. In the latter case, this effort has already been made and essentially all local subgroups of the sporadic groups have been determined.

Concerning the Sylow $p$-subgroups $P$ of a simple $K$-group $X$, we are usually interested only in the rank of $P$ or perhaps the structure of $J(P)$; but sometimes, particularly when $p=2$, we need more detailed structural information. Carter and Fong [62] have given a complete description of the Sylow 2-subgroups of the classical groups over fields of odd characteristic $q$, from which any needed properties can be read off. The corresponding results for $p \neq 2$ or $q$ were obtained by A. Weir [318]. These descriptions turn out to be very similar to the easily computed structures of the Sylow $p$-subgroups of $A_n$. The Sylow structures of most of the remaining simple $K$-groups have been determined by the combined work of a great many authors, usually in the course of various classification theorems. In particular, a description of the Sylow $p$-subgroups of the groups in $Chev(p)$ involve the explicit commutator formulas of the Steinberg relations.

## G. Fusion

One needs a considerable amount of information about the conjugacy classes of elements of prime order in simple $K$-groups. One often wants a precisely described set of representatives for, say, the conjugacy classes of involutions. For example, if $X = A_n$, the involutions $(12)(34), (12)(34)(56)(78), \ldots$ constitute such a set. On the other hand, if $X \in Chev(p)$, $p$ odd, one can obtain representatives for the classes of involutions contained in a Cartan subgroup from the action of the Weyl group. However, there may be additional conjugacy classes of involutions (these have been determined in each case). In [21], Aschbacher and Seitz have computed all conjugacy classes of involutions of $X$ in the case that $Y = F^*(X) \in Chev(2)$ (they have also determined the centralizers in $Y$ of each such involution). Finally, with the exception of possibly $F_2$, all conjugacy classes of the sporadic groups have been determined (often as a preliminary step to computing their character tables).

We consider two general questions about fusion. First, if $X$ is a simple $K$-group, does the analogue of Glauberman's $Z^*$-theorem hold in $X$ for *odd* primes $p$? In other words, if $P \in Syl_p(X)$ and $Z \leqslant P$ has order $p$, must $Z$ fuse in $X$ to some subgroup of $P - Z$? (For odd $p$, the natural extension is to *subgroups* of order $p$ rather than to *elements* of order $p$.) As in the involution case, we say that $Z$ is *isolated in $P$ with respect to $X$* if $Z$ does not fuse to a subgroup of $X-Z$.

If $P$ is cyclic, then $Z=\Omega_1(P)$ is the unique subgroup of $P$ of order $p$, so certainly $Z$ is isolated in $P$ with respect to $X$. Hence the question is of interest only when $P$ is noncyclic.

THEOREM 4.250. *Let $X$ be a simple K-group with noncyclic Sylow p-subgroup $P$, $p$ odd. If $P$ contains a subgroup of order $p$ which is isolated in $P$ with respect to $X$, then $m_p(X)=2$ and one of the following holds*:

(i) $X\cong U_3(p)$.
(ii) $p=5$ *and* $X\cong Mc$.
(iii) $p=3$ *and* $X\cong G_2(q)$, $q\neq 3^n$, *or* $J_2$.

In the study of groups of component type, the notion of a "tightly embedded" subgroup, due to Aschbacher, plays a fundamental role [11].

DEFINITION 4.251. A subgroup $H$ of even order of a group $X$ is said to be *tightly embedded* in $X$ provided $H\cap H^x$ has odd order for every $x\in X-N_X(H)$.

This concept is a direct generalization of strong embedding, which requires $H\cap H^x$ to have odd order for every $x\in X-H$. [Thus if $H$ is tightly embedded in $X$ and $H=N_X(H)$, then $H$ is strongly embedded in $X$.]

Obviously any subgroup of $X$ of order 2 is tightly embedded in $X$. Aschbacher's classical involution theorem is essentially a characterization of the groups of Lie type of odd characteristic by the property of possessing a tightly embedded subgroup with quaternion Sylow 2-subgroups (cf. D14 following Section 1.2). Furthermore, Aschbacher and Seitz have studied simple groups $G$ which possess a tightly embedded subgroup of 2-rank at least 2, under suitable additional assumptions that arise naturally in the study of simple groups of component type, and have determined the exact possibilities for $G$ under these conditions [21, 22]. All these results will be described in detail in the sequel.

Ultimately their analysis reduces to the determination of those finite groups $X$ with $F^*(X)$ quasisimple and containing a tightly embedded subgroup $H$ of 2-rank at least 2 satisfying a number of technical side conditions. In particular, $H$ must have (noncyclic) elementary abelian Sylow 2-subgroups and a normal 2-complement. For brevity, let us just say that $H$

is *restricted* if it satisfies all these various conditions. An example of such a restricted tightly embedded subgroup $H$ is the subgroup $\langle(12)(34),(13)(24)\rangle \cong Z_2\times Z_2$ in $A_n$.

The following theorem lists a few of Aschbacher–Seitz's results on restricted tightly embedded subgroups.

THEOREM 4.252. *Let $X$ be a group such that $Y=F^*(X)$ is a simple $K$-group that contains a restricted tightly embedded subgroup $H$. Then we have:*

(i) *Either $Y\in Chev(2)$, $Y\cong A_n$ for some $n$, or $Y\cong M_{12}$, $M_{24}$, $J_2$, $He$, $Suz$, $Ru$, or .1.*
(ii) *If $Y\cong A_n$, $n\geqslant 10$, then in the natural representation of $A_n$, $H$ is conjugate to $\langle(12)(34),(13)(24)\rangle$.*
(iii) *If $Y\in Chev(2)$, then a Sylow 2-subgroup of $H$ induces inner automorphisms on $Y$.*

The effect of (iii) is to reduce the analysis [when $Y\in Chev(2)$] to the case that $X=Y$. Aschbacher–Seitz obtain rather complete information in this case concerning the possible structures of (restricted) tightly embedded subgroups.

## H. Signalizers

Let $X$ be a simple $K$-group and $P$ a $p$-subgroup of $X$. One needs various properties of $P$-signalizers in $X$—i.e., the collection of $P$-invariant $p'$-subgroups of $X$—primarily when $P$ is either elementary abelian or a Sylow $p$-subgroup of $X$. For $p=2$, two questions are of special interest:

(a) Does $P$ normalize a nontrivial $q$-subgroup of $X$ for various odd primes $q$?
(b) Are any two maximal $P$-invariant $q$-subgroups of $X$ conjugate by an element of $N_X(P)$ [or even by an element of $C_X(P)$]?

For odd $p$, the primary questions of interest are:

(c) What is the 2-local $p$-rank $m_{2,p}(X)$?
(d) Is $m_p(X)>m_{2,p}(X)$?
(e) If $P$ lies in a 2-local subgroup of $X$ and $m_p(P)=e(X)$ [the maximum of $m_{2,q}(X)$, taken over all odd primes $q$], can one describe the embedding of $P$ in $X$?

Here are a few of the illustrative results.

THEOREM 4.253. *Let X be simple of Lie type defined over GF(q). Then we have*

(i) *If $P \in Syl_p(X)$ and $q=p^n$, then P has trivial signalizers.*
(ii) *If p is odd, $q=2^n$, and $m_{2,p}(X)=e(X)$, then p divides $q^2-1$.*

THEOREM 4.254. (i) *If $X=A_n$ and $P \in Syl_2(X)$, then a maximal P-invariant subgroup of K of odd order is either trivial or of order* 3; *and, in the later case,* $n \equiv 3 \pmod 4$.

(ii) *If X is a sporadic simple group and $m_{2,p}(X) \geqslant 3$, p odd, then $m_p(X) > m_{2,p}(X)$.*

## I. Modular Representations

We have already indicated in some detail the type of properties of representations of quasisimple *K*-groups on vector spaces over $GF(p)$ needed for local analysis.

## J. Small Groups

In Sections 4.14 and A–I of this section we have tried to give a picture of the "general" theory of simple *K*-groups. However, in some classification problems, particularly those dealing with low rank groups, one also needs a "special" theory of simple *K*-groups. For example, in the classification of groups with dihedral, quasi-dihedral, or wreathed Sylow 2-subgroups, one needs a "dictionary" of specialized properties of the groups $L_2(q)$, $L_3(q)$, $U_3(q)$, $q$ odd, and $A_7$ (in most cases of the type described in Sections A–I above).

## K. The Theory of *K*-Groups

As we have repeatedly observed, properties of simple and quasisimple *K*-groups are used to establish properties of the local subgroups of the simple group under investigation. The systematic development of this entire relationship amounts to a "theory (both general and special) of arbitrary *K*-groups," which is the very essence of local analysis. In Section 4.1, we have given a good indication of the special theory in the case of solvable groups.

Because of the rather haphazard way the theory of $K$-groups developed over the years, the subject is not presently in a satisfactory state. Certainly one of the important first tasks of the "postclassification" era will be to develop a coherent, systematic treatment of the entire theory of $K$-groups.

# Bibliography

This bibliography is not intended as a complete list of references on finite simple groups, but is limited to articles and books referred to in the text.


1. J. L. Alperin, Sylow intersections and fusion, *J. Algebra* **6** (1967), 222–241.
2. ——Sylow 2-subgroups of rank 3, in E. Shult, T. Gagen, and M. Hale, eds., *Gainesville Conference on Finite Groups*, North-Holland, Amsterdam; American Elsevier, New York, 1973, pp. 3–5.
3. J. L. Alperin, R. Brauer, and D. Gorenstein, Finite groups with quasi-dihedral and wreathed Sylow 2-subgroups, *Trans. Amer. Math. Soc.* **151** (1970), 1–261.
4. ——Finite simple groups of 2-rank two, *Scripta Math.* **29** (1973), 191–214.
5. J. L. Alperin and D. Gorenstein, The multiplicators of certain simple groups, *Proc. Amer. Math. Soc.* **17** (1966), 515–519.
6. S. Andrilli, Existence and uniqueness of O'Nan's simple group, Ph.D. thesis, Rutgers University, 1979.
7. E. Artin, *Geometric Algebra*, Wiley-Interscience, New York, 1957.
8. M. Aschbacher, Finite groups with a proper 2-generated core, *Trans. Amer. Math. Soc.* **197** (1974), 87–223.
9. ——Finite groups generated by odd transpositions: I, II, III, IV, *Math Z.* **127** (1972), 45–56; *J. Algebra* **26** (1973), 451–459; **26** (1973), 460–478; **26** (1973), 479–491.
10. ——A condition for the existence of a strongly embedded subgroup, Proc. *Amer. Math. Soc.* **38** (1973), 509–511.
11. ——On finite groups of component type, *Illinois J. Math* **19** (1975), 78–115.
12. ——Tightly embedded subgroups of finite groups, *J. Algebra* **42** (1976), 85–101.
13. ——A characterization of Chevalley groups over fields of odd characteristic, *Ann. of Math.* **106** (1977), 353–468.
14. ——On finite groups in which the generalized Fitting group of the centralizer of some involution is symplectic but not extra-special, *Comm. Algebra* **4** (1976), 595–616.
15. ——On finite groups in which the generalized Fitting group of the centralizer of some involution is extra-special, *Illinois J. Math.* **21** (1977), 347–364.
16. ——A pushing up theorem for characteristic 2 type groups, *Illinois J. Math.* **22** (1978), 108–120.
17. ——On the failure of Thompson factorization in 2-constrained groups (to appear).
18. ——A factorization theorem for 2-constrained groups (to appear in *Proc. London Math. Soc.*).

19. ——*GF*(2)-representations of finite groups (to appear).
20. ——Some results on pushing up in finite groups, *Math. Z.* **177** (1981), 61–80.
21. M. Aschbacher and G. Seitz, Involutions in Chevalley groups over fields of even order, *Nagoya Math. J.* **63** (1976), 1–91.
22. ——On groups with a standard component of known type, *Osaka J. Math.* **13** (1976), 439–482.
23. B. Baumann, Endliche Gruppen mit einer 2-zentralen Involution, deren Zentralisator 2-abgeschlossen ist, *Illinois J. Math.* **22** (1978), 240–261.
24. G. Baumslag, Problem areas in infinite group theory for finite group theorists, *Proc. Sympos. Pure Math.*, **37** (1980), 217–224.
25. H. Bender, Endliche zweifach transitive Permutationsgruppen, deren involutionen keine Fixpunkte haben, *Math. Z.* **104** (1968), 175–204.
26. ——Transitive Gruppen gerader Ordnung, in denen jede Involution genau einen Punkt festlasst, *J. Algebra* **17** (1971), 527–554.
27. ——On groups with abelian Sylow 2-subgroups, *Math. Z.* **117** (1970), 164–176.
28. ——On the uniqueness theorem, *Illinois J. Math.* **14** (1970), 376–384.
29. ——Goldschmidt's 2-signalizer functor theorem, *Israel J. Math.* **22** (1975), 208–213.
30. ——A group-theoretic proof of the $p^a q^b$-theorem, *Math. Z.* **126** (1972), 327–338.
31. ——Über die grössten $p'$-normal Teiler in $p$-auflösbaren Gruppen, *Arch. Math.* **18** (1967), 474–478.
32. ——On groups with dihedral Sylow 2-subgroups (to appear).
33. H. Bender and G. Glauberman, Characters of finite groups with dihedral Sylow 2-subgroups (to appear).
34. N. Blackburn, On a special class of $p$-groups, *Acta. Math.* **100** (1958), 45–92.
35. ——Generalizations of certain elementary theorems on $p$-groups, *Proc. London Math. Soc.* **11** (1961), 1–22.
36. E. Bombieri, Thompson's problem ($\sigma^2=3$), *Invent. Math.* **58** (1980), 77–100.
37. A. Borel and J. Tits, Élements unipotents et sousgroupes paraboliques de groupes reductifs: I, *Invent. Math.* **12** (1971), 97–104.
38. A. Borel *et al.*, *Seminar on Algebraic Groups and Related Finite Groups*, Lecture Notes in Mathematics 131, Springer-Verlag, Berlin, 1970.
39. R. Brauer, On groups whose order contains a prime to the first power: I, II, *Amer. J. Math.* **64** (1942), 401–440.
40. ——On the structure of groups of finite order, in: *Proc. Internat. Congr. Math.* Vol. 1 (1954), pp. 209–217, Noordhoff, Groningen, North-Holland, Amsterdam.
41. ——Zur Darstellungstheorie der Gruppen endlicher Ordnung: I, II, *Math. Z.* **63** (1956), 406–444; **72** (1959), 25–46.
42. ——On finite Desarguesian planes: I, II, *Math. Z.* **90** (1966), 117–151.
43. ——Some applications of the theory of blocks of characters of finite groups: I, II, III, *J. Algebra* **1** (1964), 152–167; 307–334; **3** (1966), 225–255.
44. ——On simple groups of order $5\cdot 3^a\cdot 2^b$, *Bull. Amer. Math. Soc.* **74** (1968), 900–903.
45. R. Brauer and P. Fong, A characterization of the Mathieu group $M_{12}$, *Trans. Amer. Math. Soc.* **122** (1966), 18–47.
46. R. Brauer and K. Fowler, On groups of even order, *Ann. Math.* **62** (1955), 565–583.
47. R. Brauer and C. Sah, eds., *Finite Groups, A Symposium*, Benjamin, New York, 1969.
48. R. Brauer and M. Suzuki, On finite groups of even order whose 2-Sylow subgroup is a quaternion group, *Proc. Natl. Acad. Sci. USA* **45** (1959), 1757–1759.
49. R. Brauer, M. Suzuki, and G. Wall, A characterization of the one-dimensional unimodular groups over finite fields, *Illinois J. Math.* **2** (1958), 718–745.
50. A. Bryce, On the Mathieu group $M_{23}$, Ph.D. thesis, Monash University, 1969.
51. N. Burgoyne, Finite groups with Chevalley-type components *Pacific J. Math.* **72** (1977), 341–350.

52. ——Elements of order 3 in Chevalley groups of characteristic 2 (unpublished).
53. N. Burgoyne and P. Fong, The Schur multipliers of the Mathieu groups, *Nagoya Math. J.* **27** (1966), 733–745; correction, **31** (1968), 297–304.
54. N. Burgoyne and C. Williamson, Centralizers of semisimple elements in Chevalley-type groups *Pacific J. Math.* **70** (1977), 83–100.
55. ——Semisimple classes in Chevalley groups (unpublished).
56. N. Campbell, Pushing up in finite groups, Ph.D. thesis, California Institute of Technology, 1979.
57. R. Carmichael, *Introduction to the Theory of Groups of Finite Order*, Ginn, Boston and New York, 1937.
58. H. Cartan and S. Eilenberg, *Homological Algebra*, Princeton University Press, Princeton, New Jersey, 1956.
59. R. Carter, *Simple Groups of Lie Type*, Wiley-Interscience, New York, 1972.
60. R. W. Carter, Centralizers of semisimple elements in finite groups of Lie type, *Proc. London Math. Soc.*, **37** (1978), 491–507.
61. ——Centralizers of semisimple elements in the finite classical groups, *Proc. London Math. Soc.* **42** (1981), 1–42.
62. R. Carter and P. Fong, The Sylow 2-subgroups of the finite classical groups, *J. Algebra* **1** (1964), 139–151.
63. A. Chermak, Finite *BN*-pairs of rank 2 and even characteristic having a nontrivial Cartan subgroup, *J. Algebra* **62** (1980), 170–202.
64. ——On certain groups with parabolic-type subgroups over $Z_2$ (to appear).
65. ——Large triangular amalgams whose rank-1 kernels are not all distinct (to appear).
66. C. Chevalley, Sur certains groupes simples, *Tohoku Math. J.* **7** (1955), 14–66.
67. ——Seminaire Chevalley, *Classification des Groupes de Lie Algèbriques*, Vol. 2, Paris, 1956–58.
68. J. Conway, A group of order 8,315,553,613,086,720,000, *Bull. London Math. Soc.* **1** (1969), 79–88.
69. ——Three lectures on exceptional groups, in: G. Higman and M. Powell, *Finite Simple Groups*, Oxford, 1969; Academic Press, London, 1971, pp. 215–247.
70. J. Conway and S. Norton, Monstrous moonshine, *Bull. London Math. Soc.* **11** (1979), 308–339.
71. J. Conway and D. Wales, Construction of the Rudvalis group of order 145,926,144,000, *J. Algebra* **27** (1973), 538–548.
72. B. Cooperstein, An enemies list for factorization theorems, *Comm. Alg.* **6** (1978), 1239–1288.
73. ——*S*- and *F*-pairs for groups of Lie type, *Proc. Sympos. Pure Math.* **37** (1980), 249–253.
74. C. Curtis, Central extensions of groups of Lie type, *Crelle J.* **220** (1965), 174–185.
75. ——Irreducible representations of finite groups of Lie type, *J. Für Math.* **219** (1965), 180–199.
76. ——Modular representations of finite groups with split $(B, N)$-pairs, Chapter B, in: A. Borel *et al.*, *Seminar on Algebraic Groups and Related Finite Groups*, Lecture Notes in Mathematics 131, Springer-Verlag, Berlin, 1970.
77. P. Dembowski, *Finite Geometries*, Springer-Verlag, New York, 1968.
78. P. Dembowski and O. Wagner, Some characterizations of finite projective space, *Arch. Math.* **11** (1960), 465–469.
79. U. Dempwolff, On the extensions of an elementary group of order $2^5$ by $GL(5,2)$, *Rend. Sem. Mat. Univ. Padova* **48** (1972), 359–364.
80. L. Dickson, *Linear Groups*, Dover, New York, 1958.
81. J. Dieudonne, Sur les groupes classiques, *Actualites Sci. Ind. No.* 1040, Hermann, Paris, 1948.

82. ——La géométrie des groupes classiques, *Ergebnisse der Mathematik und ihrer Grenzgebiete*, Neue Folge, Heft 5, Springer, Berlin, 1955.
83. A. Dold and B. Eckmann, *Proceedings of Second International Conference on the Theory of Groups*, Lecture Notes in Mathematics, Springer-Verlag, Berlin and New York, 1973.
84. L. Dornhoff, *Group Representation Theory: Part B, Modular Character Theory*, Marcel Dekker, New York, 1972.
85. B. Fein, W. Kantor, and M. Schacher, Relative Brauer groups III (to appear).
86. W. Feit, On a class of doubly transitive permutation groups, *Illinois J. Math.* **4** (1960), 170–186.
87. ——A characterization of the simple groups $SL(2,2^{\alpha})$, *Amer. J. Math.* **82** (1960), 281–300; correction, **84** (1962), 201–204.
88. ——*Representations of Finite Groups*, Yale University Lecture Notes, 1969.
89. ——On integral representations of finite groups, *Proc. London Math. Soc.* **29** (1974), 633–683.
90. W. Feit, Some consequences of the classification of finite simple groups, *Proc. Sympos. Pure Math.* **37** (1980), 175–182.
91. W. Feit, M. Hall, and J. Thompson, Finite groups in which the centralizer of any nonidentity element is nilpotent, *Math. Z.* **74** (1960), 1–17.
92. W. Feit and G. Higman, On the nonexistence of certain generalized polygons, *J. Algebra* **1** (1964), 114–131.
93. W. Feit and J. Thompson, Solvability of groups of odd order, *Pacific J. Math.* **13** (1963), 775–1029.
94. ——Finite groups which contain a self-centralizing subgroup of order 3, *Nagoya Math. J.* **21** (1962), 185–197.
95. D. Fendel, A characterization of Conway's group .3, *J. Algebra* **24** (1973), 159–196.
96. B. Fischer, A characterization of the symmetric groups on 4 and 5 letters, *J. Algebra* **3** (1966), 88–98.
97. ——Distributiv Quasigruppen endlicher Ordnung, *Math. Z.* **83** (1964), 267–303.
98. ——Finite groups admitting a fixed-point-free automorphism of order $2p$: I, II, *J. Algebra* **3** (1966), 98–114; **5** (1967), 25–40.
99. ——Finite groups generated by 3-transpositions, University of Warwick (preprint).
100. ——Finite groups generated by 3-transpositions, *Invent. Math.* **13** (1971), 232–246.
101. ——Evidence for the existence of a new (3,4)-transposition group (unpublished).
102. P. Fong and G. Seitz, Groups with a $(B, N)$-pair of rank 2: I, III, *Invent. Math.* **21** (1973), 1–57; **24** (1974), 191–239.
103. P. Fong and W. Wong, A characterization of the finite simple groups $PSp(4, q)$, $G_2(q)$, $D_4^2(q)$: I, II, *Nagoya Math. J.* **36** (1969), 143–184; **39** (1970), 39–79.
104. R. Foote, Finite groups with maximal 2-components of type $L_2(q)$, $q$ odd, *Proc. London Math. Soc.* **38**, 422–458.
105. ——Component type theorems for finite groups in characteristic 2 (to appear).
106. G. Frobenius, *Über einen Fundamentalsatz der Gruppentheorie*, Berlin-Sitz. (1903), 987–991.
107. R. Gilman, Components of finite groups, *Comm. Alg.* **4** (1976), 1133–1198.
108. R. Gilman and D. Gorenstein, Finite groups with Sylow 2-subgroups of class two: I, II, *Trans. Amer. Math. Soc.* **207** (1975), 1–101; 103–125.
109. G. Glauberman, A characteristic subgroup of a $p$-stable group, *Canad. J. Math.* **20** (1968), 1101–1135.
110. ——Central elements in core-free groups, *J. Algebra* **4** (1966), 403–420.
111. ——On the automorphism group of a finite group having no nonidentity normal subgroups of odd order, *Math. Z.* **93** (1966), 154–160.
112. ——Subgroups of finite groups, *Bull. Amer. Math. Soc.* **73** (1967), 1–12.

113. ——Isomorphic subgroups of finite $p$-groups: I, II, *Canad. J. Math.* **23** (1971), 983–1022; 1023–1039.
114. ——Failure of factorization in $p$-solvable groups, *Quart. J. Math.* **24** (1973), 71–77.
115. ——On groups with quaternion Sylow 2-subgroups, *Illinois J. Math.* **18** (1974), 60–65.
116. ——On solvable signalizer functors in finite groups, *Proc. London Math. Soc.* **33** (1976), 1–27.
117. ——Factorizations in local subgroups of finite groups, Expository Lectures, CBMS Conference, University of Minnesota, Duluth, 1976.
118. G. Glauberman and R. Niles, Pushing up theorems for finite groups (to appear).
119. D. Goldschmidt, A conjugation family for finite groups, *J. Algebra* **16** (1970), 138–142.
120. ——A group-theoretic proof of the $p^a q^b$-theorem for odd primes, *Math. Z.* **113** (1970), 373–375.
121. ——Solvable signalizer functors on finite groups, *J. Algebra* **21** (1972), 137–148.
122. ——2-signalizer functors on finite groups, *J. Algebra* **21** (1972), 321–340.
123. ——2-fusion in finite groups, *Ann. of Math.* **99** (1974), 70–117.
124. ——Strongly closed 2-subgroups of finite groups, *Ann. of Math* (2) **102** (1975), 475–489.
125. ——Pushing up $A_n$ and $\Sigma_n$ (unpublished).
126. ——*Modular Character Theory*, Publish or Perish, Inc., Decatur, Georgia, 1980.
127. ——Automorphisms of trivalent graphs, *Ann. of Math.* **111** (1980), 377–406.
128. D. Gorenstein, On the centralizers of involutions in finite groups, *J. Algebra* **11** (1969), 243–277.
129. ——The flatness of signalizer functors on finite groups, *J. Algebra* **13** (1969), 509–512.
130. ——*Finite Groups*, Harper and Row, New York, 1968; 2nd ed., Chelsea, New York, 1980.
131. ——On finite simple groups of characteristic 2 type, *Inst. Hautes Etudes Sci.* **36** (1969), 5–13.
132. ——The classification of finite simple groups: I, *Bull. Amer. Math. New Series* **1** (1979), 43–199.
133. D. Gorenstein and K. Harada, Finite groups whose Sylow 2-subgroups are the direct product of two dihedral groups, *Ann. of Math.* **95** (1972), 1–54.
134. ——On finite groups with Sylow 2-subgroups of type $A_n$, $n=8,9,10,11$, *Math. Z.* **117** (1970), 207–237.
135. ——Finite groups of low 2-rank and the families $G_2(q)$, $D_4^2(q)$, $q$ odd, *Bull. Amer. Math. Soc.* **77** (1971), 829–862.
136. ——Finite groups whose 2-subgroups are generated by at most 4 elements, *Mem. Amer. Math. Soc. No.* **147** (1974), 1–464.
137. D. Gorenstein and M. Harris, A characterization of the Higman–Sims simple group, *J. Algebra* **24** (1973), 565–590.
138. ——Finite groups with product fusion, *Ann. of Math.* **101** (1975), 45–87.
139. D. Gorenstein and R. Lyons, Non-solvable signalizer functors on finite groups, *Proc. London Math. Soc.* **35** (1977), 1–33.
140. D. Gorenstein and J. Walter, On the maximal subgroups of finite simple groups, *J. Algebra* **1** (1964), 168–231.
141. ——The characterization of finite groups with dihedral Sylow 2-subgroups, *J. Algebra* **2** (1964), 85–151, 218–270, 354–393.
142. ——Centralizers of involutions in balanced groups, *J. Algebra* **20** (1972), 284–319.
143. ——The $\pi$-layer of a finite group, *Illinois J. Math.* **15** (1971), 555–564.
144. ——Balance and generation in finite groups, *J. Algebra* **33** (1975), 224–287.
145. J. A. Green, The characters of the finite general linear groups, *Trans. Amer. Math. Soc.* **80** (1955), 402–447.
146. R. Griess, Schur multipliers of finite simple groups of Lie type, *Trans. Amer. Math. Soc.* **183** (1973), 355–421.

147. ——Schur Multipliers of some sporadic simple groups, *J. Algebra* **20** (1972), 320–349.

148. ——On the subgroup structure of the group of order $2^{46} \cdot 5^9 \cdot 7^6 \cdot 11^2 \cdot 13^3 \cdot 17 \cdot 19 \cdot 23 \cdot 29 \cdot 31 \cdot 41 \cdot 47 \cdot 59 \cdot 71$ (to appear).

149. ——The structure of the "monster" simple group, in W. Scott and F. Gross, eds., *Proceedings of the Conference on Finite Groups*, Academic Press, New York, 1976, pp. 113–118.

150. ——On a subgroup of order $2^{15}|GL(5,2)|$ in $E_8(C)$, *J. Algebra* **40** (1976), 271–279.

151. ——Schur multipliers of the known finite simple groups: II, *Proc. Sympos. Pure Mathematics* **37** (1980), 279–282.

152. ——The friendly giant (to appear).

153. J. Grover, Covering groups of groups of Lie type, *Pacific J. Math.* **30** (1969), 645–655.

154. M. Hall, *The Theory of Groups*, Macmillan, New York, 1959.

155. M. Hall and D. Wales, The simple group of order 604,800, *J. Algebra* **9** (1968), 417–450.

156. P. Hall, A note on soluble groups, *J. London Math. Soc.* **3** (1928), 98–105.

157. ——A characteristic property of soluble groups, *J. London Math. Soc.* **12** (1937), 190–200.

158. ——On the Sylow systems of a soluble group, *Proc. London Math. Soc.* **43** (1937), 316–323.

159. ——On the system normalizers of a soluble group, *Proc. London Math. Soc.* **43** (1937), 507–528.

160. P. Hall and G. Higman, The $p$-length of a $p$-soluble group and reduction theorems for Burnside's problem, *Proc. London Math. Soc.* **7** (1956), 1–41.

161. K. Harada, A characterization of the simple group $U_3(5)$, *Nagoya Math. J.* **38** (1970), 27–40.

162. ——On finite groups having self-centralizing 2-subgroups of small order, *J. Algebra* **33** (1975), 144–160.

163. ——On the simple group $F$ of order $2^{14} \cdot 3^6 \cdot 5^6 \cdot 7 \cdot 11 \cdot 9$, in W. Scott and F. Gross, eds., *Proceedings of the Conference on Finite Groups*, Academic Press, New York, 1976, pp. 119–195.

164. M. Harris, Finite groups with Sylow 2-subgroups of type $PSp(6, q)$, $q$ odd, *Comm. Algebra* **2** (1974), 181–232.

165. D. Held, The simple groups related to $M_{24}$, *J. Algebra* **13** (1969), 253–296.

166. C. Hering, W. Kantor, and G. Seitz, Finite groups having a split $(B, N)$-pair of rank 1, *J. Algebra* **20** (1972), 435–475.

167. D. Higman, Finite permutation groups of rank 3, *Math. Z.* **86** (1964), 145–156.

168. D. G. Higman, Coherent configurations: I, II, *Geometrica Dedicata*, **4** (1975), 1–32; **5** (1976), 413–424.

169. D. Higman and C. Sims, A simple group of order 44,352,000, *Math. Z.* **105** (1968), 110–113.

170. G. Higman, Suzuki 2-groups, *Illinois J. Math.* **7** (1963), 79–96.

171. G. Higman and J. McKay, On Janko's simple group of order 50,232,969, *Bull. London Math. Soc.* **1** (1969), 89–94.

172. G. Higman and M. Powell, *Finite Simple Groups*, Oxford, 1969; Academic Press, London, 1971.

173. C. Ho, On the quadratic pair whose root group has order 3, *Bull. Inst. Math. Acad. Sinica* **1** (1973), 155–180.

174. ——Quadratic pairs for 3 whose root group has order greater than 3: I, *Comm. Algebra* **3** (1975), 961–1029.

175. ——Chevalley groups of odd characteristic as quadratic pairs, *J. Algebra* **41** (1976), 202–211.

176. ——On the quadratic pairs, *J. Algebra* **43** (1976), 338–358.

177. ——Quadratic pairs for odd primes, *Bull. Amer. Math. Soc.* **82** (1976), 941–943.

178. D. Holt, Doubly transitive groups in which the stabilizer of two points is dihedral, *Quart. J. Math.* **27** (1976), 267–295.
179. ——Transitive permutation groups in which a 2-central involution fixes a unique point, *Proc. London Math. Soc.* **37** (1978), 165–192.
180. M. Hopkins, On the Ree groups of characteristic 3, Ph.D. thesis, University of Illinois, 1978.
181. B. Huppert, *Endliche Gruppen*, Vol. I, Springer-Verlag, Berlin and New York, 1967.
182. M. Isaacs, *Character Theory of Finite Groups*, Academic Press, New York, 1976.
183. N. Ito, On a class of doubly transitive permutation groups, *Illinois J. Math.* **6** (1962), 341–352.
184. N. Iwahori, Centralizers of involutions in finite Chevalley groups, in: A. Borel *et al.*, *Seminar on Algebraic Groups and Related Finite Groups*, Lecture Notes in Mathematics 131, Springer-Verlag, Berlin, 1970, pp. 267–295.
185. ——*Finite Groups, Symposium*, Japan Soc. for Promotion of Science, Tokyo, 1976.
186. N. Jacobson, *Lie Algebras*, Wiley-Interscience, New York, 1962.
187. Z. Janko, A new finite simple group with abelian 2-Sylow subgroups and its characterization, *J. Algebra* **3** (1966), 147–186.
188. ——Some new simple groups of finite order: I, *First Naz. Alta Math. Symposia Math.* **1** (1968), 25–65.
189. ——The nonexistence of a certain type of simple group, *J. Algebra* **18** (1971), 245–253.
190. ——A characterization of the Mathieu simple groups: I, II, *J. Algebra* **9** (1968), 1–19, 20–41.
191. ——A new finite simple group of order 86,775,571,046,077,562,880 which possesses $M_{24}$ and the full cover of $M_{22}$ as subgroups, *J. Algebra* **42** (1976), 564–596.
192. Z. Janko and J. Thompson, On a class of finite simple groups of Ree, *J. Algebra* **4** (1966), 274–292.
193. Z. Janko and S. K. Wong, A characterization of the Higman–Sims simple group, *J. Algebra* **18** (1971), 245–253.
194. C. Jordan, Recherches sur les substitutions, *J. Math. Pures Appl.* (2) **17** (1872), 351–363.
195. Victor G. Kac, A remark on the Conway–Norton conjecture about the "Monster" simple group, *Proc. Nat. Acad. Sci. USA* **77**, No. 9(1) (1980), 5048–5049.
196. W. C. Kantor and G. Seitz, Some results on doubly transitive groups, *Invent. Math.* **13** (1971), 125–142.
197. K. Klinger, Finite groups of order $2^a 3^b 13^c$, *J. Algebra* **41** (1976), 303–326.
198. L. Kovacs and B. Neumann, eds., *Proceedings of the First International Conference on the Theory of Groups* (Canberra, 1965), Gordon and Breach, New York, 1967.
199. J. Leech, Some sphere packings in higher space, *Canad. J. Math.* **16** (1964), 657–682.
200. ——Notes on sphere packings, *Canad. J. Math.* **19** (1967), 251–267.
201. W. Lempken, The Schur multiplier of $J_4$ is trivial, *Archiv der Math.* **30** (1978), 267–280.
202. J. Leon and C. Sims, The existence and uniqueness of a simple group generated by $\{3,4\}$-transpositions, *Bull. Amer. Math. Soc.* **83** (1977), 1039–1040.
203. J. Leon and D. Wales, Simple groups of order $2^a 3^b p^c$ with cyclic Sylow $p$-subgroups, *J. Algebra* **29** (1974), 246–254.
204. D. Livingstone and B. Fischer, The character table of $F_1$ (to appear).
205. G. Lusztig and P. Deligne, Representations of reductive groups over finite fields, *Ann. of Math.* **103** (1976), 103–161.
206. R. Lyons, Evidence for a new finite simple group, *J. Algebra* **20** (1972), 540–569.
207. A. MacWilliams, On 2-groups with no normal abelian subgroups of rank 3 and their occurrence as Sylow 2-subgroups of finite simple groups, *Trans. Amer. Math. Soc.* **150** (1970), 345–408.
208. D. Mason, Finite groups with Sylow 2-subgroup, the direct product of a dihedral and wreathed group and related problems, *Proc. London Math. Soc.* **33** (1976), 401–442.

209. G. Mason, Finite groups of order $2^a 3^b 17^c$: I, II, III, *J. Algebra* **40** (1976), 309–339; **41** (1976), 327–346; 347–364.
210. E. Mathieu, Memoire sur le nombre de valeurs que peut acquerir une fonction quand on y permut ses variables de toutes les manières possibles, *Crelle J.* **5** (1860), 9–42.
211. ——Memoire sur l'étude des fonctions de plusieures quantités, sur la manière de les formes et sur les substitutions qui les laissent invariables, *Crelle J.* **6** (1861), 241–323.
212. ——Sur la fonction cinq fois transitive des 24 quantités, *Crelle J.* **18** (1873), 25–46.
213. Matsuyama, H., Solvability of groups of order $2^a p^b$, *Osaka J. Math.* **10** (1973), 375–378.
214. P. Mazet, Sur le multiplicateur de Schur du Groupe de Mathieu $M_{22}$, *C. R. Acad. Sci. Paris* **289** (1979), 659–661.
215. P. McBride, Nonsolvable signalizer functors on finite groups (to appear).
216. J. McKay, Computing with finite simple groups, Second International Conference on the Theory of Groups, pp. 448–451.
217. J. McKay and D. Wales, The multipliers of the simple groups of order 604,800 and 50,232,960, *J. Algebra* **17** (1971), 262–272.
218. ——The multiplier of the Higman–Sims groups, *Proc. London Math. Soc.* **3** (1971), 283–285.
219. J. McLaughlin, Some subgroups of $SL_n(F_2)$, *Illinois J. Math.* **13** (1969), 108–115.
220. ——A simple group of order 898,128,000, in: R. Brauer and C. Sah, eds., *Finite Groups, A Symposium*, Benjamin, New York, 1969, pp. 109–111.
221. R. Niles, Pushing up finite groups, *J. Algebra* **57** (1979), 26–63.
222. ——$(B, N)$-pairs and finite groups with parabolic-type subgroups, (to appear).
223. S. Norton, Construction of Harada's groups, Ph.D. thesis, University of Cambridge, 1976.
224. ——The construction of $J_4$, (to appear).
225. S. Norton, The uniqueness of $F_1$ (to appear).
226. M. O'Nan, Automorphisms of unitary block designs, *J. Algebra* **20** (1972), 495–511.
227. ——A characterization of $U_3(q)$, *J. Algebra* **22** (1972), 254–296.
228. ——A characterization of $L_n(q)$ as a permutation group, *Math. Z.* **127** (1972), 301–314.
229. ——Normal structure of the one-point stabilizer of a doubly transitive permutation group, *Trans. Amer. Math. Soc.* **214** (1975), 1–42, 43–74.
230. ——Some evidence for the existence of a new simple group, *Proc. London Math. Soc.* **32** (1976), 421–479.
231. ——Local properties of sporadic groups (unpublished).
232. D. Parrott, On the Mathieu groups $M_{22}$ and $M_{11}$, *J. Australian Math. Soc.* **11** (1970), 69–81.
233. ——A characterization of the Tits simple group, *Canad. J. Math.* **24** (1972), 672–685.
234. D. Parrott and S. K. Wong, On the Higman–Sims simple group of order 44,352,000, *Pacific J. Math.* **32** (1970), 501–516.
235. N. Patterson, On Conway's group, .0 and some subgroups, Ph.D. thesis, University of Cambridge, 1974.
236. K. W. Phan, A characterization of the finite simple group $U_4(3)$, *Australian J. Math.* **10** (1969), 77–94.
237. K. Phan, On groups generated by special unitary groups: I, II, *J. Austral. Math. Soc.* **23** (1977), 67–77, 129–146.
238. R. Ree, A family of simple groups associated with the simple Lie algebra $F_4$, *Amer. J. Math.* **83** (1961), 401–420.
239. ——A family of simple groups associated with the simple Lie algebra $G_2$, *Amer. J. Math.* **83** (1961), 432–462.
240. ——Classification of involutions and centralizers of involutions, in: L. Kovacs and B. Neumann, eds., *Proceedings of the First International Conference on the Theory of*

*Lie Groups* (Canberra, 1965), Gordon and Breach, New York, 1967, pp. 281–301.

241. A. Reifart, On finite groups with large extra-special subgroups: I, II, *J. Algebra* **53** (1978), 452–470; **54** (1978), 273–280.
242. P. Rowley, Characteristic 2-type groups with a strongly closed 2-subgroup of class at most two (to appear).
243. ——Finite groups which possess a strongly closed 2-subgroup of class at most two: I, II, (to appear).
244. A. Rudvalis, Evidence for the existence of a simple group of order 145,926,144,000 (unpublished).
245. I. Schur, Untersuchen über die Darstellung der endlichen Gruppen durch gebrochenen linearen Substitutionen, *Crelle J.* **132** (1907), 85–137.
246. ——Über die Darstellungen der symmetrischen und alternierenden Gruppen durch gebrochenen lineare Substitutionen, *Crelle J.* **139** (1911), 155–250.
247. W. Scott and F. Gross, eds., *Proceedings of the Conference on Finite Groups*, Academic Press, New York, 1976.
248. G. Seitz, Properties of the known simple groups, *Proc. Sympos. Pure Math.* **37** (1980), 231–238.
249. G. Seitz, The root groups of a maximal torus, *Proc. Sympos. Pure Math.* **37** (1980), 239–242.
250. G. Seitz, Generation of finite groups of Lie type (to appear in *Trans. Amer. Math. Soc.*).
251. E. Shult, On a class of doubly transitive groups, *Illinois J. Math.* **16** (1972), 434–455.
252. ——On the fusion of an involution in its centralizer (unpublished).
253. E. Shult, T. Gagen, and M. Hale, eds., *Gainesville Conference on Finite Groups*, North-Holland, Amsterdam; American Elsevier, New York, 1973.
254. C. Sims, Graphs and finite permutation groups: I, II, *Math. Z.* **95** (1967), 76–86; **103** (1968), 276–281.
255. ——The existence and uniqueness of Lyons' group, in: E. Shult, T. Gagen, and M. Hale, eds., *Gainesville Conference on Finite Groups*, North-Holland, Amsterdam; American Elsevier, New York; 1973, pp. 138–141.
256. ——The existence and uniqueness of O'Nan's group (unpublished).
257. F. Smith, Finite groups whose Sylow 2-subgroups are the direct product of a dihedral and semidihedral group. *Illinois J. Math.* **17** (1973), 352–386.
258. ——Finite groups whose Sylow 2-subgroups are the direct product of two semidihedral groups, *Illinois J. Math.* **17** (1973), 387–396.
259. ——A characterization of the Conway simple group .2, *J. Algebra* **31** (1974), 91–116.
260. ——Transitive permutation groups in which a 2-central involution fixes a unique point (unpublished).
261. S. Smith, Large extraspecial subgroups of widths 4 and 6, *J. Algebra* **58** (1979), 251–280.
262. ——A characterization of orthogonal groups over $GF(2)$, *J. Algebra* **62** (1980), 39–60.
263. ——A characterization of finite Chevalley and twisted groups of type $E$ over $GF(2)$, *J. Algebra* **62** (1980), 101–117.
264. R. Solomon, Finite groups with intrinsic 2-components of type $\hat{A}_n$, *J. Algebra* **33** (1975), 498–522.
265. ——Maximal 2-components in Finite groups, *Comm. Algebra* **4** (1976), 561–594.
266. ——2-signalizers in groups of alternating type, *Comm. Algebra* **6** (1978), 529–549.
267. R. Stanton, The Mathieu groups, *Canad. J. Math.* **3** (1951), 164–174.
268. B. Stark, Another look at Thompson's quadratic pairs, *J. Algebra* **45** (1977), 334–342.
269. R. Steinberg, Variations on a theme of Chevalley, *Pacific J. Math.* **9** (1959), 875–891.
270. ——Automorphisms of finite linear groups, *Canad. J. Math.* **12** (1960), 606–615.
271. ——Générateurs, relations, et revêtements de groupes algèbriques, Colloque sur la théorie des groupes algèbriques, *C.B.R.M.*, Brussels, 1962, 113–127.

272. ——Lectures on Chevalley groups, Lecture Notes, Yale University, 1967–1968.
273. ——*Endomorphisms of Linear Algebraic Groups*, *Mem. Amer. Math. Soc.* **80** (1968).
274. B. Stellmacher, On graphs with edge-transitive automorphism groups (to appear).
275. M. Suzuki, The nonexistence of a certain type of simple group of odd order, *Proc. Amer. Math. Soc.* **8** (1957), 686–695.
276. ——Finite groups with nilpotent centralizers, *Trans. Amer. Math. Soc.* **99** (1961), 425–470.
277. ——Applications of group characters, *Proc. Sympos. Pure Math.* **6** (1962), 101–105.
278. ——On a class of double transitive groups: I, II, *Ann. of Math.* **75** (1962), 105–145; **79** (1964), 514–589.
279. ——On characterizations of linear groups: III, *Nagoya J. Math.* **21** (1962), 159–183.
280. ——A characterization of the 3-dimensional projective unitary group over a finite field of odd characteristic, *J. Algebra* **2** (1965), 1–14.
281. ——A simple group of order 448,345,600, in: R. Brauer and C. Sah, eds., *Finite Groups, A Symposium*, Benjamin, New York, 1969, pp. 113–119.
282. ——Characterization of linear groups, *Bull. Amer. Math. Soc.* **75** (1969), 1043–1091.
283. ——Characterization of some finite simple groups, *Internat. Cong. Math. Nice* **1** (1971), 371–373.
284. ——A characterization of the orthogonal groups over finite fields of characteristic two, in: N. Iwahori, *Finite Groups Symposium*, Japan Soc. for Promotion of Science, Tokyo, 1976, pp. 105–112.
285. J. Tate, Quotient groups of finite groups, *Topology* **3** (1964), supp. 1, 109–111.
286. J. Thompson, Normal $p$-complements for finite groups, *Math. Z.* **72** (1960), 332–354.
287. ——2-signalizers of finite groups, *Pacific J. Math.* **14** (1964), 363–364.
288. ——Factorizations of $p$-solvable groups, *Pacific. J. Math.* **16** (1966), 371–372.
289. ——Nonsolvable finite groups all of whose local subgroups are solvable: I–VI, *Bull. Amer. Math. Soc.* **74** (1968), 383–437; *Pacific J. Math.* **33** (1970), 451–536; **39** (1971), 483–534; **48** (1973), 511–592; **50** (1974), 215–297; **51** (1974), 573–630.
290. ——Toward a characterization of $E_2^*(q)$: I, II, III, *J. Algebra* **7** (1967), 406–414; **20** (1972), 610–621; **49** (1977), 163–166.
291. ——Notes on the $B$-conjecture (unpublished).
292. ——Fixed points of $p$-groups acting on $p$-groups, *Math. Z.* **86** (1964), 12–13.
293. ——Quadratic pairs (unpublished).
294. ——Simple 3′-groups, in *Symposia Math.* XIII, Academic Press, London, (1974), pp. 517–530; balance unpublished.
295. ——A simple subgroup of $E_8(3)$, in N. Iwahori, *Finite Groups Symposium*, Japan Soc. for Promotion of Science, Tokyo, 1976, pp. 113–116 (cf. Reference 129).
296. ——A conjugacy theorem for $E_8$, *J. Algebra* **38** (1976), 525–530.
297. ——The uniqueness of the Fischer–Griess monster, *Bull. London Math. Soc.* **11** (1979), 340–346.
298. ——Some numerology between the Fischer–Griess monster and elliptic modular functions, *Bull. London Math. Soc.* **11** (1979), 352–353.
299. F. Timmesfeld, Eine Kennzeichnung der linearen Gruppen über $GF(2)$, *Math. Ann.* **189** (1970), 134–160.
300. ——A characterization of the Chevalley and Steinberg groups over $F_2$, *Geometrica Dedicata* **1** (1973), 269–321.
301. ——Groups generated by root involutions: I, *J. Algebra* **33** (1975), 75–135.
302. ——Groups with weakly closed T.I. subgroups, *Math. Z.* **143** (1975), 243–278.
303. ——Finite simple groups in which the generalized Fitting group of the centralizer of some involution is extra-special, *Ann. of Math.* **107** (1978), 297–369.
304. J. Tits, Théorème de Bruhat et sous-groupes paraboliques, *C. R. Acad. Sci. Paris* **254** (1962), 2910–2912.

305. ——Groupes simples et geometries associees, *Proc. Internat. Congr. Math. Stockholm* (1962), pp. 197–221.
306. ——Algebraic and abstract simple groups, *Ann. of Math.* **80** (1964), 313–329.
307. ——*Buildings of Spherical Type and Finite* $(B, N)$ *-Pairs*, Springer-Verlag, Berlin and New York, 1974.
308. W. Tutte, A family of cubical graphs, *Proc. Cambridge Philos. Soc.* **43** (1947), 459–474.
309. ——On the symmetry of cubic graphs, *Canad. J. Math.* **11** (1959), 621–624.
310. D. Wales, Simple groups of order $p \cdot 3^a \cdot 2^b$, *J. Algebra* **16** (1970), 1883–1890.
311. ——Simple groups of order $7 \cdot 3^a \cdot 2^b$, *J. Algebra* **16** (1970), 575–596.
312. ——Simple groups of order $13 \cdot 3^a \cdot 2^b$, *J. Algebra* **17** (1979).
313. ——Simple groups of order $17 \cdot 3^a \cdot 2^b$, *J. Algebra* **17** (1971), 429–433.
314. J. H. Walter, Finite groups with abelian Sylow 2-subgroups of order 8, *Invent. Math.* **2** (1967), 332–376.
315. ——The characterization of finite groups with abelian Sylow 2-subgroups, *Ann. of Math.* **89** (1969), 405–514.
316. H. Ward, On Ree's series of simple groups, *Trans. Amer. Math. Soc.* **121** (1966), 62–89.
317. ——On the triviality of primary parts of the Schur multiplier, *J. Algebra* **10** (1968), 377–382.
318. A. Weir, Sylow $p$-subgroups of the classical groups over finite fields with characteristic prime to $p$, *Proc. Amer. Math. Soc.* **6** (1955), 529–533.
319. H. Wielandt, Primitive Permutationsgruppen von Grad $2p$, *Math. Z.* **63** (1956), 478–485.
320. ——*Finite Permutation Groups*, Academic Press, New York, 1964.
321. E. Witt, Die 5-fach transitiven Gruppen von Mathieu, *Abl. Math. Hamburg* **12** (1937), 256–264.
322. W. Wong, Determination of a class of primitive permutation groups, *Math. Z.* **99** (1967), 235–246.
323. ——On finite groups whose 2-Sylow subgroups have cyclic subgroups of index 2, *J. Austral. Math. Soc.* **4** (1964), 90–112.
324. H. Zassenhaus, Kennzeichnung endlicher linearer Gruppen als Permutationsgruppen, *Abh. Math. Sem. Hamburg Univ.* **11** (1936), 17–40.

# Index